TO HELL & HIGH WATER

Walking in the Footsteps of Henry Lawson

First published 2012

Big Sky Publishing Pty Ltd
PO Box 303, Newport, NSW 2106, Australia
Phone: 1300 364 611
Fax: (61 2) 9918 2396
Email: info@bigskypublishing.com.au
Web: www.bigskypublishing.com.au

Cover design and typesetting: Think Productions

Images: With the exception of the Henry Lawson images, all photograph credit: Sean Morahan, Jane Crowle, Barrie Bryan, Gregory Bryan.
Cover photograph: Sean Morahan. Henry Lawson image courtesy of Alexander Turnbull Library, Wellington, New Zealand.

National Library of Australia Cataloguing-in-Publication entry (pbk)
Author: Bryan, Gregory.
Title: To hell and high water : walking in the footsteps of Henry Lawson / by Gregory Bryan.
ISBN: 9781921941788 (pbk.)
Subjects: Bryan, Gregory.
Bryan, Gregory--Travel.
Lawson, Henry, 1867-1922.
Australia--Description and travel.
Australia--History.
Dewey Number: 920.710994

National Library of Australia Cataloguing-in-Publication entry (ebook)
Author: Bryan, Gregory.
Title: To hell and high water [electronic resource] : walking in the footsteps of Henry Lawson / by Gregory Bryan.
ISBN: 9781921941832 (ebook)
Subjects: Bryan, Gregory.
Bryan, Gregory--Travel.
Lawson, Henry, 1867-1922.
Australia--Description and travel.
Australia--History.
Dewey Number: 920.710994
Printed in China through Bookbuilders

TO HELL & HIGH WATER

Walking in the Footsteps of Henry Lawson

www.bigskypublishing.com.au

GREGORY BRYAN

To Jennifer
who stays behind for me
and, of course,
To Henry,
who went ahead.

TABLE OF CONTENTS

ACKNOWLEDGEMENTS

There are many people deserving of acknowledgement for their roles in helping to bring this book to fruition. I appreciate everybody at Big Sky Publishing for the quality of your work. Particular thanks to Diane for support and leadership.

Thank you to Mrs Yvonne Swaffer for sharing with me her family history about the Hungerford region, including the Royal Mail Hotel and the Henry Lawson poem in her family. Thank you to Graeme Foster for permission to reproduce that poem in this book.

Thank you to John Stephenson and Tony Marsh for kindly allowing us to stay at Warroo and Kia-Ora when we passed through. Your many kindnesses helped more than you know.

Thank you to Moc and Sheree Parker at the Royal Mail Hotel. They are wonderful hosts who made us feel welcome and helped us to fully enjoy our time in Hungerford. Thanks also to Scott and Annette Parker at the Warrego Hotel in Fords Bridge and Lochie Ford at the Port O' Bourke Hotel in Bourke.

Thank you to Dean Hutchinson, the policeman/paramedic in Hungerford. Without him, I would have been stopped in my tracks.

Thank you to the University of Manitoba. My Lawson trek and the writing of this book was a product of research study leave, without which I would never have had the time to complete this project. My thanks also go to the good people at Tenison Woods College who provided me with a computer and office space, allowing me to spread out my things and write. Julian Tenison Woods is buried near Henry Lawson at the Waverley Cemetery in Sydney. If there are things that happen beyond the grave, I would like to think that Henry occasionally gets together with Julian for midnight strolls around the cemetery. I fancy they would enjoy one another's company.

Special thanks must be given to the kind people of Leeton and Grenfell for their warm welcomes and assistance with my research. Particular thanks to Dot Eurell of the Leeton Family and Local

History Society. Dot generously shared her time and local expertise and patiently drove me all over town to visit Henry's old haunts. Thanks to George Weston for getting me inside Henry's cottage to sit where Henry sat. Thanks also to George for fighting the good fight and leading the efforts to save Henry's cottage from demolition. Thanks to Gai Lander. I was humbled to have been a guest of the Grenfell Henry Lawson Festival. As President of the committee, Gai looked after me and ensured I had a wonderful time visiting Henry's birthplace. Gai and her festival committee were terrific hosts.

Moz and Jane deserve the highest praise and sincere thanks for all their work. They toiled tirelessly to support Baz and me through to the end. I will ever be grateful for all that they did for me. I know that I would not have succeeded without them. There are times in this book when I am critical and unappreciative. Although I am embarrassed by these feelings, I have left them in the book to more accurately show the person that Moz and Jane were dealing with and, by so doing, to illustrate what a grand job they did. Jennifer, Bronwyn and Tegwen acted as the support crew on the first walk. They also deserve thanks. Without them, I would not have got through the first walk and without that first walk; there would have been no second walk.

Finally, Baz, mere words cannot say enough to convey how much I appreciate you. I may have been 'hard work' for you out there on the Track, but you were a joy during times when I felt no joy. Thanks for your loyalty. I am so glad you were out there with me and with the ghost of Henry.

FOREWORD

In the Footsteps of Henry Lawson

I am humping my bluey far out on the land,
And the prints of my bluchers sink deep in the sand.

- On the Wallaby, *When I was King and Other Verses*, 1905

I'd knocked about the Bush too long, and run against too many strange characters and things, to be surprised at anything much.

- The Babies in the Bush, *Joe Wilson and His Mates*, 1901

'How are you travellin', Gregga?' my brother asks.

'Good,' I lie.

I tilt my head further forward, looking only at the hard-baked ground beneath my butchered feet.

'Do you need anything, Greg?' Jane asks.

'No, thanks,' I lie again. However, I do need something. I need help. I need a way out of this mess that I have gotten myself into.

It is early on the second afternoon of this 15-day ordeal. I have set out with the ridiculous notion of walking from Bourke to Hungerford and then, as if that is not enough, my plan is to turn around and walk back—just like my favourite author, Henry Lawson, did in the summer of 1892-1893. I have only a day-and-a-half behind me and I have already realised I am in way over my head.

'Do you want a cold necktie?' Jane asks.

'No, I'm good.' Three lies within the space of a minute. I have to get out of here. I look only at the ground as I shuffle away from the others. I know that if one of them catches my eye, I am gone. I am losing my fight to keep my emotions intact and under control. I hide my face and the pain that must be deeply etched there. I avert my eyes from my three companions' worried looks, knowing that if they see the self-doubt hidden within my eyes that I am going to burst into a torrent of tears. I am on the edge of breaking down completely. To do so will end everything right here, where I stand, 50 km from Bourke and with 400 km of my planned trek not done. I must get away to be alone. I do not say goodbye. Fighting my pain, I concentrate on increasing the speed of my shuffled escape.

My tears are a millimetre below the surface. But I am not a crier. I can think of only one other time in recent memory where I broke down and cried. Back then, overwhelmed by the task of completing my graduate studies, I almost completely lost my mind and my grip on perspective. A broken mess, I desperately tried to hang on by a thread. But then when even that thread broke, with it, I broke too. I am on the edge of a similar breakdown now, standing at the precipice on broken, unsteady, blistered feet. I feel as if there is nothing but a black hole before me. I feel completely powerless and I see no way out. Here, in the Outback New South Wales emptiness, I recognise the same thread now. It is torn and tattered. It is as thin as paper. I cling to it like a madman, knowing that if the thread breaks, I break too and descend into madness.

'It is so hot,' I say, trying to divert my thoughts. The mercury is pushing beyond the mid-forties. It is about 115 degrees in the old Fahrenheit scale. 'One-hundred-and-something-scary in the shade,' Henry Lawson wrote of his time out here. The heat is something scary alright. Less than a week beforehand, I kissed my Canadian wife goodbye in temperatures of 20 below zero. My body is struggling with this rapid change in temperature—with acclimatisation to the furnaces of hell. My fellow Henry Lawson re-enactment walk tramps—my brother, Baz, and my old school mate, Moz—told me of a phone conversation they shared before I left Canada and returned home to Australia.

'If it was me, I would have wanted to spend some time in Australia beforehand,' Baz said, 'getting used to the heat.'

'Greg has never been one to make things easier for himself,' Moz responded. 'He has never been one to look for shortcuts or easier going.'

I am looking for an easier route now though. The Track is littered with large stones, some as big as the fists I clench against my pain. Trying to pick my way through the stones is tedious and dangerous. With a false step I will twist an ankle. With the wrong footfall the sharp edges of a stone digs deep into a blister, slicing at the tender skin like a knife through butter. I move to the roadside. It is sandier here, with less stones, and smaller ones at that. I know, however, that snakes lie in the roadside grass: Taipans—perhaps the deadliest snakes in the world. As much as I must be careful, I am too tired to think. I am too tired even to care, let alone to be careful.

I place one foot in front of the other. Left and then right. Left and then right. As difficult as it all is, I tell myself my journey is really just this simple: Left and then right. Left and then right. Left, right. Left, right. Left, right, left, right, left, right, left.

I feel myself starting to regain control of my emotions. I am a machine. Not thinking, not caring. Not feeling. I push on—on toward Hungerford.

I am committed to making it to Hungerford and back, come hell or high water. Come hell or high water. Hell or high water. Hell or...

DAY ONE

Henry Lawson's Legacy

I've fought it through the world since then,
And seen the best and worst,
But always in the lands of men
I held Australia first.
I wrote for her, I fought for her,
And when at last I lie,
Then who, to wear the wattle, has
A better right than I?

- The Wattle: No Better Right Than I, *Skyline Riders and Other Verses*, 1910

Still be noble in peace or war, raise the national spirit high;
And this be our watchword for evermore:—'For Australia—till we die.'

- 'For Australia,' *For Australia*, 1913

It is 3.45 in the morning. Unlike many people in Australia at this time, I am not awake because I am celebrating the advent of the New Year. It is the first morning of the year but I am awake because I am scared stiff about what lays before me.

I initially awoke at 3.00 am and, after getting my bearings and realising that I was not at home but, rather, in a room in the old Port O' Bourke Hotel, in Bourke, New South Wales, I wandered out of the room I am sharing with my brother and then down the hallway to the toilet.

Despite not getting to bed until 11.00 pm, here I am now, almost a full hour after awakening and still not able to get back to sleep. I am anxious and scared. I know the feelings. I have felt this way before. Back in my younger days, when I was filled like all silly youth with a misguided sense of immortality, I tackled a couple of ultra-marathons. Then, ten years ago in my early-thirties, I decided to lose weight and have another crack at an ultra distance. I remember my feelings on the morning of the run, knowing that I was about to embark on a world of hurt. I was terrified of the pain before me. Yet on that day, my pain was not going to extend beyond the twelve hours of the event. This morning however, I set out on a journey that will extend across fifteen days.

I lie in bed and watch the hands of my watch tick inexorably toward 4.00 am.

We are planning to leave at 6.00 am and I realise that, with what lays before me, having a good night's sleep would be a gift from the Heavens.

Yet time marches on and I do not sleep.

I have a recurring vision that dances across my mind like an eddy of dust on the barren Whim Plain that I must cross not once, but twice, to fulfil my goal. In my vision, the Devil delights in my suffering. I know that the Devil is out there on the Track waiting for me, as Henry Lawson would put it, to make me 'pay for my sins Out Back'.[1]

It is because of Henry, Australia's greatest writer, that I am here in Bourke. In the summer of 1892-1893, Henry hefted a swag onto his back and, together with his mate, Jim Gordon, and a Russian by the name of Ernest de Guinney, he strode out for distant Hungerford, approximately 220 km away. Upon arrival in Hungerford, Henry was decidedly unimpressed with what he found. Just a day later, he and his mate turned their heads back toward Bourke, completing the two-way journey in about five weeks.

The months during and immediately after his time in and around Bourke were among the most productive months of Henry's writing life.[2] However, for the remainder of his life his writing continued to be influenced and inspired by his experiences on the Track between Bourke and Hungerford, across the Queensland border.

When I was introduced to the work of Henry Lawson as a schoolboy, I fell under his spell. His writing immediately influenced my way of thinking and it influenced my way of looking at my country. As I was taught details of Henry's life, I learned about his tramp to Hungerford and back to Bourke. I decided then and there that one day I would like to walk in Henry's footsteps. As such, it seems like I have been waiting for this moment all of my life. Yet here I lay in bed, scared to death of what I am about to undertake.

I feel inadequate to the challenge.

I realise after more than an hour that I am not going back to sleep and so I rise from the bed and ruefully move back down the hallway toward the showers.

'Bugger it anyway,' I say to myself.

I will have two companions on the Track with me and another companion in a support vehicle. I know that while they have their own motivations for being here, they are principally here to assist me in fulfilling one of my life's dreams.

My old primary school mate, Sean 'Moz' Morahan is asleep in the first room that I pass on my way to the showers. At 6 foot 8 in the old scale (204 cm), he is a mountain of a man and, having tackled one way of the Track with him a year-and-a-half earlier, I know that he has a huge stride that can eat up the kilometres in a hurry. Eighteen months ago, after I successfully defended my Doctoral dissertation, Moz and I walked from Bourke to Hungerford. Despite my travails on that walk, it occurred to me before reaching Hungerford that Henry had made a round trip and he had also done it in the heat of the summer. Despite that blister-filled journey being physically the most difficult thing I had ever done, I realised what I really wanted to do was to do what Henry had done: I wanted to walk from Bourke to Hungerford and then turn around and walk all the way back to Bourke too. Furthermore, although mid-winter temperatures reached 30 degrees Celsius on my first tramp through the desert landscape, like Henry, I wanted to do it in summer. One thing that my one-way walk had taught me was that, with his long legs, Moz could walk a long way in a hurry.

My other companion is my older brother Barrie or, as I have called him all of my life, Baz. I remember telling the woman who became his wife that Baz was my hero. My own wife recently asked if I have ever been jealous of Baz and I told her that I have not. I have always been his biggest fan, happy to sit in the background celebrating his many achievements. I have always thought that, regardless of how good I am at something, Baz can do it better.

I know that he can do this Outback tramp too. In his mid-forties, he is in great shape. He and I flew into Sydney on the same morning—me from Canada and he from Adelaide. Moz picked me up from the international terminal first then we drove to the domestic terminal where Baz was waiting. I had not seen him since my last visit home a year-and-a-half earlier. As Moz and I drove up to where he waited by the kerbside, I was conscious of the fact that I had not lost those several kilograms that I planned to lose before this tramp. Baz, on the other hand, is in better shape than even I expected, without a spot of fat anywhere. Moz pulled to the kerbside and I opened the car door to step out and greet my brother. As I stepped from the car onto the footpath and went to straighten up, a bolt of pain shot through my body. My back has been giving me hell for a couple of weeks now. It comes and goes like this all the time—a couple of weeks of pain and then, if I am lucky, no more pain for a month or two. My back went out again a couple of weeks ago and 30-plus hours sitting on aeroplanes and waiting in uncomfortable airport lounges for connecting flights has done nothing to make my back feel any better. As I stepped from the car, Baz saw the pain dart across my face and my failure to straighten to a fully upright position.

'You're already buggered?' Baz joked as we shook hands in greeting.

'Oh, my back's stuffed,' I said. 'It comes and goes. It's mostly been coming lately.'

'You're off to a good start,' Baz said.

'It's gonna get worse before it gets better,' I prophesied, knowing that with the Lawson tramp coming, I was on a sure thing.

Jane Crowle is the other member of our team. She is Moz's life partner. She has sacrificed her annual leave to come out here to provide support to we three walkers. She will be driving Moz's car and checking on us from time to time to make sure that we have what we need.

As I step into the shower at the Port O' Bourke Hotel, I wish that, like Baz, Moz, and Jane, I was still sleeping. Baz will be fine. He is in great shape. He will eat up the kilometres without feeling it. Moz will be fine. Once he unwinds that long stride of his, the kilometres will disappear beneath his feet. My two fellow walkers will be fine. I am worried about myself. I know from last time—when I went only one way—this is going to be hell.

≈

It is almost 6.00 am when I enter the dining room for breakfast.

'Good afternoon,' old Candy Mick says as I enter the room. Mick is well into his eighties and he hobbles around like he is in his hundreds—or has just walked in from Hungerford. His hips are stuffed. Baz and I started calling him Candy after John Steinbeck's *Of Mice and Men* character of the same name. Steinbeck's Candy hobbled around the farm doing odd jobs to help out where he could. Mick does the same thing here in the Port O' Bourke pub, including laying out the bread and cereal with the breakfast dishes.

I pop some bread into the toaster.

'I thought you were leaving at five,' old Mick says.

I figure it is too complicated to explain that I never told him that. When we had chatted the day before, *he* had actually suggested that 5.00 would be a good time to start.

'We'll try to get away early,' I said in response to his enquiry about what time we would want breakfast. 'Before it gets too hot.'

'About five or so?' Candy Mick asked.

'Sometime early, hopefully.'

Moz wanted a 6.00 am start and I was onboard for that. Baz floated the idea of a later start, reasoning that, regardless of when we start, it is going to get hot. It is always hot in Bourke. Henry wrote 'that when the Bourke people die [and go to hell] they send back for their blankets.'

'We might be better to stay in our air-conditioned rooms as long as possible.' Baz made sense but Moz and I preferred the earlier start. We are anxious to get going. The sooner we start, the sooner we stop for the day. We have been out on the Track before and we know what we are in for. Our experience held sway and a 6.00 start was settled upon.

'Bloody rookie,' I said to Baz. It is my title for him out here. Anything that he says or does that I do not agree with I dismiss as a product of his inexperience. He is my older brother but he has never walked to Hungerford. I think now about his reasoning; a nice cool, air-conditioned room with a comfortable bed. A seven, eight, or even nine o'clock start. Bloody rookie.

I spread Vegemite across my toast and look out the window; the sun is bright already. We should be on the way but, having been awake since 3.00, I cannot believe I will be the last one ready to leave. Still, I will not be rushed. I will eat my breakfast peacefully, brush my teeth and then meet them outside—nice and leisurely. My aim is to keep cool despite the heat already scorching the windows.

≈

Even 80 years ago, a popular sentiment was that Henry's Australia—that country portrayed in his writing—was an Australia of the past, an Australia of memory. Were that all, Henry would still have done his country an inestimable service.[3] He recorded for posterity Australia as she was through the dark Depression days of the 1890s, the dreamy days of gold rushes to Western Australia, the days of turn-of-the-Century excitement, the pride-filled days of Federation, and those heady days and militant tones of a world rumbling towards war.

Yet, Henry's service to his country was not merely to record the way that she was. Writing at a time when Australians still looked longingly and for approval towards Mother England, Henry gave us our own Mother Country. He gave us not just the right, but the reason, to take pride in ourselves and our country and, as the grateful Miles Franklin said of her mentor, 'Henry Lawson gave us this kingdom for our own [and] wove it so that we could fold it around us with the comfort of a blanket.'[4]

Through Henry, *mateship* became 'the watchword of Australia'—a principle to guide our behaviour and our treatment of one another. 'She'll be right, mate,' we say, and we mean it, for we will watch out for each other and provide whatever assistance it is that our mate requires. Described as 'the apostle of mateship,'[5] Henry left a legacy of mateship and courage in the face of difficulties that continues to instruct and inspire Australians almost 100 years after his death. For these things, he is deserving of the lavish praise heaped upon him by one contemporary: 'He is the greatest gift the gods have so far given to Australia.' Unfortunately, many Australians no longer read Henry. I read him constantly and I love what he wrote. It is what has inspired me to be out here at Bourke, ready to walk through hell.

≈

As we stand on the veranda of the Port O' Bourke Hotel, we are a mixture of obvious excitement and nervous energy. This pub stood at the time of Henry's visit to Bourke, albeit under the name, Royal Hotel. As one of the town's main Union pubs, there is little doubt that Henry would have been a frequent visitor.

I take the opportunity to express my gratitude to my companions and to place an apology in advance. 'Before we start, I just want to thank the three of you for being here,' I say. 'I know that over the next couple of weeks, I am not going to say very much and so I want to say "thank you" right now, while I still feel like talking.'

Given my experiences of tackling the Track 18 months earlier, I am conscious of the fact that the difficulties will likely see me withdraw behind a shield to try to shut things out, including the people around me. My previous tramp with Moz was blister-filled hell. 'Nothing can prepare you for this trek,' I warned Baz. 'Imagine the worst it could possibly be, and then multiply that by ten!'

Before we begin, Moz, Baz, and I line up for a photograph outside the pub. We want to preserve an image of how we look at the start. We are all freshly showered and clean, knowing that we will soon be sweating beneath the sun and that the toil of the next several days will leave us dirty and grimy.

After the photos, it is time to begin. It is 6.27 am, 1 January 2011. We are on our way.

I run a few steps ahead.

'There, now I've at least temporarily been out in front,' I joke. I expect that with Baz's fitness and Moz's huge stride, my time ahead of the others will be limited. Jane has decided to walk out to the bridge with us too and she also looks fit, so there will be no relief for me there either.

Initially, there is lots of chatter as our excitement bubbles at finally starting this big adventure. Moz and I are two of just four people ever to successfully complete the one-way re-enactment walk. No one has ever retraced Henry's entire Bourke-to-Hungerford-and-back trek. After many months of build up and preparation, the journey is finally underway.

≈

Over the initial kilometres, the four of us generally walk together. I ask Moz to take a photo of me as I walk past the site of the old Carriers' Arms Hotel—the setting for some of Henry's best Bourke-inspired stories. As I walk by the now decrepit building, I imagine thirsty drinkers sitting on the veranda alongside the former hotel proprietor, Watty Braithwaite, 'known as "Watty Broadweight," or, more familiarly, "Watty Bothways".' Henry said, 'Watty was considered the most hopeless publican and his customers the hardest crowd of boozers in Bourke.' A large, round figure, Watty used to settle into a comfortable chair on the pub veranda to snooze and be entertained by his increasingly drunken clientele. The Salvation Army would preach and bang their drums and sing outside the hotel while Watty sat with 'an indulgent or fatherly expression on his fat and usually emotionless face. And by and by he'd move his head gently and doze. The banging and the singing seemed to soothe him, and the praying, which was often very personal, never seemed to disturb him in the least.'

For action and excitement, Watty's pub was the place to be. 'Most things that happened in Bourke happened at Watty's pub, or near it.' Henry said:

> *If a horse bolted with a buggy or cart, he was generally stopped outside Watty's, which seemed to suggest, as Mitchell said, that most of the heroes drank at Watty's...there was something in Watty's beer which made men argue fluently, and the best fights came off in his backyard. Watty's dogs were the most quarrelsome in town, and there was a dog-fight there every other evening, followed as often as not by a man-fight. If a bushman's horse ran away with him the chances were that he'd be thrown on to Watty's veranda, if he wasn't pitched into the bar; and victims of accidents, and sick, hard-up shearers, were generally carried to Watty's pub.*

I imagine that, if such things exist, the ghosts of Watty and other characters who appear in Henry's Bourke-based poems and stories—the likes of the Giraffe, One-Eyed Bogan, Barcoo Rot, Gentleman Once, and Jack Mitchell—are interested to see us pass by. I expect Henry's ghost comes in for some good-natured ribbing about we fools and our desire to walk where he walked.

I notice a man in a green t-shirt behind us. He is walking in our direction. 'There's some silly bugger behind us who looks like he is planning to walk to Hungerford,' I joke.

Our spirits are upbeat and positive; however, it is impossible to ignore the fact that, despite not yet being 7.00 am, the sun already carries a sting. I notice the sweat soaking through Moz's blue shirt as we pass the Exhibition Centre and head into the mythical Back o' Bourke—the great Outback heart of the country. Still today, people talk of going Outback to find oneself in the emptiness. 'Yer wanter go out back, young man, if yer wanter see the country. Yer wanter get away from the [railway] line,' Henry was told. But as Henry found, there is always somewhere further out. 'You could go to the brink of eternity so far as Australia is concerned and yet meet an animated mummy of a swagman who will talk of going "out back".'

I come across a colourful Eastern Ringneck parrot dead by the roadside. I stop to take a photograph, knowing from my last tramp that it will be merely the first of many dead birds and animals that I will see on the Track. We will be traipsing through harsh country, where Death hunts greedily at every turn.

We reach the Polygonum Swamp, approaching the bridge across the Darling River, and are shocked to see the water level. Hundreds of trees stand in the still blue water, their thick foliage and strong branches a testament to the fact that their trunks have not long been submerged. Henry said that, depending on the season, the Darling River can be described as 'either a muddy gutter or a second Mississippi'. With the floodwaters coming down from Queensland, there is no doubt that, today, the Darling is no muddy gutter. When Henry saw it, the Darling 'was a narrow streak of mud between ashen banks, with a barge bogged in it.'

As we approach the bridge, a young couple from Ningaling Station, near Hungerford, pass us as they drive toward Bourke. They see us but do not stop. I later learn that Will, the driver, turns to his girlfriend, Narelle, and says, 'Look at these mad bastards walking home from the pub'. Being early in the morning on New Year's Day, Will and Narelle assume we have been partying hard. After all, that is what I assume about the bloke behind us in the green t-shirt. He shadows us from the distance of about a hundred metres until, about five kilometres out of town; I do not see any more of him. Perhaps, he is just like Henry's horseman 'who looked like a drover just returned from a big trip, [who] dropped into our dusty wake and followed us a few hundred yards' until we lost our shadow when 'a friend made wild and demonstrative signals from a hotel veranda—hooking at the air in front with his right hand and jobbing his left thumb over his shoulder in the direction of the bar—so the drover hauled off and didn't catch up to us any more.'

Crossing the bridge, we see a long snake squashed on the road. Henry was said to have 'a holy horror of snakes.' As fascinating as they are, I am not keen on snakes either. The morning before, while having breakfast alongside Candy Mick, I sought reassurance about the danger of running into trouble from snakes.

'Have you been to the jetty to see the river?' Candy Mick asked.

'Yeah, we had a look yesterday,' I replied. 'Gee, the water's high.'

'It's high alright.'

'There were lots of frogs,' I said, recalling the loud croaking. There must have been thousands of frogs hidden in the shadows.

'That's right,' Mick replied.

'Probably a few snakes too?' I posed it as a question, cautiously venturing forward.

'That's for bloody sure!' Mick responded, leaving no room for doubt.

After a short while, Candy Mick and I were joined by a wheat harvester who was working long hours in the fields.

'Did you see any snakes out there?' I asked.

'Yeah. I've seen *lots* of them.'

It was time for me to stop asking about snakes.

≈

At 7.30 am, we cross the road and walk toward the old North Bourke Bridge. Built in 1883, it is now just a walking bridge but, as the first built across the Darling River, at one time travelling stock and wagons overflowing with wool would have crossed here constantly. This is the same bridge Henry and his companions crossed as they set out toward Hungerford in late December 1892.

I am under the impression that Jane is turning back at the bridge and so I wait for her to say her goodbyes. When none are forthcoming, I take the initiative, say goodbye and push on. There is much ground to cover today and I am anxious to keep moving.

Across the bridge I pass the Bourke Bridge Inn and, as I am now in front, I am the first to see a Bearded Dragon sunning itself on the grass. I have never seen a Bearded Dragon before and, surprised that it is being so compliant; I take out my camera and snap a few photos. The lizard seems comfortable with me getting close-ups and it is only with seeming reluctance that it eventually decides that the paparazzi is too intrusive and moves to a more secluded spot.

After Jane begins her walk back to the pub and Moz's car, Moz, Baz and I push forward. My eyes are drawn skyward when I hear the raucous nasal screech of cockatoos. I watch as several Red-tailed Black-Cockatoos pass overhead with their slow, lazy flight. Like us, they seem desirous to conserve energy in the early morning heat.

We have a long, flat, straight stretch of road before us as we negotiate Walkdens Plain and then Seventeen Mile Plain. The plains are devoid of trees but with the rain and sunshine they have had out here, Mitchell grasses stand thick and tall in the paddocks. The bitumen road is hot and hard on the feet. I would prefer not to walk on the road but the roadside is thick with grass, even extending above knee height. We see hundreds and hundreds of grasshoppers bouncing in the grass. I cannot help but wonder about the number of snakes that might be hidden there too. As a result, I stick to the hard but clear road surface.

'How far does the bloody bitumen go?' Baz wants to know. 'I didn't come out here for this. I can walk on a road at home.'

Moz and I begin to tell Baz about what we call the red carpet—the soft, fine red sand that starts up ahead about 30 or 40 km out of

Bourke. We will be off the bitumen before then but until we reach the red carpet the road will either be sealed or will consist of gravel or large, sharp rocks.

We continue to press forward, the sun growling louder. Moz often takes the lead while Baz and I walk together, me occasionally breaking the silence to clear my throat. I wish I was feeling better. I have been fighting with illness for weeks now and coughing up phlegm.

We watch as Moz stops ahead of us and bends from the waist as if he is about to throw up. Baz and I wonder what he is doing. After a while, we see Moz do the same thing again and, a short time later, again. He does not appear to vomit but he looks as if he is about to.

'Moz is struggling,' Baz says.

'Nah. He'll be alright,' I reply, thinking back to the way Moz had covered the distance in 2009. He surprised me with his strength and endurance then. He is made of the right stuff.

As Moz continues to stop and bend over, hands on knees, Baz becomes more concerned. 'He's struggling, Gregga. He's struggling.'

'He'll be okay. He's a strong bugger.' I am blinded by my belief in Moz's ability to keep going and so I dismiss Baz's concerns. A little after 9.30, we reach the first water stop. The first 15 km are behind us. With our trowels in hand, Baz and I dig up our buried water. We spent New Year's Eve burying water and food every 10 or 15 km along the Track. Digging up the water today is no easy task under the blazing sun. We are soon sweating profusely. There is no shade out here on the treeless plain and I wonder why we buried the water so deeply. We are, however, rewarded when the buried water with which we replenish our hydration pack bladders is noticeably cooler than the water we were carrying. The cooler the water on the shadeless plain is, the better.

We rebury what water we do not need and leave the first water stop. We see trees in the distance far in front of us, beyond Seventeen Mile Plain. I long for some shade to provide respite from the angry sun.

'I've been doing some calculations,' Baz says, 'regarding the amount of time we have been walking and the amount of time it might take us to walk the whole distance at a rate of about five kilometres per hour. We have completed about three percent of the journey.'

I find this announcement uplifting because it seems like a sizeable bite from the overall task. Eating an elephant may be accomplished one bite at a time, but we have already gobbled up three percent of the beast.

The flies are thick. There are hundreds of them hitching a ride with me but, fortunately, they seem content to crawl around on my hat and my back and generally stay out of my face. It is only when I stop and turn my head to look behind me that they cloud before my eyes.

A breeze picks up and, with no trees to block its path, it offers some partial respite from the heat. How we long for the shady trees in the distance though.

≈

Having been born and raised in the Bush, Henry's experiences endowed him with an appreciation of the difficulties of life in the country. Although he did not always enjoy his time in the Bush, he did admire the people. Henry predicted that Australia one day would have to fight to suppress foreign invaders. He used to say to his brother-in-law that it would be the men from the Bush who would save the country.[6]

In terms of beauty and splendour, Henry did not always have glowing tributes to write about this country out here, because really, at every turn, what it is trying to do is to kill you. What Henry did write about in glowing terms were the people who survived and, indeed, occasionally thrived despite the harshness. They are the people Henry praised and to whom Henry sang tribute. As one friend said, Henry 'truly did understand and love his people...Lawson is, and ever will be, Australian of the Australians.' Another contemporary described Henry as 'the first articulate voice of our nationhood'—the one who helped us to see and appreciate who and what we are as people and to be proud of that—to take pride in being Australian. In shaping our identity as a unique People, Henry helped us to recognise our own self-worth and, for that, rather than fading like an old photograph, our indebtedness to Henry grows with each passing year.[7]

≈

It is a matter of walk, walk, walk.

The trees seem never to draw nearer.

The sun climbs higher in the sky and its intensity is incredible. Just a week ago, I was in Canada and the temperature was around minus 20 degrees Celsius. Out here in the sun, with no shade whatsoever on the treeless plain, the mercury races towards its high point for the day of 47 degrees. I have never felt so much heat. I wish that, like Baz, I could be in shorts. Having just emerged from the Canadian snow though, I know that I will burn like paper. I am wearing long tracksuit pants and a long-sleeved shirt. Beneath my Canadian Tilley hat I have draped a bandana, extending down over my neck and ears. I know that sunburn will stop me in my tracks and so I must stay protected.

The bitumen eventually gives way to a gravelly surface baked hard in the sun. The pebbles and rocks crunch beneath our feet and we sound like an army on the march. My feet are tiring and I feel the first suggestion of blisters. Blisters were the bane of my existence on the tramp eighteen months ago. I am desperately hoping to avoid them. Last time, in the quest for authenticity, I wore a pair of elastic-sided boots. I figure that Henry probably wore elastic-sides himself. By the time I reached Hungerford, my feet were bloated and distorted by blisters and that one-way journey had been the most physically demanding task I had ever undertaken. My wife, Jennifer, quite rightly put much of my suffering down to my blisters and, hence, to my footwear. She believes that things will be better this time around with, as she put it before I left Canada, 'a sensible pair of shoes.'

I bought a new pair of hiking shoes while I was in Canada. They are Merrell shoes and have an air cushion sole. I am under the impression that I have sufficiently worn them in over the months preceding this tramp. At the same time, I want them to retain that cushy new shoe feel too, and so I have not worn them too much. I thought what I had as I left Bourke this morning was a nice balance of comfortable, cushioned new shoes that had been worn in to such a degree that blisters were not likely to be a problem. I was wrong.

My shoes are rubbing on the inside of the pads of each heel. I had blisters there last time too. In order to reduce the risk of blisters, I spent more than an hour after my shower this morning powdering my feet and applying bandaids and bandages. Recalling the spots that had been particularly troublesome last time, I applied moleskin adhesive bandages as a preventative measure. I also put in place a couple of small rubber 'sleeves' over each of my little toes. Those two toes were butchered last time, rubbed red raw before eventually becoming infected. As I walk along reflecting on my feet, the toe sleeves at least seem to be working well. My little toes are giving me no trouble at all, but I am bothered by what seems to be forming on the inside of my heels and the general foot sore feeling that I have as I walk on the hard road.

Moz is in front and at 10.45am, he reaches three or four spindly little gums and decides that will be a good place to rest. Baz pulls in next to him and by the time I get there my first thought is that there is bugger all shade left for me. I find a spot that seems to offer a little respite from the direct glare of the sun but there are burrs and prickles aplenty.

'Prick of a spot,' I say to myself, but beggars can hardly be choosers.

I throw down my swag and lie on the prickly ground, using my swag as a pillow to cushion my head. I take off my hat to try to cool down a little but I leave my bandana draped across my face. Moz is on his back but his head and shoulders are propped on his backpack which is, in turn, propped against the trunk of a small tree. His knees are bent up in front of him as if he is trying to fold himself into an envelope. He looks anything but comfortable. His face reminds me of a Muscovy drake, all red and puffy. He looks as hot as hell and, frankly, he looks exhausted.

'I wonder what was going through Custer's mind when he realised that he'd led his men into a slaughter,' I think to myself, recalling one of Mel Gibson's lines from the movie, *We Were Soldiers*.

Baz and Moz would not be out here if it were not for me. I am too buggered to say anything out loud and, anyway, I know that if I do, Baz will respond with, 'Custer was a pussy.'

We rest for almost an hour before we decide to move on. The high sun is noticeably hotter as we leave what shade the trees offer. All three of us seem to have lost a lot of energy during our stop. Seventeen Mile Plain is as flat and devoid of trees as its predecessor, Walkdens Plain.

The difference is that the sun is now higher and hotter. My feet feel worse for the rest and my blisters are more painful.

We see kangaroos on both sides of the Track.

The track that seems arisen up
Or else seems gently slopin',
And just a hint of kangaroos
Way out across the open.

Some kangaroos pause and gaze at us curiously. Others immediately turn and flee, as if embarrassed to be in any way associated with what they perceive to be a fiasco. 'What are these fools doing out here?' they seem to ask.

Baz and I see many Emus on the plain too. Moz comments on our eyesight. Invariably, we spot the Emus before he does. It becomes almost an unspoken contest between Baz and me to see which of us can be the first to spot the next Emu. The more tired I get however, the more tired of the game I become.

By noon, Moz is out in front again and Baz is a little ahead of me. Baz pushes out a fart and leaves it behind for me to walk through. He will be 45 in a few days but as he looks back at me walking through his flatulence he still has the cheeky grin of a 12 year old. It is the same grin that Henry might have had in mind when he commented that 'the average Australian boy is a cheeky brat with a leaning towards larrikinism.' And flatulence, I might add, Henry.

The half an hour since leaving our rest stop has been a tough one. Still, we figured we had 21 km behind us before we stopped and so we are three-fifths done for the day. Only 14 more kilometres to go and we can call the day over. Fourteen kilometres is a bloody long way when the temperature is 47 degrees though.

The grey gravel road gives way to the first of the red sand. The Track is still rocky but the walking is now easier on the feet. The sun is almost directly overhead and the shadow that I cast does not even extend beyond my footfalls.

We come to a cattle grid across the Track. 'These things get to be a pain in the arse when you have a few days behind you and you're buggered,' I tell Baz as we negotiate this one with relative ease.

We notice that Moz starts to struggle again, bending over presumably either to stretch out his back or to dry retch. In this heat, we are all struggling and so I do not bother to ask him about what he is doing. Baz starts to move ahead of Moz. He must be cooler in his shorts than Moz and me in long pants. I only hope he is sensible enough to make sure he does not get sunburnt.

It is unbelievably hot when, mid-afternoon, I catch up to Baz, seemingly asleep under the shade of a big tree. He has a towel spread beneath him in the dirt. I am happier with this spot than our morning rest stop. This one is shadier and clearer and so I throw down my swag and backpack and join Baz. I take off my shoes and socks and see I have two nasty blisters forming and, as I expected, they are right at one of the spots I had bandaged this morning. Much to my frustration, the bandages have not held in place. It is annoying for me to know that I picked the right spot but that I was let down by failure of the bandage to adhere. 'Then again,' I think as I look at my sweaty feet, 'I would almost have had to nail it in place for the bandage to stay there under these conditions.'

One who has not carried a load in 47-degree heat cannot understand just how hot I feel. As Henry said, 'No one who has not been there can realise the awful desolation of Out Back.' Were fire-breathing dragons to confront me with fanged jaws and roar in my face, I could not feel hotter.

It has been a tough couple of hours and Moz is in a bad way when he joins us. When Jane appears in the car, Moz goes to the vehicle and sits inside with the air-conditioner running. He remains there for 40 minutes or more. He invites me to join him but I do not feel right about using the car in that way. Furthermore, I think it is a bad sign for Moz. Using the car to cool down seems to me to be an admission that he is in trouble and cannot stand it out here. Mentally, I think it is a knock to him and, while it is cooling him down, I see it as an admission of weakness.

Jane hands me something to eat and I scoop some tuna out of a tin with rice crackers. It goes down well, despite the flies' determination to share my lunch. The juice from the tuna oozes onto my hands and I feel sticky and dirty but I know this is just the beginning. I will get worse before I get better.

I lay down and I try to sleep but the heat will not let me. All the same, it feels good to be off my feet.

Baz begins to stir a little after 3.30 and slowly and reluctantly I climb back on my feet. As we fill our water bottles and backpack hydration bladders with water from the back of the car, Baz sees an Emu close in front of the car.

'There's an Emu,' he says.

I am filling my water. That is my task for now. In this heat, I do not have the energy for anything else.

'Look, Greg, he's right there,' says Baz, pointing.

I am too tired to tell him I do not care.

'Look, Greg. See how close he is?'

I am too tired to tell him I have reached the stage where it is just another bloody Emu.

When Baz sees that I am not going to lift my head, he gives up on me and moves off in the direction of the Emu.

We rested for more than two hours but it seems likely this will be necessary to avoid the worst of the sun during the early hours of each afternoon. I find it hard to get going again though. There are still nine kilometres to go before we reach the place we chose the day before as our first camping spot. It is going to be a tough slog for me to get there on this hateful, spiteful Track.

I catch up to Baz, feeling guilty about ignoring him while he had tried to point out the Emu.

'There was an Emu back there, eh?' I ask.

'Yeah,' says Baz, 'it was right in front of the car—just a few metres away.'

The day before was New Year's Eve and I had asked Moz about the fireworks show they put on where he lives in Sydney. 'I don't know what you think,' Moz had said, 'but I think if you've seen one firework, you've seen them all.'

I think about Baz's Emu and I turn to him and say, 'Seen one Emu, seen 'em all.' I do not believe it for a minute. It is not the way that I look at the world but I want Baz to be aware that I am not going to have the energy to look at every Emu that crosses our path.

≈

Living overseas as I do might be one reason my love of Henry's writing grows and grows. 'For an Australian to read him in another country,' one literary analyst wrote, 'is to breathe the air of home.'[8] I inhale a long, deep breath. As much as I am struggling, it is nice to again be able to breathe Australian air.

≈

Moz has rebounded well and he is again in front. On the voice recorder that hangs on my chest from a strap on my backpack, I record a description of Moz as a trooper and a fighter. He looked terrible this morning but he is now moving well.

The unsealed road is dry and dusty and consists mostly of fine red sand, covered in places with loose rocks. Two vehicles drive by and kick up enormous clouds of dust that envelope me completely. I cannot see Baz in front of me or Moz in front of him. I am reminded of my dream of three nights earlier—my first night back in Australia. In my dream, I was stuck in a thick fog and unable to find my way out. I was lost and alone and helpless. In the dust and heat today, I feel that way again.

The road goes back to sealed bitumen and I curse Mans' determination for so-called progress. On foot, an unsealed road is a much kinder surface than the black top bitumen. The heat radiates through my shoes and the soles of my feet cook. The wispy cirrus clouds low on the horizon behind me do nothing to block the sun, which has by now passed to the other side of my body but still blazes like a furnace. I keep forcing myself to drink. I cannot allow my body to dehydrate.

I am pleased to find Moz and Baz resting up ahead. Baz stands in some shade by the roadside but Moz is prostrate in the long grass. I notice a channel of water in the grass and move across to soak my bandana. Dozens of unseen frogs croak to one another, conveying the news of my arrival. The grass is almost to my thigh. With the long grass, the heat, the water, and the frogs, it is a dangerous place to stop. It must be full of snakes but I am too exhausted to care. I throw my pack and swag in the grass and collapse on top of them.

'That's a hell of a snaky spot you've got,' Baz says.

'I'm just hoping that any self-respecting snake will see us as such easy prey that they'll see no challenge in it for them to bite us.' I am more hopeful than realistic but too tired to move or to care. It is only the first day and already I am worn 'dirty and careless and old' and, as much as I try to resist it, 'my lamp of hope' has gone dim.

Moz rouses enough to sing a ditty. 'We're here. We're queer. We don't want any more snakes!' It is an adulteration of a line from the television programme, *The Simpsons*, and he uses it to let the snakes know of our presence and, hopefully, to encourage any nearby snakes to move away.

I wring my bandana out and let the refreshing water dribble onto my neck.

Baz rests for a moment but does not join us in the long grass. I tell him that he need not feel obligated to wait with us. If it is best for him, he should push on at his own pace. Baz checks that we are okay and asks if there is anything we need. He gives us both a refreshing squirt in the face from his hydration bladder.

Baz leaves Moz and me in the grass, flat on our backs, roasting despite the dappled shade. As he leaves, Baz drops a laminated piece of red paper onto my chest. On it, there is a type-written quote from Shakespeare's Saint Crispin's Day speech. I read it to Moz: 'We few, we happy few, we band of brothers; for he today that sheds his blood with me shall be my brother.'

There is no doubt that there is a brotherhood out here that extends beyond just me and Baz. It also embraces Moz and, although not with us right now, even Jane as well. We are a team all working toward the goal of getting three of us to Hungerford and back.

Moz and I lie in the grass for about forty minutes. As I lie there, I am already questioning the wisdom of what I have gotten myself in for. It is still Day One. How can I possibly survive 15 days of this? I cannot see or hear Moz in the long grass. He is lying somewhere above my head. For all I know, he might be dead. How can *he* possibly survive 15 days of this?

We are in a mess. We have only managed about half an hour of travel since our long afternoon stop and here we are again, flat on our backs melting in the heat.

I hear a car approaching. In my confused state, it seems to me that the sound is coming from the Hungerford direction. Fortunately, Moz correctly discerns the noise to be coming from the direction of Bourke. Jane had driven back into Bourke and Moz perhaps recognises the sound of his own vehicle. He pops up from the grass so Jane sees us and does not drive past. She has with her some Powerade drinks.

'Thank you, Jane.'

The cold drink tastes wonderful as it slides down my parched throat.

≈

Before leaving us, Jane says we have less than four kilometres to go until we reach our first night camp. There is no way to know how far ahead Baz might be. Even when I am back on the Track, I cannot see him in the distance. I suspect he might already be at the campsite. He is a pocket rocket of energy and he is travelling very well. I later learn that he fails to recognise the campsite and continues past. It is not until Jane catches up to him in the vehicle that he learns he has walked about a kilometre-and-a-half past the spot we had picked to stop for the night. Seeing as we have already dug a fire pit, collected firewood, and buried water the day before, he must turn back. When he tells me later that he overshot the campsite, I am so exhausted that I say to him, 'If I walked three kilometres extra, I'd bloody well shoot myself.'

Right now, the thought of going one step further than I have to is not even comprehensible. I just try to keep plugging away and not let Moz get too far from me. Fortunately, Moz is only travelling in short bursts. He goes for a kilometre or so and then stops and rests and allows me to catch up. Then we push on again and he strides out in front and then stops and rests again and allows me to catch him. It is slow progress but progress nonetheless.

Rather than cooling down in the evening, it seems to be getting hotter. I am ever so glad when I see Moz and Jane's tents set up ahead.

At 6.45 pm, Moz and I stagger into camp. Day One is over. Thank God for that. Thirty-five kilometres are behind us. As exhausted as I am, even I can raise a smile at that.

≈

The sun is a long time in going down and, even when it does, the temperature seems not to drop. As darkness descends, Moz and Jane are in separate tents while Baz and I are on our blankets beneath the stars. My swag consists of two blankets—one of which I have spread over as clear and prickle-free a spot as I can find. I am aware that with the heat, I will not need the other blanket for warmth but I figure it might be necessary to keep the mosquitoes off me.

Boy, am I right.

The mozzies arrive in droves. The constant buzzing around my head drives me crazy. The mozzies are so loud they sound like huge squadrons of helicopters, descending to feed from me. Yet, when I pull the blanket over my body, I cook. It is so hot that I cannot bear to have the blanket over me but the mozzies are so hungry that I am losing my mind.

When Henry was out here, he said he 'spent most of [the] night hunting round in the dark and feeling on the ground for camel and horse droppings with which to build fires and make smoke round our camp to keep off the mosquitoes.' As I lay in the darkness, it is not just because of the absence of camel and horse droppings that I have an overwhelming sense of being poorly equipped for this task. My blanket is too hot. My shoes give me blisters. My bandaids and bandages are not doing the preventative job I had hoped for. There is no way to escape the mosquitoes. I have no insect repellent. I am feeling overweight and out of shape. I am so tired. My body craves sleep. I watch the hands of my watch climb towards 10.00 and then 11.00 pm. Then it is midnight. Then it is 1.00 am. Now it is 2.00 am. I am still wide-awake.

I try to distract myself from the constant drone of the mosquitoes. The blanket of stars overhead is awe-inspiring. I think it is the most stars that I have ever seen. I pick out the Southern Cross and, knowing that he is also still awake, I ask Baz for confirmation that I have the right cluster.

I cannot stop urinating. I also am too tired to rise and hobble to a more discreet spot away from our camp. With Jane in her tent, I hope she cannot hear me peeing as I kneel on my blanket and pee to one side. I do this over and over again—seven or eight times in four or five hours of trying to get to sleep. I console myself in believing that at least my peeing seems to be an indicator that I am not dehydrated.

Earlier, I noticed Jane had some mosquito repellent. I suspect she is asleep but I am going crazy in the storm of mosquitoes.

'Jane,' I say into the darkness.

'Yes?' she answers. I cannot tell if she was asleep or not.

'Do you still have that insect spray?'

'It's here in the tent,' she tells me. Despite the hour, she seems not to be irritated.

I am too buggered to stand up and move away to pee but, at this point, I would walk a hundred kilometres to get insect spray. I haul myself onto unsteady feet and slide them into my thongs. I hobble towards Jane's tent and she kindly passes the repellent to me.

'Thanks a lot,' I say and I mean it as much as any thanks I have ever given. I spray myself all over and give an extra spray on and around my head.

Baz is listening closely. Like me, he is bombarded by mosquitoes.

'Can I have some of that, Gregga?' He seems almost to be begging. I know how he feels. I toss the spray bottle toward him and I hear him drowning himself in bug spray.

Another hour passes. The repellent keeps the mozzies from landing on me. But they still fly in great swarms about me, searching for a way to my blood. The noise of their buzzing does not stop for a second.

I hear Baz's voice in the darkness. 'Moz...is the car locked?'

'Yes.' Moz sounds groggy. I am certain he was asleep.

'Where are the keys?' Baz sounds desperate.

'In my pocket,' Moz replies.

'Can I have them?'

'What are you doing, Baz?' I ask.

'The mozzies are driving me crazy. I'm getting in the car.'

It is 3.00 am and I have not slept a wink. What a horrible end to a horrible first day. What a horrible start to what I can only expect will be a horrible second day.

If sheltering in the car is good enough for Baz, it is good enough for me. 'Wait up, Baz. I'm coming too.'

DAY TWO

Henry Lawson's Parents

I am an Australian Bushman (with city experience)— a rover, of course, and a ne'er-do-well, I suppose.
I was born with brains and a thin skin—worse luck!

- The Babies in the Bush, *Joe Wilson and His Mates*, 1901

It is when the sun goes down on the dark bed of the lonely Bush, and the sunset flashes like a sea of fire and then fades, and then glows out again, like a bank of coals, and then burns away to ashes—it is then that old things come home to one. And strange, new-old things too, that haunt and depress you terribly, and that you can't understand.

- 'Water Them Geraniums', *Joe Wilson and His Mates*, 1901

Sitting in the front passenger's seat of the car, I still have difficulty falling asleep because I am unable to stretch out. At least I have eluded the army of mosquitoes. Baz is snoring beside me and silently I curse him. Considerably later, I too eventually drift off to a shallow sleep of only an hour to an hour-and-a-half's duration.

I awaken at 6.00 am and know that I must get moving. We have another 33 km to tramp to reach Fords Bridge. Oh, how I long for the Warrego Hotel at Fords Bridge and the bed that awaits.

I prepare my feet and then cover any exposed skin with sunscreen. I fill my water bladder and two spare water bottles and squeeze them into the pockets of my backpack. I am only dimly aware of the others as they begin to stir. I am caught up in my own dread of what the day might bring.

It is 7.10 when I step onto the Track, going only a few paces before I hack up a colourful wad of phlegm. The sun is already high and hot. So far, there is only one blister bothering me on each foot, but they are major blisters. The inside of the ball of each heel is agony and as I take my first, uncertain steps for the day, my feet scream in pain. Baz takes the lead early on. Three Emus race up the road in front of him.

Although Henry was born at the small New South Wales town of Grenfell, at the start of the gold rush that lured his parents to the area, the place was originally known as Emu Creek. The settlement was renamed in honour of John Grenfell, a Gold Commissioner who was shot by bushrangers and killed in the year before Henry's birth. Henry was born on 17 June 1867 and legend has it that he was born in a tent on the diggings during a flooding storm.

The sky is a beautiful clear blue this morning. When I am not at home I miss the blue of the Australian sky. There is no sky like it—no colour like it—anywhere else in the world. It is a blue that I imagine a painter could spend a lifetime trying to create, only to die frustrated and disappointed. 'I cannot do it. It is beyond me,' I imagine even Leonardo or Michelangelo saying. I wonder at my own limitations and what things are beyond me. Van Gogh cut off his ear. I do not think that taking a razor blade to my feet will help me out of this situation but it might not make things much worse either.

I extract from my pocket the two portraits of Henry Lawson that will accompany me for every step of the journey. One portrait is a copy of a photograph of Henry taken in 1893, later in the year that he completed his Outback trek. Although the photo was taken in New Zealand, I expect that it is a good reflection of what Henry's appearance would have been as he set out from Bourke en route to Hungerford. The other portrait is actually a series of eight photographs believed to have been taken in 1915 by William Johnson. Henry would have been a little older than I am now. I am 43 years old. In the photos Henry would be 47 or 48 years of age. Using what was termed a 'multiplying camera,' Johnson was able to take eight separate portraits of Henry on the one negative. Although the technology is beyond me, I like the different moods reflected in the images. I have long considered this series of photographs to be my favourite Lawson portrait. Henry is laughing in some of the images and, despite his troubled life; I like to think of Henry in this cheerful mood.

As I look at the portraits, I wonder what Henry would make of me walking along the dusty Track in his footprints. Perhaps he would simply ask, 'Did you not read what I wrote?' Enough warning flowed from Henry's pen for anyone to know that this trek is one to be avoided. 'It's not glorious and grand and free to be on the track,' Henry wrote. Then again, he did follow that sentence with the words, 'try it'. I am trying it and, like Henry, I am hating it. I expect that Henry would be surprised and more than a little amused that somebody would be inspired to go to all of this trouble in honour of him.

My hands are swollen and my fingers are tingly. Baz is having the same problem. We suspect it is a product of our backpacks and the straps inhibiting our circulation. As a result of my swollen hands and wrists, I dangle my watch from one of the straps of my pack. It keeps the watch in a position from which it is easy for me to keep track of the passage of time, yet is not uncomfortable on my wrist. I also have my camera in its case and my voice recorder device in its case dangling from the shoulder straps of my pack. It is comfortable and makes things easily accessible, but with the straps of my backpack, plus the straps to the camera and the voice recorder, I have more 'straps' than an old-time Marist Brothers school.

≈

Henry's father was Niels Hertzberg Larsen. Niels went by the name, Peter. The surname was anglicised to Lawson some time after Peter left his home in Norway at the age of twenty-one. Peter was a sailor who arrived in Melbourne during the peak of the 1850s Victorian gold rush. Caught up in gold fever, he 'wandered off in the quest of gold' and left his seafaring days behind him.

On 7 July 1866, at age 33, Peter married 18-year-old Louisa Albury. While the newlyweds were fossicking on the New South Wales goldfields, Henry was born. Peter was known as 'a most industrious man [who] toiled manfully as gold seeker, bush farmer, and builder for his young Australian wife and four children.' In an autobiographical work, Henry described his father as 'short, and nuggety and fair' but Henry conceded, 'He died before I began to understand and appreciate him.' Nonetheless, Henry retained fond memories of Peter and when speaking of his father throughout his life, Henry spoke with love and affection.[1]

Henry remembered his father 'was always working, or going somewhere with an axe or a pick and shovel over his shoulder, and coming home late.' In a short story entitled 'A Vision of Sandy Blight,' Henry wrote, 'Father was always gentle and kind in sickness.'

When Louisa eventually decided that a life in the Bush was not for her, she took Henry and his two brothers and sister, left Peter, and moved to live in Sydney. At the time of his parents' separation, Henry—the eldest child—was sixteen. Peter died when Henry was twenty-one.

≈

By 9.00 am, Baz is well in front and Moz is somewhere behind me, struggling in the early heat. A local farmer named Rod Thompson stops to chat. Like almost everyone driving past on the previous day, Rod wonders if we are in trouble and in need of help. He tells me to watch for the ruins of a pub that existed at Lauradale during Henry's time and he tells me also of a shanty that stood just around the bend. There were many more spots to stop and buy a drink when Henry passed this way. Nowadays, the only watering hole between Bourke and Hungerford is at Fords Bridge and I cannot wait to get there later today.

Baz waits for me and we reach Sutherlands Lake together. There are hundreds of birds on the lake and it is an idyllic, lovely scene. I pick my way through the long grass to the lake's edge. It is unbearably hot. I fall to my knees and then lay flat on the ground, with the upper part of my torso and my head resting in the water. I pull my body further into the water and completely submerge my head. I lay there with my head under water, my hands sinking into the gooey muddy bottom. The water is dirty and it occurs to me that it is no doubt full of leeches, but it is cooling and refreshing. I surprise myself by how long I keep my head under. I do not feel any need to breathe. I feel as if I have become a Water Being with no need for air. Given our state of exhaustion, when Moz arrives, he thinks I am trying to drown myself—to take the easy way out.

Eventually, I rise to the surface and as I get to my feet, the water runs down my body. I feel cool but also light-headed. In fact, I soon feel queasy and realise that I overdid it under the water. I wander to where Jane has pulled up in the car and I retrieve one of my blankets and set off in search of shade. I find some small trees and spread the blanket over the red ground. Ignoring the prickles, I lie down feeling sick and in danger of vomiting all over myself and my blanket. The others join me on the blanket in the shade. They proceed with breakfast.

'What do you want to eat, Gregga?' Baz asks.

I am fighting the urge to throw up and so I decline to answer. My eyes are closed and I keep my mouth closed too, just in case.

We rest for a long while but all too soon it is time to move again. The day is only getting hotter. We bid farewell to Jane and then we trudge on wearily. There is not much traffic but whenever a car or a truck does pass, clouds of dust envelop us.

We hear vehicles rumbling along the Track from a long way back. Out here, intrusions on the silence are easily detected and we know when a vehicle is approaching long before we see it. Most people stop for a chat, but some simply wave as they drive past, covering us in dust.

'It is the second one that will kill you,' I say to Baz. Buried in dust, if a second vehicle were to approach, the driver would be hard-pressed to see us. Furthermore, because we can still hear the first car moving away from us for a long time, it is difficult to hear the second vehicle approaching from behind. Fortunately, out here, even a lone vehicle is rare. There are few second vehicles.

≈

Henry's mother, Louisa, was a woman ahead of her time. Amongst other things, she was a poet, publisher, social reformer, and leading figure in Australia's women's suffrage movement. Louisa was the second daughter in a family of ten girls and two boys. She was born in New South Wales on 17 February 1848 to English-born parents. She married at 18 and gave birth to Henry when she was 19 years old. 'Someday,' Louisa wrote when she was almost 50 years of age, 'I hope to be strong enough to look the public in the face as Louisa Lawson, and not as the mother of a man. A man writes me up, a man takes my photograph, and I appear as the mother of a man.'[2]

After leaving her husband and moving to Sydney, Louisa bought the *Republican* newspaper. In 1888, she founded the *Dawn*. Both publications were used as vehicles to push social causes and advocate reform. Louisa was described as 'a remarkable woman, with many graces of character, destined to take a conspicuous and honourable part in public affairs.'

Henry referred to his mother as '*The Chieftainess*'. He had an antagonistic relationship with her,[3] claiming at one time that she was 'a selfish, indolent, mad-tempered woman [who] was insanely jealous on account of my "literary success".' This criticism is interesting in light of a description by one person who knew both Henry and Louisa well and said that in many ways mother and son were identical in character.[4]

≈

At noon I am alone in the high sun. My feet are blistered to hell, my back is sore and my lips are dry and cracked. I am back on the bitumen and the heat from the road feels like it will melt my feet. The others are in the distance ahead. Occasionally, I catch a glimpse of them as they crest a rise in the road out in front of me. 'There are no mountains out West,' Henry wrote, 'Only ridges on the floors of hell.' I can see Satan dancing across this blazing inferno and tempting passersby with relief from suffering. Yesterday's blisters are getting worse and I have two new ones. As I press against my toes with each step, a razor blade seems to cut through my feet. My feet feel as if they are being torn to shreds.

After a while, I pass where Moz has stopped to rest in the shade. He is dripping with sweat. I am struggling but he is struggling too. They say that misery loves company but I feel no inclination to stop and chat. Although this is a team endeavour and our success will be heavily influenced by one another, it is as much an individual pursuit as anything else. We depend on one another but no one can take the steps for us. One has to do that for himself and, as I hobble along in the heat, that is the hardest thing—to keep taking steps, to keep pushing forward.

I had told myself that I would push on until 11.15 and stop for a rest. When it got to 11.15, I decided to keep going until 11.30 am. When I am still going at noon, I decide to keep going until 12.30 pm. I crest another hill and see Baz sitting in the shade on the roadside. The water and food we buried for lunch cannot be too far ahead—perhaps only a few kilometres. I decide to keep pushing on to the lunch stop.

'Bazzaaaaaaaa,' I say as I walk past. I say it in a way that he will know I am alright and that, for now at least, if he is worrying about me, he need not. Moz is the one he should worry about at present.

Baz smiles as I walk past. He knows I am fine. I have not said much to him or Moz or Jane since about mid-morning yesterday. If I can force out an upbeat 'Bazzaaaaaaaa,' I must be feeling a bit better.

Jane has the car out in front of me somewhere, presumably at the spot we have chosen for our lunch stop. I scan the road ahead for a sign of the car shining in the sun.

A car approaches from the Fords Bridge direction. Two young men sit inside.

'How are you?' the driver says when he stops.

'Yeah, I'm fine.'

'Sorry about covering you in dust yesterday,' the young driver says. I would be surprised if he is 20 years old.

'No worries,' I say. 'It can't be helped.'

'I didn't see you until I was almost on top of you.'

Great. That makes me feel much better.

'Where are you going?' the driver asks.

'I'm walking to Hungerford with my brother and a mate.'

'How far's that?' the driver wants to know.

'Um, about 220 kilometres or so,' I reply. I do not tell him about the return leg. I have stopped thinking about the return journey. It is not that I have counted it out. It is just that I have stopped thinking about it. Get to Hungerford first. That is my job for now and, even then, I am primarily thinking only of getting to Fords Bridge today. Bit by bit is how I will get this done.

'Is it hot enough for you?'

I laugh. 'You bet.'

We chat for a bit and they offer me a drink. They tell me that there is a car parked over the next rise, 'Only 200 metres or so down the road.' It must be Jane.

I thank them for taking the time to stop.

I reach the lunch stop first, arriving at 1.00 pm. Baz gets in about ten minutes later, right as the temperature hits its peak for the day of 46 degrees in the sun. 'Gregga, you're travelling better than I expected this morning.'

'You seem always to be travelling well, Baz.' When it comes out of my mouth, I realise it sounds almost resentful. 'You bloody rookie.'

Baz shakes his head as if to suggest he might be making it appear easier than it really is for him.

Moz eventually arrives and we all have a bite to eat—tuna and rice crackers—before stretching out on the ground. I am on the bedroll that Baz packed for me and I have a blanket for a pillow. I roll onto my side and stuff a towel between my knees because that relieves some of the ache in my back. I rest my hat and bandana over my face and neck. The best that we can get from trees out here is only mottled shade. I remove my shoes to try to cool down my feet. I keep my socks on because doing so will help to keep my bandaids and bandages in place. I prop my feet on my backpack because I do not want them dragging on the ground, picking up dirt and grit that will rub and form yet another blister. I keep my pannikin mug and a water bottle close at hand.

I steal a glance at Moz lying on his back on a red fleece blanket four or five metres from me. He is using his backpack for a pillow. Because of the mosquitoes here in the shade, he has a Citronella coil burning beside him. He still has his mosquito net over his head. His blue shirt is unbuttoned and the yellow t-shirt underneath is stained with sweat. Given his shirt has long sleeves, I wonder why he does not discard the t-shirt.

I close my eyes and try to sleep.

≈

When Henry was nine years of age, his life changed dramatically. 'I remember we children were playing in the dust one evening,' Henry wrote, 'and all that night I had an excruciating ear-ache and was unspeakably sick on my stomach. Father kept giving me butter and sugar, "to bring it up", which it eventually did.' Henry said that 'it was the first and last time I had the ear-ache.' Unfortunately, however, the next day Henry was 'noticeably deaf'. Over the next five years, his hearing continued gradually to deteriorate. By the time he was 14 years old, Henry was suffering major hearing loss. Although he later sought treatment in Sydney and Melbourne, nothing could be done to repair the damage. Already quiet and shy, Henry withdrew further into his shell.

At 3.00 pm, Baz begins to stir and I reluctantly take it as my cue to start to move too. Baz is wearing pants that can zip off at the legs and convert to shorts. He wore them as shorts all morning but now he attaches the legs so as to get more protection from the afternoon sun. I drag my shoes back onto my protesting feet. I roll up my bedroll and carry it and my blanket to the car, hobbling with each painful step. To stop and to rest feels good, but to try to get going again after a rest is a nightmare. When I stop, it feels as if my feet contract and curl into tight balls. Getting back onto them again is agony. I ferry the empty lunchtime waste to the car and place it in the plastic bag that we are using for rubbish. I begin to refill my water bottles and my hydration bladder.

'How are you travellin', Gregga?' Baz asks, noticing my stiff, wooden steps. In contrast, he seems not only to be doing well but, indeed, to be trouble-free.

'Good,' I lie. My feet are butchered.

'Do you need anything, Greg?' Jane asks.

'No thanks,' I lie again.

'Do you want a cold necktie?' Jane asks. There are three cooling ties in the esky. They feel good on the back of the neck, especially when you first wrap them on. They are cold and refreshing.

I have got to get away. 'I'm good.'

I feel horrible. I am in so much pain that I am filled with doubts about what I am doing out here. I feel I am on the verge of a breakdown and so I have to get away from the others. I do not want to talk to them. I keep my head bowed and shuffle away, my eyes hidden behind my sunglasses. I know that if one of them catches my eye, I am going to burst into tears. I do not want for the others to see the pain that is in my eyes or the doubt that must surely be etched into my face. I must be on my way, away from them. I must be alone. I do not say goodbye. I increase the speed of my shuffled, painful escape, knowing that tears are barely below the surface. I am broiling in the sun. In my battered state, with my feet already blistered to buggery, all that I can do is to keep shuffling forward. I doubt my ability to keep up with my companions.

There is the stink of death on the hot wind and I know that somewhere near there is a dead kangaroo or Emu or some other animal. If they have not already started, it will not take long for the desert scavengers to tuck in and start dining.

≈

I know that I have not been talking very much to the others. I knew this would be the case though and so I apologised in advance. I learn from Moz later that Jane has said to him, 'Greg never says anything.' Another time, Jane asks, 'Does Greg not like me?' She does not understand. I have no energy for talking. All that I have, I must give to moving down the Track. This is not a place for small talk or niceties.

Fortunately, Moz has been out here before. He knows. He explains to Jane that one does not have the energy to think, let alone to talk.

'But Baz talks to me,' Jane counters.

'Baz is a freak,' Moz replies.

Jane has taken a shine to Baz. She likes him a lot more than me. I can tell and I do not mind for a second. I understand. Hell, I like Baz a lot more than me. I am his biggest fan. 'Join the queue, Jane,' I think.

Anyway, try as I might, I am not without feeling and I suspect Jane is wondering about me. I am sure she is wondering how it could be possible for her boyfriend to have such a high opinion of me—this withdrawn, sullen and morose pile of pelican shit. I really appreciate what Jane is doing for us—for me—in devoting her holidays to serve as support for this endeavour. I try to ensure this is obvious to her in those few instances where I do have the energy to talk.

When Jane next stops and asks me if there is anything I need, I refill my water.

'I really do appreciate everything you are doing, Jane. I want you to know that.'

Jane does not miss a beat. 'Show me by finishing the bloody thing.'

Nice one. The truth is, I like Jane. I am not out here to make friends though. Unlike Baz, I do not have the energy for that. I understand if she thinks I am a sombre and gloomy prick. I press on, determined to finish the bloody thing—for myself and maybe now for Jane too.

After a while, Baz catches up to me. Like always, he wants to talk. I do not. He asks a few questions and occasionally I grunt a one-word response. Mostly we walk together in silence. After some time and by way of apology I say, 'I am an unresponsive, uncommunicative prick, aren't I?'

'That's for bloody sure.'

I do not mind. We are good. Baz is starting to understand.

Henry would understand:

On sunset tracks they ride and tramp,
Till speech has almost died,
And still they drift from camp to camp
In silence side by side.

I pick my nose and marvel at the strangely coloured extraction. It is almost an orange colour, not unlike the wax one fishes out of his ears if he neglects to clean them for a few months.

'Bloody hell my boogers are an odd colour,' I say. That is a conversation starter for you. Who says I am an uncommunicative prick?

'Yeah, I noticed a big hunk of snot back on the road this morning. I knew no one else out here would be clearing his throat like that.'

I have been hacking up phlegm all day.

I push on through the afternoon, spitting out big gobs of snot every once in a while. I find the colour intriguing. I have never seen phlegm that colour before. I wonder if it is just a product of my flu or if it might be a warning of dehydration. I try to ensure that I keep my fluid levels up. For all I know, I might be over-hydrated.

When next I see her, I ask Jane how far I have travelled but I doubt the accuracy of her response. Not for the first time, I wonder about the things she is telling me about the distances I have covered. All I need is the truth. Do not water it down. Do not dress it up. Just give me the truth and I will deal with it. Do not make me second guess, because that I cannot deal with. It bothers me that I have to second guess and figure out if she is trying to play psychological mind games.

As I struggle along in the heat, I think about how this is the first time I have been granted research study leave from my university, where I am a professor of literacy education and children's literature at the University of Manitoba in Winnipeg, Canada. 'This is very odd research that I have chosen to do,' I say out loud, but only to the emptiness around me. 'Very, very odd.'

≈

At 4.15, we are all within 30 or 40 metres of one another, but we are not talking. We are on the red carpet and enjoying the soft sand beneath our feet. Each step leaves distinct tracks in the sand. The sand is so soft and in some places so deep that you have to pick the right line to walk along or your feet sink and the walking is tough. There is also the possibility of filling your shoes with sand and that is a recipe for more blisters. We carefully pick a line where vehicles have packed the sand a little harder and flatter. We enjoy momentary respite from the sun when a white cloud passes in front of it but, all too soon, the cloud is chased away and the sun blazes on the left side of our faces again. I notice our assorted headwear. Moz still wears his insect net over the top of his floppy white hat. Baz also has a floppy white hat, but his is over the top of a legionnaire-style cap with a peak out front and a long flap trailing down his neck. Like the others, I also have two pieces of headwear, with my floppy white Tilley hat atop the bandana that is covering my neck and the sides of my face.

We have a long, gradual climb up a rise to the horizon. The roadside contains tussocks of coarse grass and then back from the road on each side, a smattering of sparse, thirsty trees. Moz is in front but he labours up the rise, sweat running in torrents down his long body. Nearing the crest, he places his hands on his hips but keeps walking. From my viewpoint, he appears short of breath as he disappears over the crest.

When I top the rise, I see him standing in the shade of trees to the left of the road. I walk past and so does Baz. Moz rests only for a moment and then he is back on the wide-open road, striding out. He rebounds well from his breaks and I start to think of him as a pogo stick. He can appear spent, but then after a break he gets those long legs going again and quickly bounces back into a rhythm that gobbles up the kilometres. He catches up to me and we walk with our heads down, neither saying a word. There is nothing to say, but plenty to do. We trudge on and Moz slowly draws away and is soon well in front again.

I love the aesthetics of the red sand. It is so 'Australia' to me. It is the Outback. It is the blood that pumps through the heart of the country.

'It's beautiful, eh, Baz?' I venture.

'Yep. It sure is.'

I continue to bake and my blisters ache. 'But I bet some of the most beautiful women in Hollywood are real bitches.'

'You don't have to go to Hollywood to find that out.' Baz seems to be speaking with the voice of experience. A smile creases my face.

Baz increases his speed and I again find myself at the back of the pack.

After a short while, I find Moz laying on the side of the road, his backpack for a pillow. He has his knees bent in what is his characteristic way of resting. If I do not have anything on which to put my feet, I rest in the same way. It stops sand and grit and other such things from getting into your shoes if you keep the soles of your shoes flat on the ground. I lie on my back in the dirt and I gaze up at the clouds. 'Rain, rain, come and stay,' I sing. 'Be as hot as hell another day.'

Moz does not respond.

In the position in which I lie in the dirt, my left arm does not fall flat to the ground. The arm is propped up at the elbow—propped up because elbows do not bend in the manner necessary for my arm to fall flat. My position reminds me of a famous photo from the 1890 massacre at Wounded Knee on the Pine Ridge Indian Reservation in South Dakota. An old Miniconjou Sioux Indian chief, Spotted Elk (known by white soldiers as Chief Big Foot), lies dead in the snow, his left arm propped up at the elbow like mine, despite his still heart. It is not a comfortable position, but I am too tired to move.

I start to make out shapes in the clouds. I see a fierce cougar, its mouth open and sharp fangs bared. I see several snakes in the sky, each one more sinister than the one before it.

Moz gets to his feet and I slowly rise too. 'I'll walk for half an hour and then stop and wait for you, Grog,' he tells me. For as long as I can remember, his nickname for me has been Grog. The nickname is a good one considering the connection with Henry through 'Bogg of Geebung,' where Henry wrote: 'The local larrikins called him "Grog," a very appropriate name, all things considered.' My nickname is, however, ironic seeing as I do not drink anymore. In this heat, I cast my mind back to my younger, wilder days and find myself thinking how nicely a cold beer would go down.

'Okay, thanks.'

I have pins and needles running down my left arm. I am aware this can be a symptom experienced by heart attack victims, but I reassure myself the discomfit in my arm is merely a product of the way that I was laying.

Moz strides out for half an hour and I try my best not to fall too far behind.

I am pleased when I see him wander off to the side of the road and go down for a rest. When I catch up to him, he is the same as before—knees bent, iPod in, backpack pillow, hat and sunglasses still on, flat on his back.

We rest together for a while.

'I'll walk for another half an hour and then stop and wait for you,' he says again.

This will get us there. Slowly but surely.

Baz is travelling much better than Moz and me and so he is well out in front now.

During our next rest, Jane arrives in the car. Moz sits in the passenger seat while Jane keeps the air-conditioner running. After a while, they join me outside in the heat.

'Do you want to know how far you've got left to go?' Jane asks Moz.

In my mind, I have it figured that we must still have about eight kilometres to Fords Bridge.

'Is it less than 12 kilometres?' Moz asks, because that was how much further we still had to go when we resumed from our lunchtime stop. Lying in the dirt, exhausted, I misunderstand Moz and I think he asks, 'Is it less than four kilometres?'

'Yes,' Jane replies. My heart soars. Less than four kilometres to go? You bloody beauty. I can do that.

Four kilometres? I am really surprised. I thought it would be twice that. I am usually quite accurate with my estimations; having a good idea of my pace and keeping my eyes open for whatever road signs might exist. Doubts start to creep into my mind.

'Moz,' I venture hesitantly, 'what did you ask Jane?'

'What?' Moz does not know where I am coming from.

'Just a minute ago, what did you ask Jane about how far we have left to go?'

'Is it less than 12 kilometres?' Moz says.

'Twelve?' I feel deflated—kicked in the guts. 'Oh.'

Bugger it.

'How far is it, Jane?' I ask. Can we please stop playing stupid bloody games?

'You have about eight kilometres to go.'

Bugger it.

≈

We cannot see him. For all we know, Baz is probably in Fords Bridge already. As Jane left, I told her to be sure to tell Baz not to wait for us. He might as well get to the Warrego Hotel where he can shower and rest properly, rather than waiting in the dirt for us to appear.

Moz and I continue our pattern of the afternoon. He moves ahead and then rests while I catch up. We rest together for a short while and then push on, Moz quickly moving to the lead and then eventually resting again.

I am resting with Moz when Jane drives up. We still have maybe three or four kilometres to go to get to Fords Bridge, she tells us.

'Let's go, let's go,' she says. 'Meals go off at eight o'clock.'

I want a shower and a bed. If I do not get to eat, I don't give a stuff.

We have tramped more than 65 km and Moz and I are battered, almost broken men. The last thing I need is to listen to some cheerleader who has been sitting in air-conditioning all day.

'You won't get there sitting down,' Jane says.

You are a genius, Jane.

'You should go on and eat.' I applaud myself for not saying more.

'Oh no, I'll wait for you two.'

'You should go on and wait. We'll be there soon.'

I think she probably gets bored before she gets my hints, but she leaves anyway and that is enough.

Moz and I plug on.

I find a snake dead on the road. It is almost two metres in length and it has not been dead long. It has only been run over once, just a few inches back from the head and, although I know it is dead, it still looks lively enough that I am not keen to venture too close. Still, it is very interesting and I struggle between my fascination with this remarkable creature and my seemingly innate fear. If Moses really did write the Book of Genesis he has a lot to answer for.

The hateful Track draws from me every ounce of energy that I possess. It is all I can do to remain upright. We have about two-and-a-half kilometres to go. It has been a very tough leg since our lunch break. At 7.30 pm, Moz is about 15 m in front of me when a one tonne Ute passes him and pulls up beside me.

'Hot enough for you?' The driver asks. He and his mate are covered in tattoos and both look as rough as guts.

'That and more,' I say, resting some of my weight on the window ledge as I lean against their vehicle.

'Are you walking through the heat of the day?'

'Some.'

'Maybe you should walk at night.'

'Yeah, maybe.' It would be bloody hard to see where you are going. The snakes would be on the move too, hunting in the cool.

'Do you need anything?' the passenger asks.

'Nah, I'm okay,' I say, 'We're stopping at Fords Bridge.'

'I can give you a lift in to Fords Bridge,' the driver offers. 'I won't tell anyone.' He has a cheeky smile on his weather-beaten, whiskered face.

They look toward Moz, resting ahead.

'Is your mate alright?' the passenger is concerned.

'He's okay,' I say, 'He's just taking a rest.'

'What do you do for a living?' the driver asks.

'Me? I'm a professor.' They will think I am crazy; a mad professor.

'A scientist?'

'No, a professor of literacy.'

'You're mad.'

They might be right.

'Oh, well, I give you ten out of ten for effort,' the driver says.

The passenger offers me a cold drink of water.

'That'd be great, thanks.'

They both get out of the Ute and move to their cooler in the tray at the back of the vehicle.

'Have you got a cup?' one of them asks.

I do have a cup and I have been carrying it for two days but I tossed it in the car last time Jane visited. I was too tired to attach it back on my pack.

'I did have one but I threw it away,' I say.

The passenger finds an old Styrofoam cup that has been bouncing around in the tray. As he rinses the dust out of the cup, I reach for it. 'You can't drink that,' he says, looking at the dust swirling in the cup.

If it is cold, I can drink it. 'That's okay.'

He throws the dusty water on the parched sand at our feet and refills the cup. It is cold and lovely. As brutish as they appear in their tattered jeans and blue singlets, I could hug them both. They are rough-looking, but friendly fellas with hearts of gold. They tell me they drove through Cuttaburra Creek earlier today. That is where we had turned back when we were emplacing our water and food stores on the 31st. It is reassuring to learn that it has probably receded a little in the two days since.

'How much further to Fords Bridge?' I ask.

'You're almost there. Once you get over this next hill there'll be a long sweeping curve and then you'll be there. Two or three kilometres is all.'

'Good.'

'They have the barbeque fired up for you.'

'What time do the meals go off?'

'They'll wait for you.'

I thank them for stopping and thank them for the drink. I bid them farewell and they drive off in the direction of Bourke.

Moz and I walk side by side over the final couple of kilometres to the Warrego Hotel.

'You never told me that Baz is a freak,' Moz says to me.

I laugh. He is a bloody freak alright. I know that it is not easy, even for him, but there are times when he is making it look easy.

We cross the first bridge over the swollen Warrego River. Henry described it as merely 'an alleged river with a sickly stream that looked like bad milk.' The sunset is creating interesting light on the treetops. We hear thunder and see occasional flashes of lightning in the far distance. Twenty minutes later we reach the second bridge over the Warrego. Baz walks out to meet us and welcomes us to Fords Bridge. In response to my enquiry, he tells us he reached Fords Bridge about an hour and a quarter earlier. He is showered and looks refreshed. He has also phoned home to speak with his family in Adelaide. I am pleased for him.

'There are a couple of female Scandinavian backpackers working at the pub,' Baz says. I guess he thinks it is something to look forward to.

'Lookers?'

'Yep, but they're not blondes.'

It is almost 8.30 pm when Moz and I stagger into the back entrance of the pub. Moz and I look beaten. I feel that death would be a relief.

≈

I shower and shampoo and try to pull myself together. The toilet and shower building are full of frogs and mosquitoes. I slip my blistered feet into my thongs. Scott Parker is the hotel owner and his cold water is straight from Heaven. Scott is a little ball of muscle in his blue shearer's singlet. I reckon he could handle himself.

Jane and the others go for something more substantial, but I am happy with chips and gravy for tea. They taste great and go down well with lots of water. I am aware of the 'not blonde Scandinavian backpackers' moving about and serving our meals but I am too tired to look up. One of them asks shyly, 'Are you alright?' The concern in her voice is genuine.

'Yeah, I'm okay.' I keep my head down. Telling lies comes easy but easier still if I keep my face hidden.

After tea I shuffle painfully out the back to brush my teeth and take another pee. I hobble to the room I am sharing with Baz and then crash on top of my bed.

I hear Baz asking Scott for a Citronella coil to burn in our room.

'You'll be lucky if I manage to find one,' Scott says. 'They're all sold out in Bourke. There is not a coil left.' The mosquito situation is extreme.

Shortly afterwards, Baz comes to the room with a couple of coils that Scott has obviously managed to locate.

'Hey, Baz, what are my chances of getting you to check under my bed for snakes?'

He bends down and does a cursory check. I notice he is a little more thorough when he checks under his own bed. It is a nice, sheltered place for a snake to hang out and avoid the heat of the day before slithering out to the shower to gobble up a few frogs under the cover of darkness. If I was a snake, that is what I would be doing.

'When I was on the phone with Leeanne, she asked how we were getting on,' Baz says as he gets back to his feet.

'Yeah?'

'I told her you and Sean are doing it hard at times.'

'Yep.' I look up at the cracks in the paint on the ceiling.

'She asked if there was any talk of maybe just stopping at Hungerford and calling it quits there—seeing as it is so hard.'

This piques my interest. 'What did you say?' I am just checking.

'No, not at this stage.'

Baz has the door open and a fan going. It feels wonderful as it blows across my body. He strikes a match and lights a Citronella coil.

'Those backpackers. Do you think they're sisters?' I ask my brother.

I am asleep before he replies.

DAY THREE

Influences on Henry Lawson's Writing

Old Mate! In the gusty old weather,
When our hopes and our troubles were new,
In the years spent in wearing out leather,
I found you unselfish and true.

- To an Old Mate, *In the Days When the World Was Wide and Other Verses*, 1896

But what's the use of writing 'bush'—
Though editors demand it—
For city folk, and farming folk,
Can never understand it.

- But What's the Use, *Verses Popular and Humorous*, 1900

I sleep the sleep of the damned—the sleep of one for whom the future holds nothing but waking up to walk through hell…again. At 11.00 pm, I am jolted from my anguished sleep when something slaps the calf of my left leg. Snake! I spring off the bed as quickly as the darkness and my battered body will permit and hobble past the fan and flick the light switch.

Woken from slumber, Baz grumbles into the suddenly lighted confines of our room.

Sorry to disturb your beauty sleep, Princess, but I have a killer on the bed with me.

'There's a bloody snake on my bed.' I pant.

We both look to my bed, wondering whether it is an Inland Taipan or a King Brown—either way, highly venomous.

But there, sitting on the clean white quilt, is an innocent-looking Green Tree Frog.

'What's all the commotion?' the frog seems to ask.

I love frogs, I really do. My favourites are Green Tree Frogs too. But this bugger is pushing the friendship, hard.

'It's a bloody frog!'

It doesn't matter whether Baz says it or whether I say it. We are both thinking it. 'It's only a bloody frog!'

Bugger. Still, it will be easier to get off my bed than a Taipan. I give the frog a push and it hops from my bed onto the floor.

'That's right, go over and see Uncle Bazza.'

'What are you gonna do with it?' Baz asks.

I herd it across the room and out the door. When last I see it, it is hopping into the darkness toward the open door to Moz and Jane's room.

'Yeah, go and visit Mozza,' I say.

I switch the light off and fall back onto my bed. Baz starts snoring.

Bugger.

Where's a bloody snake when you need one?

≈

Arguably the biggest influence on Henry's career was his editor at the *Bulletin* newspaper, Jules François Archibald. Archibald co-owned the *Bulletin* with William Macleod. As editor, Archibald promoted Australian writers and artists and provided opportunities for them 'to reveal the atmosphere and spirit of Australia,...inculcating a truly Australian sentiment' rather than to portray Australia merely as 'a land for exiles to languish in and lament upon.'[1] Archibald invited all to partake of this 'Australian sentiment' and encouraged contributions from readers. One of his readers was the young Henry Lawson.

Recollecting his first published work, Henry said:

> *I wrote it and screwed up courage to go down to the* Bulletin *after hours, intending to drop the thing into the letter box, but, just as I was about to do so, or rather making up my mind as to whether I'd shove it in, or take it home and have another look at the spelling*

and the dictionary, the door opened suddenly and a haggard woman stood there. And I shoved the thing into her hand and got away round the corner....

I hadn't the courage to go near the Bulletin *office again, but used to lie awake at night and get up very early and slip down to the nearest news-agent's on Thursday mornings, to have a peep at the* Bulletin, *in fear and trembling....At last, sick with disappointment, I went to the office and saw Mr Archibald, who seemed surprised, encouraged me a lot and told me that they were holding [the poem] over for a special occasion.*

'A Song of the Republic' appeared in the *Bulletin* on 1 October 1887. Henry was just 20 years of age.

Henry dedicated his 1896 book, *In the Days When the World was Wide and Other Verses*, to Archibald. As part of the dedication, Henry included a poem entitled, 'To an Old Mate'.

I have gathered these verses together
For the sake of our friendship and you.
You may think for awhile, and with reason,
Though still with a kindly regret,
That I've left it full late in the season
To prove I remember you yet;
But you'll never judge me by their treason
Who profit by friends—and forget....

Archibald had a way of placating moody writers like Henry and diffusing volatile situations. 'You couldn't quarrel with Archibald,' Henry wrote. 'A writer would go up the *Bulletin* stairs furious and boiling over...and come down a few minutes later soothed, and even remorseful.'

Upon Archibald's death many years later, Henry was moved to reflect on Archibald's life and influence. 'The lines we wrote when our hearts were young, / Are Archibald's Monument,' Henry wrote.[2]

As he was leaving Archibald's funeral dry-eyed ('No tear is needed, nor funeral frown' because of the lasting impact of Archibald's life), a reporter took Henry's arm and asked for the names of some of the people who had been in attendance. Henry responded by identifying a number of the grateful writers who had gathered to mourn the passing of their dear editor and generous supporter. 'But there was someone else, someone very great, someone much more important than anyone of us,' Henry continued. 'Archibald was there.'

≈

I wake at 5.10 am, gather my toiletries and wander out to use the toilet and then to take a shower. There, I am greeted by more frogs than I have ever seen. Dozens and dozens and dozens and dozens of them are scattered around. There are even frogs in the toilet bowl. I try to encourage them out before I do my business and wonder where my night-time visitor has ended up. I hope he is not passing the morning down the throat of some big snake.

After my shower, I go back to my room and start attending to my feet. One of the rubber sleeves for my little toes has worn through and broken. Psychologically, this is a major blow because the toe sleeves seem to be working so well. It was my little toes that took the greatest battering last time and, with the protection of the rubber sleeves so far, they are in good shape. If my shoes have been rubbing so much that they would wear through the rubber sleeve, I wonder how butchered my little toe is going to be without the sleeve.

I have some major blisters. The ones at the base of the second toe on each foot are a real problem. They are in such an awkward spot for me to get at. I am the least flexible person at the best of times and with my legs having covered almost 70 km in the past two days, I have no chance of getting them to bend far enough so that I can gain access to the underside of the front ends of my feet. It is easier to get to the blisters on the inside of each heel. These are big, ugly bubbles and there also appears to be substantial bruising just below the blisters. How on Earth can I expect to walk on these feet for the 150 km left to Hungerford? In purchasing a good pair of hiking shoes, my hope had been that I would avoid blisters. For my feet to be so badly blistered so early into this endeavour is a real kick in the teeth.

I spend more than an hour with powder, bandages, bandaids, and

Vaseline before carefully wrestling on two pairs of socks. My feet are such a mess that I decide to abandon the Merrell hiking shoes. They are obviously not worn in enough. I know that my old pair of Nike runners are well worn in because I have had them for years.

I had bought my Merrells with plenty of room inside in order to accommodate bloated, swollen feet. As I pull on the old Nikes, they are noticeably tighter, particularly around the toes. So far, my toes are in reasonable shape and I fear that a change of shoes will create a whole new series of blisters.

As I tie my shoelaces, I notice a mouse in the room. Great—another reason for a hungry snake to slither inside.

I wander into the kitchen and help myself to breakfast. I have three pieces of toast with Vegemite and then I down a banana and several glasses of water. Scott comes in for a chat and I ask him about a photograph of three children that adorns the wall. He tells me that he has two sons, 19 and 17 years of age, and a 12-year-old daughter. The eldest son is working in Brisbane. He is about to join the army.

'What do you think of him joining the army, Scott?'

'Well, I'm proud,' Scott says. He hesitates, choosing his words. 'I'm proud that he wants to do it.' Another pause. 'I just hope he's a better shot than they are.' He smiles, but I know it is not just a joke.

His younger son attends boarding school in Toowoomba, 900 km away. Toowoomba is the scene of some of the worst of the flooding up north at present.

'I hope that he's not caught in the flooding,' I say.

'No, no,' Scott replies. 'Fortunately, both my sons are home for Christmas.'

'How often do you get to see your youngest boy?' I ask.

'About four times a year.'

Scott tells me that his daughter will also go to boarding school in Toowoomba with the start of the new school year. My oldest daughter is 11 years old. I think of my wife Jennifer. 'Gee, that's going to be tough for your wife,' I say, 'with her only daughter moving away.'

'Well, yeah,' Scott says, 'but it is hard for kids out here, with no friends to play with.'

Scott talks with pride about how his daughter recently finished sixth in the high jump for her age group at the School Sport Australia championships in Bendigo.

'She's just training with me on some old mats out the back,' Scott says. 'It'll be good for her to have some proper training facilities when she moves away.'

'Yeah, but I'm sure she'll miss training with Dad.'

Scott laughs. 'Yeah, I guess so.' He pauses, deep in thought. His wife will not be the only one to miss her daughter when she moves away. 'Anyway, I have work to do. Good luck today, Greg.'

After Scott leaves, Moz and Jane enter the kitchen. Jane complains that their room is full of mosquitoes and she has a number of bites on her legs. There were no mozzies in our room. Maybe my mate, the Green Tree Frog, ate them all.

≈

I am first on the Track at 8.00 am. And yes, I know I awoke and arose from the bed at 5.10—that's just how long it takes to get going. Toilet, shower, feet, pack, breakfast, water, sunscreen. It all takes time, particularly on stiff, pained feet. I would love to start earlier in the mornings and get down the road before the sun, but it takes so long to get ready that the sun always wins.

Despite the distant thunder and lightning the previous evening, there is no sign of rain this morning. The sky is cloudless and blue—a rich azure directly overhead, and then toning down through various blues to the softer sky blue towards the horizon. It is the type of sky that one dreams of through a long Canadian winter. On another day with a different agenda, one would consider it beautiful. Today, the clear morning sky is an ominous threat of extreme heat.

At different times, Henry worked in New Zealand and in England. Like me, when Henry was abroad, he suffered badly from homesickness. Missing home, Henry wrote:

But I have lost in London gloom
The glory of the day,
The grand perfume of wattle bloom
Is faint and far away.

Through a long cold English winter, Henry and his wife, Bertha, would long for the sun and imagine being back under the blue Australian sky. Their miserable loneliness was alleviated when they found a florist selling Australian wattle. Henry bought it all and he and Bertha brightened their small London flat by decorating it with the golden sprigs.[3]

≈

I look back and see that Moz and Baz start about three or four minutes behind me. Jane is with them, out for exercise—perhaps company too. It cannot be easy for her out here but she seems to be doing well and her help has been invaluable.

I carry a soft blue towel in my hands. It is the colour of the sky toward the horizon. I also have my thongs pushed into my backpack. I know from when we buried our supplies that we have a water crossing coming up at Green Creek only six or seven kilometres from the pub. The floodwater coming down from Queensland has left water over the road. Henry described the state of the rivers and creeks out here as being 'vaguely but generally understood to depend on some distant and foreign phenomena to which bushmen refer in an off-hand tone of voice as "the Queenslan' rains", which seem to be held responsible, in a general way, for most of the out-back trouble.'

Just after 9.00, I notice an animal skull on the roadside and stop to take a photograph of the sun-bleached skull with the long boar tusks. I then reach a gulley of water running along the side of the road and dunk my hat and bandana and enjoy the cool water running down from my head. I notice a long snake on the road, about 15 m in front of me. I move forward for a closer look as the snake attempts a wriggled escape toward the longer grass and water. I get a good look at the uniform greyness of the skin and its black, unblinking eyes. It is a heavily-built snake with a small head. I am no snake expert but my first thought is 'Taipan'. Upon reflection and after consulting a number of field guides, it was probably a King Brown, otherwise known as a Mulga Snake. We are getting deeper into the Mulga country. The Mulga Snake is considered very dangerous and is said to have the largest recorded output of venom of any snake in the world. The Inland Taipan, or Fierce Snake, is considered the world's deadliest snake, producing venom more toxic than any other snake, a staggering 50 times as potent as the venom of India's feared Cobra.

As the others approach, I call out, 'Baz! Snake!' Now, I know I have previously expressed that I am not too keen on snakes; however, Baz and I discussed how we would like to see them, provided they are on clear ground and pose no danger. Snakes are amazing creatures and fascinating to watch. The snake disappears before Baz reaches me so I indicate the general area where it went. Baz approaches tentatively, peering into the grass and the water to see if he can catch a glimpse but the snake is gone.

Baz picks up the boar skull I photographed and takes a photo, then extracts the tusks and gives me one as a memento to keep. I place it in my pack to carry with me.

A little before 9.30, we get to where Green Creek has spilled across the road. This will be the only water crossing for the day and a short one at that. It must be only a 50 m stretch from dry road to dry road. After spending so long in getting my feet ready this morning, I am worried that all of my work is about to be undone. It is with reluctance that I begin to untie my shoelaces.

Baz can see my reluctance. 'I can piggy back you across,' he offers.

I imagine the scene—my brother carrying me down the road early on Day Three. Little Bazza—the runt of the family. I know that he will do it too. He is not just saying it. He will do anything it takes to get me to complete this trek and realise my dream. But I think how ridiculous it will look. I also think that I am out here to walk to Hungerford, not to be carried.

I pull off my shoes and slip on my thongs. I leave my socks on, thinking they will help to keep my bandages in place in the water. I roll my pants up to my knees and step into the water, which is like bathwater. The road here is sealed and so it is an easy walk across the pool. I keep my eyes open for any snakes that might be out for a morning swim.

Jane does not do the crossing. 'Good luck,' she calls, beginning the long lonely walk back to the car. She is a trooper alright.

Moz and Baz enter the water too. The sunshine sparkles off the ripples that their steps create as they wade through.

Once across, it is time to dry our feet and get our shoes on again. I sit in the middle of the road to peel off my wet socks. I am deflated to have my fears confirmed—my bandages and bandaids are soaked and refuse to stick.

I give voice to my frustrations.

'I haven't heard you swear like that for 20 years,' my brother tells me.

I have used almost all of the moleskin bandages that I brought with me. I thought that I had enough to last me the entire trip. It is early on Day Three. What am I going to do with my blisters? My visit to hell just got worse.

I use my towel to dry my feet.

'That string of expletives,' Baz says, 'are you going to include that in your book?'

'That'll be the title,' I suggest, grumpily.

Moz is lying on his back and he seems amused. I however, am in no mood for joking.

'If a car comes along, I'm not moving,' I say.

'Yeah, they can go around,' Baz replies, missing my point.

'No. They can run right over the top of me.'

I have my shoes and socks on but I am still seated in the middle of the road when a vehicle appears, coming toward us from the direction of Fords Bridge. The car slows down to enter the water. I guess I want to live after all. I slowly get to my feet and hobble to safety while Baz carries my pack to the side of the road. We see that it is Scott and the two 'Scandinavians,' headed out to a property under Scott's management. They smile and wave, but do not stop.

'They're actually not Scandinavian,' Baz says. 'I found out one's from France and one's from Germany.'

'I guess they're not sisters then, eh?'

Baz and Moz have both taken off their shirts and dunked them into the water. I do the same and agree with Baz that putting it back on, it feels like an ice vest. When a slight breeze pushes the wet fabric against my hot skin, I even feel cold. It feels lovely.

It only took a few minutes to actually wade across the water, but it has taken about an hour from when we arrived at the water's edge until we are walking the Track again. In that hour, the sun has only got higher and hotter. Moz and Baz move ahead and I limp along behind.

≈

In addition to J. F. Archibald, another person who exerted an enormous influence on Henry's professional life was George Robertson of the Angus & Robertson bookstore and publishing firm. David Angus started a bookselling business in 1884 but in 1886 he entered a partnership with Robertson and in 1887 they began publishing. Robertson played a key role in the publication of several of Henry's books.

In one friend's opinion, 'No publisher ever did more for one of his clients than the late George Robertson did for Henry Lawson.' Robertson generously often paid Henry in advance—and sometimes twice—for his poems and stories. When Henry was leaving to pursue his writing career in London, Robertson went to see Henry off and to make sure he had enough money to establish himself overseas. The kindly Robertson also paid money to Henry's estranged wife after she separated from her husband.[4]

In 1910, Henry composed a poem entitled 'The Auld Shop and the New' that, Henry wrote in the dedication, was 'written specially for "The Chief," George Robertson, of Angus and Robertson, as some slight acknowledgement of and small return for his splendid generosity during years of trouble.' Robertson loved the poem and the accompanying notes for, as one friend recalled, 'Nothing that Lawson wrote made [Robertson] laugh so hilariously.'

≈

'I'm here. I'm queer. I don't want any more snakes!' Moz calls.

I had almost passed by without seeing him lying in long grass in the shade beneath the Green Creek Station mailbox. I am only a kilometre or two beyond the water crossing and I had assumed that, like Baz, Moz was way out in front of me.

'I am out of water,' Moz croaks.

I do not know how he is out of water, but he does not look well. It is only 11.00 am but Moz says, 'I have already drunk three litres today.'

I wonder to myself, 'Did you fill up at our water dump then?'

On the 31st, the day before we started walking, we had driven eight kilometres out from Fords Bridge. We filled a container with drinking water, tied it with a piece of rope and sunk it in the cold floodwater running under the road through a culvert. I had fished it out as I passed earlier this morning and filled my own hydration bladder. Moz mustn't have done the same.

One has to make sure that one carries sufficient water out here. It is easy to feel too buggered to carry a full load though. We carry between three and four litres each and refill at our water storage spots every ten kilometres or so. A full hydration bladder and a couple of extra water bottles in your pack can weigh you down like a sumo wrestler on a race horse, but it has to be done. Henry's tramping mate, Jim Gordon, said that Henry was always fastidious about ensuring his waterbag was full. 'Although at times we were walking through water,' Jim wrote, '[Henry] always insisted on keeping the bag full—"in case of accidents".'

Moz is flat on his back and looks to be in a bad way. I have emptied one of my bottles but I give him the other bottle of water from my pack.

'Do you want me to wait with you, mate?' I wonder how long it will be before Jane appears. I am feeling okay and I have a nice rhythm working at present.

Moz tells me to keep going. 'To live you must walk,' Henry wrote. 'To cease walking is to die.' I push on in the heat.

Moz rebounds like the pogo stick that he is and soon passes me again. Only a few minutes later however, I find him sheltering in the grass under a small shrub. I know that I cannot afford to give it up, but I offer him my hydration pack anyway. It is what Henry's laws of mateship demand.[5] If Moz accepts my offer I am out of water, but he looks worse off than me.

Moz declines the offer and I cannot suppress the sigh of relief that I breathe.

'I'll hang on to this bottle that you gave me and just lie here and rest and sip until I cool down.' He tells me to keep going.

'Will you be okay?'

'Yes. I just need to rest.'

'Are you sure you're okay?'

'Yes.'

I keep going. I am sure Moz will rebound and in a few minutes will be out in front of me again.

I get two kilometres down the Track and am surprised I have not seen Moz again. I am also surprised when it suddenly occurs to me what song has been running through my head for the past day-and-a-half. The song is R.E.M.'s *Losing My Religion*, but it is not the title that is surprising to me. I lost my religion a few years ago, when I decided amidst the struggle to complete my Doctoral dissertation that I have enough to worry about in this life without also worrying about another life. When a religious person descends to the black depths that I did at that time, their religion is of no value. But rather than the *title* of the song in my head surprising me, it is to suddenly realise that the line I am repeatedly singing is the one, 'I don't know if I can do it.' It is strange that I could have that song line running through my head for so many hours without being consciously aware of what I was singing. 'I don't know if I can do it.' What a dangerous line. It is a startling discovery for me. One of my favourite quotes has always been, 'Whether you think you can or can't, you're right.' In the book in which I first encountered this quote as a young teenager, the words were attributed to the great heavy-weight boxer, Muhammad Ali. I have since seen the quote attributed to Henry Ford, the car and assembly line bloke. Either way, both Ali and Ford achieved great things through strong self-belief. My choice of song suggests that, at least subconsciously, I am experiencing doubts about my ability to complete the task I have set.

Having given Moz my spare bottle of water, I soon run dry. I find a clear and shady spot and wait for Jane to appear in the support vehicle with some water. I have to stay still and cool because if I overheat with nothing to drink, I will be in trouble.

I lick my dry, cracked lips and then run a finger over them, feeling flakes of skin hanging loose, but I resist the urge to pick. The backs of my hands are beetroot from sunburn on the first day.

I sit and think and wait.

I wonder whether Moz will catch up to me before Jane arrives. It is 11.45 am and Jane cannot be far away.

When she appears, Moz is in the car with her. 'The support crew just doubled in size,' he tells me. He has a broad, toothy smile.

'I see that.'

Moz is sitting in the front passenger seat and I can hear the air-conditioner humming. 'Oh...Australia are none for 30-odd in the Test,' Moz says.

I consider Test cricket to be the ultimate sporting contest. Five days of ebb and flow, up and down. It is a battle of wills. A war of endurance. The final Test of the summer starts today and I am disappointed not to have the opportunity to watch it. Although it is not true to say as Henry did that as an 'average Australian' I am likely to have 'no ambition beyond the cricket and football field', I do love talking about Test cricket. Right now however, I think there are other conversations to have.

Moz, have you given up?

He tells me I am 80 km from Bourke. 'I'm surprised how far you got since I saw you last.'

'Yeah, I'm feeling okay—moving pretty well.'

'Do you need anything?'

'I'm out of water.'

Moz brings me one of the full water containers from the back of the car.

'Thanks mate,' I say.

'I was getting chest pains,' Moz says. I nod, but do not say anything. 'The heat was too much and I couldn't cool down and I was having chest pains.'

I am listening.

'I've lost my cold necktie,' Moz says. 'It must have fallen off.' He starts fiddling around in his pockets. He has obviously misplaced something else. 'Damn. I think I've lost my iPod too.'

'It must be back where you were lying in the grass,' I say. 'Don't worry about it, it'll be under the mailbox or where you were lying under the shrub. Jane will be able to find them.'

'I better go back. I know where I was laying.' Moz is keen to get away.

'Well, come back,' I say, 'I want to talk to you.'

Moz and Jane drive back in the direction from which they came.

What can I say to Moz?

It is horrible out here. I hate it.

But it is only the third day. We have been preparing for this for months and it is only the third day.

I am disappointed—mostly disappointed for Moz. He must have been dreaming about tackling this walk for months. From the emails we have exchanged, Baz and I are under the impression he has done far more training for this than either of us. Moz has told us of long bushwalks outside Sydney, carrying a pack on his back. I am disappointed for Moz, yet I cannot completely suppress some disappointment in Moz. It is not even lunchtime on Day Three.

≈

When Moz reappears, he tells me that both his missing cold necktie and his iPod were actually in the back of the car.

'I wasn't thinking straight,' he says.

Jane decides to drive ahead to find out how Baz is coping with the heat. She knows I want a word with her partner.

Moz sits down in the dirt beside me.

I have been wracking my brain for what I should say to him. I cannot help but think he would have made a different decision if he were not alone. In hindsight, I should have stayed with him. It was too easy for him to quit when he was baking in the heat alone.

'First off,' I begin, 'to walk 80 kilometres in this heat is crazy courageous.'

'Yeah,' Moz agrees.

'That is first and foremost...' My pause is a long one. I listen to the stillness around us. '...but I think I owe it to you not to just stop there.'

'Hmmm.' Moz waits for me to continue.

I am searching for the right words. We had talked about ensuring that we all managed to make it—not just to Hungerford, but all the way back too. We were not planning just to encourage one another, but to do whatever it took to ensure we were all successful. 'Don't let me fail,' I had written in one of my last emails, 'because I will be a long-time failed.' There were always going to be times—lots of them—when I felt like I wanted to quit, but pain is temporary and failure can be permanent. I did not want to live the rest of my life as a failure.

Baz had replied to my email with,

Come Hell or High Water, there is NO WAY IN THE WORLD that any of us will be failing...that is where the physical and mental toughness of the others MUST squash such feelings [of wanting to quit], pick the person up (literally if necessary!), arrest the momentum shift, and ensure the collective job (i.e. getting us all from Bourke to Hungerford and back) is completed!

Words. Just words. In this instance, Moz had not even allowed us to pick him up. He pulled the pin behind Baz's and my back.

Moz interrupts my thoughts. 'I was having chest pains and I decided this is not worth it.'

We are all having pains—pain all over. It is what I expected.

'Well, I think you should rest today and cool down in the car.' I go easy. There is nothing I can do now that he has severed the string by which the three of us were connected. 'You should just rest today and maybe start up again from wherever we are in the morning,' I suggest. 'Maybe you should just think about making it to Hungerford, and forget about the return journey.'

'What you are saying is just like what Jane said to me too.'

Good one Jane. I wondered what she said to him.

'See how you feel tomorrow and then decide what you'll do.'

'Yeah, I'll see how I feel,' Moz says, 'and play it by ear.'

I pause again. There was something else I was thinking about when he went back in the vehicle to search for his iPod.

'What about us—me and Baz?' I ask hesitantly. 'I know it is costing you money to be out here and, of course, you can be making money back in Sydney.' Would I want to stay out here if I had just quit? Not likely.

'No, no.' Moz is decisive. 'I committed to coming out here for two-and-a-half weeks. I knew that. I'm going to help you and Baz do this.'

Good for you, Moz. That is a big relief. I would understand if he wanted to call the whole thing off and drive back to his home in Sydney. Sitting here in the dirt, having already been through hell, but knowing that worse is ahead of me than behind me, I had decided that if Moz left for Sydney—taking the car and Jane with him—Baz and I would push on regardless. I have not thought about how we would possibly do it, but I decided that we would do it anyway. We would worry about the logistics later, but we would keep going.

I tell Moz I appreciate his attitude. I tell him Baz and I would be stuffed without him and Jane. I am telling the truth. We would be stuffed without their help, but I do not tell him that we would keep going without them anyway.

Jane returns. She has told Baz about Moz's withdrawal.

Moz and Jane decide to drive back to Bourke for petrol and, presumably, to cool down. They make sure I have everything I need and then they leave. It is after 1.00 and it is time for me to push on.

A car approaches from behind me. Two young blokes pull up. They must only be teenagers.

'Are you okay?' the young driver says. He might be 20 but is more likely to be 16 years old.

'Yeah, I'm fine.' I notice the three huge dogs in cages in the back of their vehicle.

'What are you doing?' the driver wants to know.

'I'm walking to Hungerford with my brother.' Moz is not walking to Hungerford any more.

The passenger—the younger of the two occupants—is not saying anything.

'How far's that?' the driver asks.

'It's 220 kilometres to get there,' I reply.

The passenger is silent, but he is slowly shaking his head and the look of incredulity on his face speaks volumes. His face is a mixture both of admiration and serious doubt about my sanity.

'That's pretty far,' the driver says.

'That's right.'

'It's pretty hot.' The driver is a master of understatement.

I laugh. 'That's for bloody sure.'

We talk for a while and they offer me a drink. I tell them that I appreciate their concern and thank them for stopping. They drive away and I learn later that they stop and chat with Baz and regale him with stories of 'dogging'—hunting wild pigs with the three huge dogs in the cages. They even show Baz a video they made of the dogs attacking a big old boar in what Baz describes as 'a ferocious encounter'. 'It's fun,' they tell Baz. Not for the boar, it aint, I think.

Not long after the young blokes have left with their dogs, another vehicle pulls up with another couple of young fellas inside. They tell me Baz is about two kilometres ahead of me.

In spite of a new blister, all things considered, I am feeling okay. I am walking what I describe on my voice recorder as a 'pretty section'. The road is straight and wide and consists of soft red sand, not fine, moving sand though. It is soft enough not to bother my sore feet, yet it is packed enough that I do not feel like I have to work hard to slog my way through.

At 12.15, Baz made it to our lunch spot and stopped to rest. At 1.45, I catch up and share a bite of lunch with him. There are fruit boxes to drink and tinned tuna and rice crackers to eat. Jane caught up to Baz just as he reached the lunch spot where we buried food and water; she left with him some other useful things. She will pick up the folding chair, the ground sheet, and any garbage later. Baz and I take the opportunity to talk about Moz's withdrawal. Baz tells me that he is not surprised. Earlier in the day, he had recorded in his journal, 'Greg hobbling with blisters and Sean really struggling with heat...how can they possibly continue for another 12 days of this?' Like me, Baz is disappointed for big Mozza.

Baz wants to know if Moz intends to go home to Sydney. I tell him that I asked Moz about that very thing and that Moz said he is happy to stay.

After I have eaten, I want to lay down out of the heat to rest. It is 42 degrees. The sun has moved since Baz's arrival and the shade that we have is not what I consider to be sufficient. I am roasting and, in any case, I have been growing increasingly particular about my shade.

The Mulga Trees, Black Box eucalypts, and local wattles rarely offer anything better than a mottled, spotty shade. The trees and shrubs of the Outback have adapted over millennia to life in the desert better than have I in the few days since flying from the Canadian winter of ice and snow. The linear or needle-shaped leaves are perfect for reducing moisture loss through transpiration but they offer bugger all shade to an overheated weary traveller.

Baz searches about and finds a shaded spot close in under a few shrubs. They probably stand only two or three metres high, but if I crawl right in, as close as possible to the base of a couple of the plants, it seems more shaded than anywhere else. At this time of the day the sun is very high in the sky. Baz spreads out his towel for me and I lie down and remove my shoes. I am careful to ensure that my socked feet are on the towel, forcing the upper half of my body to lie in the red dirt. I prop my feet up on one of the ten-litre containers that Baz dug from where we had it buried. Baz also gives me what remains in a two-litre bottle that was also buried. I slowly pour water over my face and chest, soak my bandana and dab at my throat. I drink and drink.

I am melting under the sun. Although I can see that clouds are building up, I cannot get cool. My sunglasses rest on my chest and I am conscious of them rising and falling with my increasingly heavy breathing.

My body is seriously overheated but, out in the middle of nowhere and with the support vehicle well over an hour's drive away in Bourke, there is nothing I can do but try to stay calm.

I notice that I have tingles down my left arm and into my hand and fingers. I am not happy with my shade. It does not provide enough protection from the sun, so I sit up, pull on my shoes and get to my feet. Baz is surprised to see me up.

'I'm going to look for another spot,' I say. 'I need more shade.'

I know that Baz picked out this spot for me and helped to set me up. By way of apology, I say, 'I'm getting to be a fussy bastard when it comes to shade.'

I think I need to be on the other side of the road. I put on my pack and take the ten-litre container and the water bottle in my hands.

'I'll check out those places over there,' I tell Baz, pointing across and down the road. 'I'll see if there's more shade on that side.'

'I'll carry the water,' Baz offers, but I tell him I am okay.

'How about you carry the food and other stuff?'

'I'll get the water.' Baz takes the heavy container out of my hand. I gather up the towel, a few loose tuna tins and crackers, and Baz's ground mat. Baz collects the garbage and puts it in a bag in the hole where we buried water. Moz and Jane will retrieve it later.

I do not wait for Baz. I move down the road and check a couple of trees, but I cannot find what I am looking for—an open, clear spot where I do not have to worry about snakes, but a spot that offers a solid enough blanket of shade that the sun is not constantly piercing through what out here is supposed to pass for leaves.

I choose another spot. It is not perfect, but it will do, I guess. I am under a Casuarina. The tree is big enough but the needle 'leaves' do not do the job. I stay only a couple of minutes then I decide it will not do. I get up and move again.

I find another spot. It offers more shade. I lie down. Baz arrives with the water and I take it from him and drink thirstily. We rest there for almost another half an hour, but I cannot settle. I want more shade yet.

I get up and move further down the road, carrying some food and the ground mat. Baz follows with the water. A mirage dances in front of me.

Baz spots some wild goats off the road, picking their way through shrubbery to the right. I am 10 or 12 m out in front of him when he tells me to come back to look. Fortunately, I can see the goats from where I stand. Big, shaggy ones. I was not walking back to Baz to see goats. I only move forward. I am not in a position to be walking over ground I have already put behind me.

Baz and I lug everything with us for another kilometre before I find what I am after—a clear, shady space to sprawl out and try to sleep. I am almost overheated and now I desperately need to rest.

Baz holds the big water container above my head and gives me a good drenching, tipping water into my face and onto my neck while I lay flat on my back. He tips so much water in my face that I struggle to breathe. I turn my head to the side and hold up my arm to signal for him to stop. He is slow to respond.

Baz goes to tip some more water over me.

'When I hold up my hand, you're supposed to stop. You know that, right?' I am just checking.

Baz is amused. 'It'd be bloody embarrassing to drown out here.'

The shade feels good and, with the drenching I receive from Baz, I feel my body temperature slowly dropping. Baz helps me to get set up with things close at hand—a little food and lots of water. I have what I need.

Baz pushes on, leaving me to continue to rest, sprawled out on the ground, with his ground mat beneath me. I hate it out here. It is horrible. My yearnings for wandering in the Outback are completely satisfied. I am over it.

Whether he was right or not, Alfred George Stephens, an editor at the *Bulletin*, said of Henry: 'Lawson hates the bush; he is blind to its beauty, deaf to the cheerful strain heard continually through the sighing of the wilderness. His six months' journey to the Queensland border in 1892...was like the journey of a damned soul swagging it in Purgatory.'[6] Well, bugger you, Mr Stephens. Were you ever out here? Beneath the mind-destroying scream of the cicadas, I hear no 'cheerful strain' of a 'sighing wilderness'.

I cannot wait to be finished. I can already hear myself when I get back to Sydney, rehearsing Henry's lines: 'I am back from up the country—very sorry that I went.' I run the lines through my head over and over and over again. Like Henry, I am dreaming of spending the rest of my life 'Drinking beer and lemon-squashes, taking baths and cooling down.'

I never want to see the Outback again, let alone to walk in it.

≈

Long before Henry arrived in Bourke, he was already a celebrity among the literati as well as the everyday people who recognised their own struggles within his stories and poems. His work had appeared in several newspapers, including Sydney's popular *Bulletin*. The poems,

'A Song of the Republic,' 'The Wreck of the Derry Castle,' and 'Andy's Gone With Cattle' all attracted enthusiastic followers. All three were published before his 21st birthday. But it was 'Faces in the Street' that set people alight with its sympathetic portrayal of the down and outs. Young Henry remained unaffected. A friend recalled, 'His "Faces in the Street" had made a sensation, and readers were watching for more....It was more than enough to turn the head of youth, even poetic youth, yet Lawson's head remained absolutely steady. He was not so much flattered as surprised.' In 1889, the men at the *Bulletin* pulled together the best work of their talented group of contributors to publish a book compilation entitled *A Golden Shanty*. 'Faces in the Street' was selected for inclusion, as was young Henry's prose story, 'His Father's Mate'.

≈

I rest until 4.30 and then take to the Track again. I have only been walking for 15 minutes when Moz and Jane return from Bourke, catching me in the kilometre between the entrance way turn offs to South Kerribree to my left and North Kerribree to my right.

'How are you poppin' up?' Moz asks, repeating a phrase he has borrowed from Henry[7] and has adopted as both his greeting to me and his form of enquiry into my welfare and potential needs.

I do not tell them that I have been dangerously close to overheating. They tell me about their time in Bourke, enjoying a cold drink in the air-conditioned comfort of the Port O' Bourke Hotel. Moz says that the proprietor of the hotel, Lachlan Ford, was very surprised to see him. Lochie did a double take. 'What are you doing here?'

I wonder how Moz felt about that. I know I would have been embarrassed as we have only been on the Track for three days since we left Lochie.

'Anyway, you are looking a familiar sight, Grog,' Moz says, changing the subject. 'Seeing you today reminds me of your style from last time we were out here together. You are like a tractor. You just keep chugging away.'

I smile and nod. 'Just keep chugging away.'

Moz and Jane drive on to find Baz and then to set up our camp for the night.

I am moving along alright—chugging away. At 5.15 pm, a big semi trailer rumbles up behind me, breathing dust like a dragon snorting smoke. The truckie brings his rig to a shuddering halt. The passenger door opens and I see a smiling woman and a couple of inquisitive kids beside the burly driver.

'Are you okay?' the woman asks.

I wander over for a brief chat and tell them what I am doing out here, alone. They seem to think I am crazy, but the broad smiles remain on their four faces.

Convinced that I am probably only crazy enough to do damage to myself and no one else, they drive off, the truck seemingly heaving a sigh as it pulls away and is soon lost again in its trail of dust. I wonder about the 'poor, tortured' bullock teams that would once have worked along this road, heaving their burdens onward, 'tugging and slipping, and moving by inches'.

With eyes half-shut to the blinding dust,
And necks to the dust bent low,
The beasts are pulling as bullocks must;
And the shining tires might almost rust
While the spokes are turning slow.

≈

Henry's flame burned brighter and brighter. In the five years before his journey to Bourke in 1892, he published several works in a variety of newspapers. His tribute to the heroic women of the Outback, 'The Drover's Wife,' won loud praise. He became known for his socialist and patriotic writing. He published several pieces singing the praises of those involved in the Eureka Stockade because he felt that Australians needed to learn more about Australia. In an 1888 piece Henry wrote: 'It is quite time that our children were taught a little more about their country, for shame's sake. Are they always to be "Colonials" and not "Australians"?'

≈

At 5.40, I approach the bridge over the swollen Kerribree Creek. The bridge is perhaps 30 or 40 m from one side to the other. There are a dozen Red Angus cows standing on the opposite side of the bridge, with more than a dozen spread down the banks and onto the grass on each side of the road. They are big and fat, obviously enjoying the lush green grass that grows about the creek. My eye is drawn to the huge bull under a lone tree to the right. He has a sack that drags to his knees. The bull's head is enormous. His body is enormous. The brisket is a wall of muscle, as are the shoulders and rear flanks. Gee, he is imposing. He does not seem at all interested in me though, and that is a good thing. He is certainly the king of his domain and I see many calves spread throughout the herd. Some of them appear only a few weeks old. When they see me, a few of the younger calves run to their mothers' sides. The running calves alert those few in the herd that had not seen me approaching. My eyes stay with the bull but he seems determined not to condescend to acknowledging me.

The way the herd is spread, I will have to walk between them. I would much rather they were all to one side. I would hate to get between a mother and her calf. Henry's 'The Story of Malachi' is running through my head, sounding alarm bells. In 'a roar, a rush, and a cloud of dust,' poor Malachi, the butt of all the station hands' jokes, was trampled to death by a cow with a newborn calf.[8]

I start to cross the bridge but only one of the cows moves down the bank to the right, seductively dancing her way toward where old Ferdinand is resting under his tree. About 10 or 11 cows remain on the bridge, including a couple of calves. As I move closer, a couple of cows turn as if thinking about moving out of my way. The other cows stand and stare, not moving a centimetre. I pause in the middle of the bridge.

'Yaaah,' I yell, waving my arms in the air. The cow in the front bats her eyelids but offers me no other movement. 'Yaah!' Nothing.

I take a few steps forward.

The front cow takes a piss, a torrent spraying over the hot, sealed road.

'Bugger off!' I yell.

I walk further forward and engage in a staring match with the lead cow.

'Bugger off!' I yell again.

Nothing. Not even the twitch of an ear.

'Piss off, you fat bitch!'

'Sticks and stones...' the cow seems to say.

It gives me an idea. I retreat until I am back off the bridge, where I can pick up a few loose rocks. I bounce them down the road without effect. I pick up a couple more and move closer. I hurl the rocks at the cows but my arm is weak and my aim is bad. All the same, a couple of cows get bored by me and move off the bridge.

After the bridge, the road makes a long, sweeping bend to the right. There are plenty of trees if I can get to the apex of the curve but, before that, there are bugger all places I might scamper to if I need to run and hide. The best climbing tree has the bull sitting under it, dreaming of which heifer he might have sex with tonight, so I don't figure I will be running toward that tree. To the left side of the road, there are only a few wispy bushes until the tree line that runs along the road from the middle of the bend. There is a fence on the left, perhaps 60 m from the road, but there are cows on both sides of the fence. If a cow decides to run at me, I have nowhere to go. What's more, my feet are letting me know that they do not fancy trying to outrun a maddened mother cow anyway. I try to calculate the state of my feet multiplied by my determination to live. I figure that, in my current state, I can outrun a mad cow for five metres, tops. Presuming I would have a head start (in that I will not wait for the cow to be on top of me before I run), I might be able to stay in front for fifteen or twenty metres. The nearest tree that I might try to scramble up, or at least duck behind, must be 100 or 120 m from me. Bugger.

I retreat back off the bridge. There is no shade out here and, despite the build up of clouds, it is still sweltering and is very humid. I decide to go down to the creek to cool off. Perhaps while I am dunking my hat and bandana in the water, the cows will move away from the bridge.

The water is refreshing.

I climb back to the bridge, my feet occasionally slipping in the loose rocks. I venture onto the bridge and walk forward, hurling a couple more rocks and some more expletives. Some cows have moved, but three remain, blocking my way and looking determined not to allow me across.

'Who's that walking on my bridge?' The front cow's eyes say a lot that her mouth does not.

I wonder how Baz got across. The bloody cows must not have been here when he crossed. If they were, the bugger probably told them to wait for his brother. 'He's bigger and juicier than me.' I try the same tactic, telling the cows that there are three Bryan brothers and the biggest and juiciest is still behind me. The lead cow does not believe me for a minute.

I retreat again. I sit on the guardrail and watch a flock of 50 Australian Wood Ducks fly in and settle into a pool of the creek.

I scramble down to the water again and fill my hat and tip the water over my head.

I climb back and return to my position sitting on the bridge guardrail.

The lead cow takes a few steps toward me.

If only the truck that had stopped for me had been running a little behind schedule. I wish another vehicle would come along. I know the old Bushmen out here will laugh at me, but I am not above asking someone to use their car to shepherd me across the bridge.

I sit and wait.

'It is about time Moz or Jane came along to see if I need a drink or something to eat,' I tell the cows.

I wait some more.

Eventually, I see Moz's vehicle on the sweeping bend before me. I return to my feet and move forward. The vehicle reaches the cows on the bridge but they do not move. The car stops. I see that Jane is alone in the vehicle. There are times when I think Moz loves that car more than anything in the world—more than he loves Jane even. She is not going to want to dent it against some stubborn old cow. Jane edges forward slowly, but the cows barely budge. Jane pushes on, slower still. Reluctantly—very reluctantly—the cows give ground. They slowly move to one side, allowing the car through.

'Boy, am I glad to see you,' I say to Jane.

'They wouldn't move, would they?' Jane says. 'How long have you been here?'

'Thirty-five minutes or more.'

Jane shakes her head.

'Can you just shepherd me across?' I ask. 'If you keep the car between me and the cows, I'll stay close and jump in if I need to.'

'I'll just have to turn around,' Jane says.

I sit on the guardrail and wait for Jane to pass over the bridge and turn back. The three stubborn cows remain on the side of the bridge and watch as I walk by with the car as my shield.

'Don't forget that you have to pass back this way.' The lead cow's eyes indicate that she knows all about my plan to walk to Hungerford *and back*.

Jane carefully shepherds me to safety and gives me water to top up my supplies. Before she drives back towards the camp, she tells me I have another eight kilometres or so to get to the camp. That will take me an hour-and-a-half or two hours. There is nothing else for me to do but put my head down and my tail up.

The sky grows grey and the wind picks up sharply. I hear thunder rumbling in the distance behind me, somewhere near Fords Bridge. A four-wheel drive passes me, heading into the approaching storm. At 6.45 pm, a gust of wind blows the hat from my head. I watch the branches of trees swirling around in the wind and I notice the dip in the temperature. Big drops of rain fall on me. Before I know it, I am pelted by hailstones. The bigger ones are the size of my thumbnail or, with Henry in mind, as big as bullets. He wrote:

> *One day I was out in the bush lookin' for timber, when the biggest storm ever knowed in that place come on. There was hail in it, too, as big as bullets, and if I hadn't got behind a stump and crouched down in time I'd have been riddled like a—like a bushranger.*

I see lightning in the distance and watch with concern as it comes closer. The gaps between the lightning flashes and thunderclaps get increasingly smaller. Suddenly, an ear-splitting crash shakes the sky above me. Lightning flashes around me. I fear that I might get struck and try to recall what I have been taught about lightning.

My school teachers always said, 'Don't seek shelter under a tree.'

Anything I have read about lighting has warned, 'Don't get caught out in the open.'

I have trees lining both sides of the road. Alternatively, the middle of the road is out in the open. The pelting rain soaks me while the red clay from the road surface bounces up and covers the bottoms of my tracksuit pants in a layer of sticky red dirt.

My metal pannikin is waving around at the top of my pack, where I have it Velcroed onto the pack. I take off my pack and detach the cup.

Within 15 minutes, the rain eases and settles into a steady rhythm that keeps up for over an hour.

At 7.45 pm, Moz appears in his car. 'How are yer poppin' up?' he drones.

'Good.'

'Isn't this great?' he says to me, sweeping his arm around to indicate the cooling rain.

It is easy to appreciate a storm from the safety of a car. 'I'm not real fond of the lightning.'

'Oh, right.' My soaked body has served as a wet blanket over Moz's enthusiasm. After all, I am the poor bugger still out here walking the road when it is almost 8.00 pm. 'It was not raining at the camp. We didn't know you were out in a storm.'

'Here, take this please,' I say, handing Moz my cup. 'I wasn't sure if it'd be a lightning rod.'

'Your camera is getting wet,' Moz notes. 'Maybe you better give me that, too.'

My voice recorder must be wet too. I decide just to give Moz my whole pack.

Moz tells me that he, Jane, and Baz have set up three tents for tonight. He and Jane will share the biggest tent and that will allow Baz and me to each have a tent. Like a good boy scout, Moz came prepared. It will be nice to have a tent and not to have to fight the mozzies again.

'It has cooled off a lot,' Moz says. 'Back at camp before I drove out to meet you, in just ten minutes the temperature dropped from 38 to 26 degrees.'

'Wow. Ten minutes? Twelve degrees in ten minutes?'

'You only have a couple of kilometres to go now,' Moz tells me. He takes a couple of photographs of me walking in the rain, soaked to the skin, and then he drives back to the camp.

I finish the remaining two kilometres and walk into camp, arriving almost three hours behind Baz. I head straight for the small campfire. I am wet through. 'Wouldn't you know it,' I think. 'I'm almost cooked during the day and then I get wet and chilled just before I want to get to bed.' Someone has set up a folding chair beside the fire and I am ushered to that seat.

I warm up a little and then ask which tent is for me. I have the smallest tent but, they tell me, for no other reason than that that was where they tossed my pack when the rain started.

'You can have the other one if you want,' Baz offers. He is ever ready to put himself out to make me comfortable.

'This one'll do,' I say.

Jane is inside her tent and I ask her to stay inside while I change into dry clothing. I use a towel to dry my hair and, with dry clothes, I soon feel much warmer. I notice as I change though that my crotch is developing a nasty red rash. 'Crotch rot,' I call it—an unavoidable malady out here in the dust and the heat when one is walking 30 or 35 km each day. I remember the last trek out here with Moz. 'For a man who is faithful to his wife,' I said, looking at the ghastly mess about my inner thighs, 'I don't deserve to have a crotch that looks like that.'

≈

Although 'Faces in the Street' and 'His Father's Mate' had been included in the *Bulletin*'s book compilation, *A Golden Shanty*, when he came out here to Bourke, Henry had not yet had a book published that consisted only of his work. This changed when his mother decided to take matters into her own hands. In 1894, Louisa published *Short Stories in Prose and Verse*, which, as the name suggests, consisted of a collection of both poems and prose stories.

Three years later, Henry grumbled that his mother had made a lot of money from the book but that he had received very little of it. After all, Henry had complained to a mate at the time of the book's publication, 'My mother's the hardest business man I ever met.' With a mother who was a noted and well-connected figure in Sydney's literary scene; however, there is no doubt that Henry benefitted well from Louisa's influence.

≈

Moz has picked cans of food from the back of the car and he and Baz tend to them in the fire pit. Moz has chosen ravioli for me. He remembers that I liked it last time we were out on this Track. It goes down well and warms any remaining chills from my body.

'Well, that was an interesting day,' I say, thinking back on all that had taken place since we left Fords Bridge in the morning.

'How so?' Jane asks. It seems an odd question. Is it not obvious?

'The water crossing...getting held up by a herd of cows...being caught in a hailstorm... and I almost overheated this afternoon too.' I deliberately exclude mention of Moz's withdrawal this morning because I suspect Jane is fishing for me to say something about that.

'You almost overheated?'

'Yeah, I couldn't cool down.'

I crawl into my tent, buggered, but thankfully cool.

DAY FOUR

Henry Lawson and Mateship

A bushman has always a mate to comfort him and argue with him, and work and tramp and drink with him, and lend him quids when he's hard up, and call him a b—— fool, and fight him sometimes; to abuse him to his face and defend his name behind his back; to bear false witness and perjure his soul for his sake; to lie to the girl for him if he's single, and to his wife if he's married.

- That Pretty Girl in the Army, *Children of the Bush*, 1902

The man who hasn't a male mate is a lonely man indeed, or a strange man, though he have a wife and family. I believe there are few such men.

- Mateship, *Triangles of Life and Other Stories*, 1913

In the morning, Moz is cheerful and happy. The weight of the world seems to have lifted from his shoulders and I find his demeanour interesting. In his position, I would be grumpy and morose, eaten away by disappointment.

When I hit the Track at 7.10, I wonder what misery it has in store for me today. As I hobble away from camp, the blisters on my feet shoot pain through my body. Not being able to put weight to the front or the back of my feet makes walking bloody difficult and bloody painful. When Henry explained his discoveries from the Track, he concluded that the things he had written 'might be new, interesting, and perhaps startling to most Australian poets'. Then, in classic understatement, Henry added, 'We found it was also painful.'

I hope that my feet will loosen up in 15 or 20 minutes and allow me to start to stride out with a bit more surety and a lot less pain. Other than that, mentally, I feel alright as I start Day Four. It is nice to be starting the day cooler, not having had to contend with heat through the night.

Before I set off, I drank lots of water at camp. Moz also cooked me a couple of pieces of toast and spread some Vegemite over them. I walk with them now, munching as I go. At the moment, I cannot think of anything that would be a better way to start my morning than with Vegemite on toast. Good old Mozza. Before I left the campsite, I said to him, 'Do you want me to pack up the tent?'

'No,' Moz replied emphatically. 'I'm hardly going to give you something that creates extra work for you to do.' He will pack up the tent.

'Thanks a lot, mate.' I mean it. He is bloody considerate. He knows what it is like out here on the Track. He understands. It occurs to me that the fact that Moz has pulled out will increase the likelihood that I will make it. Having him to help Jane will increase the amount of assistance that I receive but, because of his inside knowledge, there will be some things that Moz is aware of that Jane might not otherwise consider. He will do anything and everything that he can for me, as a mate. Henry was like that too. He was generous, even to a fault. One mate said that Henry 'couldn't say no to a friend'. Indeed, friends did not even need to ask. One time Henry and an unemployed mate met in a park before the friend attended a job interview. Seeing the ragged state of his mate's shoes—for the man had been out of work for some time—Henry immediately removed his own new boots and the two exchanged footwear.

The sky is a soft powder blue, getting darker and brighter above me. A few wispy clouds remain from last night's storm but as the sun gets hotter, they will burn off. The Track has a slightly damp appearance, the rain having settled the dust. The shadows cast from the tallest of the area's small eucalypts and acacias stretch almost to the centre of the road, running from right to left. If there were not so much roadside grass to weave through, one could get good shade walking off the road, along the verge beside the trees. The grass is full of burrs though. I walk down the centre of the road, where it is flattest, avoiding the strain on knees and ankles otherwise caused by the road's camber.

Baz started only about five minutes behind me and before I have gone very far he catches up. He slows his pace to accommodate me and I tell him that I will warm up soon enough and be able to hit a decent stride. Until then, I can raise little more than a shuffle. Baz is up for a chat though and so he stays by my side. I tell him about my earlier realisation that Moz's withdrawal from the walk actually increases our—certainly *my*—chances of successfully completing our mission. Baz agrees. Moz's inside knowledge will be helpful.

Baz and I walk alongside one another for two hours, talking and catching up. He notices that I am obviously feeling okay because I am willing to chat. He certainly knows that is not always the case. In an email I receive from Baz in February, after we have recovered from our ordeal, Baz writes to me, 'You were "hard work" at times in those first few days when you were really struggling and when I was lucky to get a grunt of acknowledgement out of you (let alone a conversation!).' Cheeky bugger. Anyway, I give him the conversation he wants this morning.

We talk about home, Mum and Dad and about growing up. When she was younger, Mum could really fire up. And just as Henry heard from his mother, 'You'll drive me mad,' we certainly heard it from ours.

Your father's coming home to-night;
You'll catch it hot, you'll see.
Now go and wash your filthy face
And come and get your tea.

Moz and Jane drive ahead to a spot a couple of hours' walk down the road. When Baz and I get there, the first thing Moz says is, 'How are yer poppin' up?' I tell him I am fine.

We eat and then enjoy a rest. With all the moisture from last night's storm, the air is humid and sticky. Sweat cascades down my body. Moz says he was told the storm was huge in Fords Bridge last night. They apparently received 150 points of rain—over 50 millimetres. As bad as it was on me, it was worse behind me.

Moz and Jane plan to drive ahead to check the depth of the water crossing at Cuttaburra Creek. 'We'll see you at lunchtime,' Moz says as they leave at ten o'clock.

Baz and I return to the Track. I have a long, difficult time getting going again after having stopped for almost an hour so I tell Baz not to wait for me. The way my feet feel, I am all talked out.

The sun is shining brightly now and, although the mammal life has largely retreated to the cover of shade, the cold-blooded lizards are enjoying the warmth after receiving a drenching the previous evening. I see many Eastern Bearded Dragons and the odd Shingleback or Stumpy-Tail.

Baz is a couple of kilometres out in front of me. As I walk along, head bowed, I come across a large arrow scratched into the dirt. My curiosity is satisfied when I see that the arrow points directly at a lizard at the side of the road. It is a Central Netted Dragon, close to the eastern perimeter of its range. A field guide tells me that, when approached, a typical Central Netted Dragon 'swiftly retreats to the safety of its burrow,' yet this one does not move as I get to within three or four metres to take photos. It has not moved an inch in the half an hour since Baz passed. Out here, it is too hot to move. Yet, I push on.

At 11.14, I pass a road sign indicating it is 110 km back to Bourke and 105 on to Hungerford. I am over halfway to Hungerford. This knowledge lifts my flagging spirits. I decide to push on until 11.30 and to then rest.

I note that I always seem to be in the sun. Although a cloud offers some shade, the shadow of that cloud seems to be just ahead of me. I see an area ahead that is in shadow, but when I get to it, the shadow has moved ahead and I am in full sun. I see where the shadow now falls and move ahead to that spot, only to find that by the time I get there, it is in full sun and the shadow has moved further ahead. It is an endless game of shadow chasing and I never win.

By 11.30, I am in need of a rest. The previous five minutes have been bad. The arm broke from my sunglasses, which was a kick in the teeth I could not afford. Out here, sunglasses are essential. They remove the blinding glare from the hot road and, with the sun shaded; psychologically they make it seem cooler than it really is. A minute later, while I am still coming to grips with my broken sunglasses, a burr starts to cut into the big toe on my right foot making it necessary to sit down and remove my shoe. In doing so, I got sand down my back while I was tugging at my shoe and as soon as I stood up, some of the sand fell down into my underwear. I did not want that. My bum is no oyster and having sand rubbing between my cheeks is not about to create a pearl. I know from experience on the Track last time that it will create a nasty and painful rash. 'Butt rot,' I call it—a close cousin to crotch rot. I pull my pants down and try to brush my bum clean.

≈

At 12.30 pm, a pig shooter named Terry Bates stops and asks if I need help. I tell him that I am out here with my brother walking to Hungerford. He looks around for my brother but, by this time, I expect he is a few kilometres out in front of me.

'What, did he just leave you?' Terry asks in surprise.

I explain that, out here, you just have to go when you can go. It does not work to try to stick together.

'What are you walking to Hungerford for?'

'We're walking in Henry Lawson's footsteps.'

'You're effing mad!'

He is probably right. I laugh and tell him that I have already walked to Hungerford once before.

'You're effing mad,' Terry repeats. 'Why are you doing it again?'

'Well, because Henry did it in the summer. I figured I needed to do it in the summer too.'

'You know that Henry did most of his walking at night, don't you?'

It is certainly possible. There is no doubt from his writing, though, that Henry did a lot of walking in the full sun. He does not write much about walking at night. 'Hmm. Possibly,' I concede.

'You'll have to come back again and do it at night.'

I laugh. I am not coming back.

Terry proceeds to tell me about some photographs that he claims to have seen of Henry Lawson on the Track. 'Not many people know about it,' Terry says. 'Not many people have seen them, but I have.' He tells me that he saw the photos at a place called Warroo Station. I misunderstand him because I think he tells me the station is south of

Hungerford, on the road to Wanaaring. We will not be going that way and so I do not worry about getting details.

'What are you going to do when you get to Hungerford?' Terry asks.

I am not telling him about the return journey. 'I'll just look at all the beautiful women there, I guess.' It is the first thing that comes to mind.

Terry laughs, thinking of Hungerford and its permanent population of seven or eight individuals. 'You're gonna be disappointed.'

I laugh too.

'Hungerford...' Terry says, 'Old Henry said it was the arse end of the world.'

I laugh again. Henry did not think very highly of Hungerford and the more I speak with Terry, the more I suspect he might share that opinion.

I ask Terry if he knows Andrew Hull. Hully is a Bourke poet who completed the walk to Hungerford with his mate, Tonchi, a few years before Moz and I completed our trek.

'I offered Hully a ride in the truck when he was out here,' Terry says. 'He wouldn't take it...do *you* want a ride?'

I laugh again. I am here to walk...or stagger, I guess.

'If I was you, I'd jump in the truck for ten kilometres.'

≈

Henry had a well-developed sense of humour, albeit one playfully described as 'quaint and twisted'. One mate said that Henry's sense of humour 'ran deep' and was 'his chief characteristic'. Henry and his mates loved nothing more than to engage in 'pranktical jokes' on one another. One Christmas Eve, Jack Moses and Henry were walking home through downtown Sydney. Under his arm, Moses carried a live rooster that would serve as the following day's roast dinner. Amongst the hustle and bustle of crowds rushing to get home to start their Christmas celebrations, Henry discretely cut the string binding the rooster's legs and gave the bird a shove, sending it flapping its way through the pedestrian and vehicular traffic. Some days later, Moses received a summons to appear in court for disturbing the peace and causing damage to property. Moses spent a whole day at the courthouse waiting to be called before the judge. At day's end, a clerk of the court informed Moses that his name had been called and that he had been fined for failing to appear. Moses protested that he had waited all day but had not heard his name called. The clerk said that he was authorised to collect the ten shilling fine on behalf of the court. Later that evening, Moses found Henry and the so-called clerk in a pub making merry on the proceeds of the supposed fine. This was typical of Henry. His escapades were described as 'hilarious business, with a touch of enjoyable unlawfulness thrown in.'

On another occasion, one of Henry's friends was accosted on a dark night. With a revolver stuck into the nape of his neck, the man was forced to make his way homeward to provide the desperate criminal with food and supplies. Imagine the surprise of the family when the head of the house arrived home pale and shaken with his good friend Henry Lawson standing behind him with a tobacco pipe jabbed into the man's neck.

For good and for bad, Henry was a Peter Pan figure—a child who never grew up—and remained 'a child with a singing soul in a world of business men.' This characteristic undoubtedly created difficulties for Henry but also created fun for his friends. 'What kids they were, Lawson and his associates, in those careless days!' said his mate, Jack Brereton, and Henry agreed, 'We were not content with common jokes...we aspired to some of the higher branches of the practical joker's art.'

≈

I ask Terry about how to get to Lake Eliza. Henry stopped there and wrote a poem about the experience. Terry tells me what to watch for. 'Up ahead, where the power line crosses the road, you walk in to the left there and follow the power line.'

I would like to get there, but I doubt I will have the energy to go off the Track.

Terry believes we will not get Moz's vehicle through Cuttaburra Creek. He also cautions that there is a worse crossing about eight kilometres past the Cuttaburra at a place called Back Creek. Although not as deep, the road is washed out there and will be a dangerous crossing to attempt. To avoid the crossing himself, Terry made a detour this morning on his way to Bourke. He says that Moz will have to turn west from Yantabulla, heading toward Willara Crossing and then eventually turn north to Hungerford. He will then be able to drive back toward us on the road to Bourke. It sounds like a long way around but at least if Baz and I manage to make the water crossing on foot, Moz eventually will be able to get the support vehicle back to us.

'What were you doing in Bourke this morning, Terry?' I ask.

'I had to get cold and flu tablets because I've been feeling crook.'

Involuntarily, I take a step back from his Ute. I do not know if he notices but I figure that picking up another flu on top of my own dose of what Baz is calling the Swine Flu is hardly what I need right now.

'I had the bloody flu shot,' Terry says, 'and now this is the first time I've ever had a cold in the summer.' The longer I chat with Terry, the more he reminds me of my father. 'I ought to tell the doctor he can shove his needles up his black arse.' Terry and my father not only share the same sentiments, but they even use the same words to express those sentiments.

I have been chatting with Terry for more than half an hour. 'Anyway, I guess I better not hold you up any longer,' I say, thinking that I better get moving or I will never see Baz again.

Terry bids me farewell and moves away. I am surprised to see him pull off to the right-hand side of the road only about a kilometre out in front of me. As I eventually draw closer, I see Baz seated in the shade by the side of the road. Terry has stopped to regale Baz with his sense of humour too.

Baz later tells me that the first topic of conversation is the previous night's rain. Terry scratches his hand in the sand and notes that, despite the day's sun, there is 'still a bit of moisture in the ground.'

Baz can barely contain his mirth as Terry scratches about where Baz relieved himself just a few minutes before Terry's arrival.

By the time I catch up, Terry is sitting back inside his vehicle. I mention to Baz that Terry said Moz will not get his car through Cuttaburra Creek. Terry then has Baz and me in stitches as he recalls an earlier time of flood, when he found himself driving behind a bloke towing a caravan through the Cuttaburra crossing. Unbeknownst to the driver, water flooded into the caravan, making it increasingly heavier and heavier until the caravan's back wall eventually surrendered to the pressure and gave way. All of the furniture and supplies within the caravan started soaking out the back end, scattering things all along the creek. The further the lead vehicle went, the more the caravan fell apart until, as Terry said, 'By the time the bloke got over the crossing, his caravan had turned into a nice little trailer.'

'You've lost all you're things,' Terry told the tourist, helpfully, as they surveyed the damage.

Incredulous that Terry had done nothing about the events that had unfolded in front of him; the bloke asked Terry, 'Well, did you pick any of it up for me?' The tourist was looking for some of that friendly assistance for which country folk are renowned.

'No.'

'Why not?'

'Well, it was no effing good to me.'

Terry is an entertaining storyteller, but I am mindful that Henry had not even been in Bourke more than a few hours when he wrote, 'I have already found out that Bushmen are the biggest liars that ever the Lord created.'

Remembering what I told him I was planning to do in Hungerford, Terry asks Baz what he is planning when he gets to Hungerford.

'Just like Lawson, we are going to turn around and walk back to Bourke.'

Terry looks at me long and hard. 'You're effing madder than I thought you were.'

'What are you eating out here?' Terry wants to know.

'Tuna and crackers.'

'Eff me,' Terry says. 'You're eating like black fellas.' Yet another sentence straight from Dad's mouth.

'Well, what do you expect we'd be eating?' I ask.

'If it was me, I'd be expecting ice-creams,' Terry replies, wiping sweat from his brow.

I look about at the emptiness surrounding us. 'There aren't a lot of ice-cream shops out here.'

'I'd be telling my bloody support crew to get off their arses and bring me an ice-cream.' I can imagine the melted mess an ice-cream would be before it reached me. Rather than licking it from a cone, I would be licking it straight from Moz's hand.

None too subtly, Terry looks directly at me and says, '*You're* gonna lose a lot of weight out here.' He emphasises the word 'you're'—no mention of Baz.

Terry updates us on the cricket score. Australia is falling further and further behind in the quest for the Ashes. 'Serves them effing right,' Terry says. 'They need to get some young fellas in the side.'

'It's Ponting's fault,' Terry says.

Baz and I are big supporters of Ricky Ponting. I am not about to sit idly by while Terry joins in the growing chorus of Ponting knockers.

'How the hell is it Punter's fault?' I demand.

'He wanted his own team. He pushed all the old players out.' He proceeds to name the old players who Punter has supposedly forced into retirement—Langer, Hayden, Warne, McGrath, Gilchrist. I do not bother to point out to Terry that his argument somehow lacks logic—he wants more young players in the side but he is also complaining about the older players supposedly being forced out.

I also fail to see why the captain of any side would willingly choose to deny himself of the services of some of the game's greatest ever players. 'That's ridiculous.'

'It's not ridiculous.'

As we continue out discussion of the Australian cricket team, Terry mentions another great Australian cricketer. 'I go pig shooting with Glenn McGrath.' McGrath is from up this way—Narromine. We drove through it on the way to Bourke and passed the statue honouring McGrath's great contribution to Australian cricket. Terry then tells us that he has also taken Brett Lee pig shooting. 'He's fast,' Terry says. 'He just jumps out and runs them down.' I think of Baz's description of the young doggers' video and the chest and throat guards the dogs wear as suits of armour to prevent them being torn to shreds by a boar's razor sharp tusks. I cannot see one of the hottest properties in world cricket wrestling wild pigs.

After almost two hours of entertaining us with his humour and his stories, Terry eventually decides he has more important things to do than to sit and talk with a couple of mad so-and-sos. He suddenly realises he has misplaced his keys. As he digs around in his pockets, I suggest he might have to get out and walk alongside us.

'You're effing mad,' he says, shaking his head as he extracts his keys from a pocket. He drives away, leaving behind clouds of dust, and an audience of two highly amused, fully entertained spectators. Terry is the type of Outback character who might just make all this suffering worthwhile.

I tell Baz that I am going to walk down the road to look for a shadier spot. I lug the big water container with me, checking any likely-looking spots where I might stop and be comfortable and sheltered from the furnace blazing in the sky. It is 38 degrees and, although layers of stratocumulus clouds are building up on the horizon, at present they do nothing to filter the rays of a sun that seems almost forever to hang high in the sky.

It is approaching 3.00 pm and, despite telling us they would be back for lunch, we have not seen Moz or Jane since the support vehicle left at 10.00 am. Given they have gone ahead to check the depth of the water over the road, we both worry about what might have happened. Moz's vehicle is his pride and joy. In the absence of a child or a pet in his life, I suspect many of the warm spots in the depths of his heart are reserved for his vehicle.

'Bloody hell, I hope nothing has happened to his car,' I say.

'Hopefully they're not stuck somewhere.'

'Where could they be?' I ask. I imagine the car stalled in a stinking, dirty swamp, the engine seized, smoke billowing from the bonnet and Moz sitting behind the steering wheel blubbering like a baby. 'Gee, I hope they're alright.'

We walk on some more, me ever searching for that elusive shaded spot. There are lots of small trees but few big ones. Few stand beyond 10 or 12 ft in height—about three to three-and-a-half metres. There are also lots of small scrubby bushes. Because the heat is so extreme, they do not offer much shade. They are made to withstand the heat and, therefore, they do not have very much foliage. Any leaves that they do have are typically thin or needle-like. The she-oaks have needles like a pine tree, while the Mulga and other acacias have elongated foliage more reminiscent of needles than leaves. Even the taller trees—the eucalypts like the Black Box trees and Desert Bloodwoods that sometimes rise two times above the height of the dominant Mulga—have elongated, almost needle-shaped leaves. The thin leaves, and the general sparsity of those leaves, means the trees out here do not offer a lot of shade at any time, but particularly when the sun is high through the middle of the day and the shade basically falls only at the trunk. One has to get right in next to the tree to get any shade. This tends to be where there are fallen sticks and pokey branches—and sometimes long grass—that make for a place that is not a comfortable one to be seated. It is also a place where it seems more likely to encounter an unseen snake. I want open spots, away from the trunks of the trees but, nonetheless, under the shade of those trees. Such spots do not exist through the several hours in the middle of the day.

I have grown fussy about shade because, having almost overheated, I know how critical it is that I am cool. And when resting, I am cooling. A rest stop should not be a time of continued or added heat. If I am getting hotter while resting, that defeats the purpose of the rest. I might as well keep moving.

As we wander down the road, Baz breaks the silence. 'Gregga, you know how a fella supposedly thinks of sex every few minutes?'

'Yeah.' I wonder where this is going. For all we know, alone out here in the middle of nowhere, we well might be the last two people left on Earth.

'Well, I have been so tired I haven't thought about sex even once out here in four days.'

I laugh. With my wife a world away in Canada and my crotch rot eating away at me, I doubt that I will ever think of sex again.

We push on, moving a couple of kilometres down the road before I find a shaded spot that I am happy with. I lie down for a good spell but, despite a bad headache, Baz decides to move on. I offer him a couple of painkillers from the little container in my pocket.

'What are they?' Baz asks.

'Buggered if I know but I have orange ones and white ones.'

'I still want to be eligible for the Olympics,' Baz jokes.

'Well, I can't guarantee that. These will drop an elephant.'

Baz seems reluctant but I force a couple of pills on him. As he leaves, Baz tells me that this is his hardest day yet. I tell him it is my best. My body is hardening. Just as Henry said, the first days on the Track are the hardest. They are 'twice as hard as any of the rest,' Henry wrote. I hope he was right.

≈

As I lay in the dirt, I keep a stick close by and occasionally give it a few taps on the ground. I do not want any snakes to inadvertently slither my way, not knowing that I am in the vicinity. I rest for a while but eventually decide that I must push on. I sit up and look to my left and am surprised to see, not 15 m from me, a black mother goat and her tiny white kid wander out of the Bush and onto the edge of the Track. The kid cannot be more than a couple of weeks old, if that. They are more surprised to see me than I am to see them and disappear back from where they came.

I get to my feet and, as I am taking my first few hobbled steps, a rabbit darts onto the Track from the right. It runs to the midpoint of the road but then, seeing me, it pauses then turns and dashes back into the Bush behind it. Gee, no one is getting where they want to go today. Then again, unlike the goats and the rabbit, I am not turning around.

After stopping to get out of the height of the sun's roar, my feet refuse to loosen up. Each step feels as if it is taken on bloody, pulpy stumps. My crotch rot is eating me alive and my butt rot is a painful mess.

I stop at a small lake to dunk my hat and my bandana—and to pour refreshing water over my head—and a vehicle pulls up. The vehicle holds two more of the roughest, toughest hombres I will ever see but, as with everyone out here, they are kind and considerate. They ask if I am alright. After they find out what I am doing out in the middle of nowhere, they offer me a drink and tell me that Baz is a few kilometres in front of me. There is a boy in the backseat. He looks on in wonder but holds his tongue. The driver asks me how far it is to walk from Bourke to Hungerford. I tell them it is about 220 km. Again, I do not tell them about the return leg. I am still not thinking that far ahead. The driver and his front seat passenger both agree that I am 'an effen mad bastard' for doing what I am doing. As with one or two others who have stopped earlier, the driver wants to know 'if it is hot enough' for me.

'You should have been here before Christmas,' the passenger says. 'It was much cooler.' Henry did not walk to Hungerford before Christmas though.

'Up until the last few days, it has been the best summer I can remember,' the driver says. 'The last few days have been hot though.' Thanks for the update.

The passenger tells me he will be back to check up on me tomorrow.

'I'll be driving the water truck from Yantabulla,' he says. 'I'll watch out for you.'

Not long after my visitors have disappeared, I pass by the spot where the telegraph wires cross over the road—the spot that Terry Bates had said was the place to turn off to Lake Eliza. As much as I would like to see it, I do not have the feet for detours at present.

The day's last several kilometres are long and hard. The featureless Track drains me and my brain shuts down. 'You might walk for twenty miles along this track without being able to fix a point in your mind,' Henry wrote. 'This is because of the everlasting, maddening sameness of the stunted trees—that monotony which makes a man long to break away and travel as far as trains can go, and sail as far as ships can sail—and farther.' Had the Track not drained me of the energy to do so, I would dream of escape.

After much toil and struggle, I get into camp just after 6.30 pm. I have left about 31 km in the dust behind me today. Just before I finish, I am pleased and relieved to see Moz and Jane, and the precious vehicle, all safe and sound. Baz has been in since 5.00, with the others arriving more than half an hour after him. Poor old Bazza spent that time mentally preparing himself for camping and sleeping out in the open without food, shelter or supplies, and with the worry about what might have happened to Moz and Jane.

The tents are set up in a row, beside a barbed wire fence. I hang some of my clothes on the fence to get a bit of air. I am pleased that my tent is at the end of the camp. It affords me more privacy as I lie in the tent in the nude. It is too hot to have the tent fly over the top but I am aware that the others, including Jane, can inadvertently look through the mesh of the tent and have their eyes affronted by the site of me, red all over with heat rash and, what's more, with a volcano of crotch rot bubbling away in my nether regions.

After, we all sit by the tiny campfire and eat tea. My tin of potatoes and bacon goes down a treat and, although I do not say anything, I wish there was another tin. It is the first time that I have wanted more. My appetite is returning as my body slowly grows accustomed to my toil. I am hardening to the task.

'Today was my worst day,' Baz says. 'All the walking and the lack of sleep must be catching up with me.'

I repeat my earlier assertion that today was my best day. Moz says that my posture was good today. I wonder at how informed his opinion is, given that I had seen neither hide nor hair of him from 10.00 am until 6.30 pm.

Earlier in the day, I told Baz about my broken sunglasses and—good old Bazza—he not only has a spare pair, but two spares from which to choose. I try them on now. Neither pair is ideal, but I decide to go with the ones that sit most comfortably on my face.

Moz draws a hearty laugh from Baz's and my tired bodies when he tells us about his solution to the difficulty he had getting to sleep the previous night. The Sydney city boy could not fall asleep because of all of the noise of the Bush. The cacophony of chirping crickets, buzzing mosquitoes, and croaking frogs had kept him awake. I point out the irony that such Bush choruses are often used for relaxation tapes that people use to help to put them to sleep.

'Not me,' Moz says, 'I had to stick wet wipes in my ears.' The wet wipes have been a Godsend for cleaning filthy hands, removing grit from between toes and wiping sweaty brows, but their use as earplugs was heretofore unknown.

'I'm hoping for a better night's sleep tonight.'

Moz and Jane tell us that they went on to Hungerford today and that they managed to drive through all of the water crossings. They tell us we have five crossings to negotiate tomorrow. I did not know there would be so many. I was hoping for only one or two, but to have to face five tomorrow is going to be a real drain. Each one will slow us down so much and add to my frustration.

Despite a vibrant sunset, I am in my tent and asleep before the sun disappears. Already, I am dreaming of completing tomorrow's fifth and final crossing and putting the floodwaters behind me.

DAY FIVE

Henry Lawson's Bush Battle with Banjo Paterson

For my beds were camp beds and tramp beds and damp beds,
And my beds were dry beds on drought-stricken ground,
Hard beds and soft beds, and wide beds and narrow—
For my beds were strange beds the wide world round.

- The Wander-Light, *Selected Poems of Henry Lawson*, 1918

From blacksoil plains burned bare with drought
Where years are sown that never grow—
From dead grey creeks of dreams and drought,
Through black-ridged wastes of weirdest woe.

- Till All the Bad Things Came Untrue, *For Australia*, 1913

It rains a little through the night and I hear Moz and Baz get up to stretch the flies over their tents. Because the tents were erected before I arrived at camp, I have no idea where I might find the fly for my tent and so the rain splashes through the mesh. It cools me down, but I have a difficult time trying to sleep with rain dripping on my face. Where it falls on my body it feels good, but I turn on my side and cover my head with the towel I have been using as a pillow. The rain never gets heavy, but it keeps up for half an hour or so. When it stops, I fall back to sleep, only to be reawakened at about 4.00 am by another light sprinkle of rain.

After the last of the rain, I fall back to sleep but I am awake again before 6.00 am. It is still too dark to get ready for the day's walking in the tight confines of my tent. I lie there until the first of the morning's light starts to chip away at darkness' defences. As my surroundings lighten, the first things I see are hundreds of mosquitoes on the mesh of the tent.

With the best of intentions of getting on the Track in a timely fashion so as to avoid some of the heat of the day, I start getting ready early but I am frustrated by how long it takes. In my cramped quarters, occasionally my back leans against the door of the tiny tent, allowing the mozzies to jab me through the mesh. Before long, I have an itchy back. I hear Baz getting ready as well. He is up and out of his tent before me.

I soon hear Baz talking about some scats located right beside our previous night's tiny campfire. Baz wonders if the droppings are from a Dingo or a fox. We are inside the state border Dingo fence, but that does not mean that there are no Dingoes here.

Baz takes to the Track at 6.45 am, but I am not ready to follow for almost another 30 minutes. During that time, I continue to struggle with bandaids and Vaseline, trying to manoeuvre my stiff legs into a position from which I can see and work on the blisters underneath my feet. In the cramped tent, my efforts are futile and, in resignation, I begin trying to work my socks over my feet without upsetting the bandaid and Vaseline work that I have done.

I hear Moz outside playing peekaboo with Jane through their tent. It is a typical demonstration of the light-hearted nature that Moz has exhibited ever since he withdrew from the tramp. Freed now due to the enormous weight that has been lifted from his shoulders, he is playful and carefree. Because I am so buggered and have nothing but misery to look forward to, the peekaboo playfulness irritates me no end. As I struggle with my second pair of socks, I keep telling myself that Moz is not doing anything wrong. He and Jane have to amuse themselves somehow. I know that the problem is all with me and that if there is anyone in the wrong, it is me. Yet, I cannot help but be irritated. I do not like feeling the way that I do, but in my pain I cannot help myself.

When I eventually emerge from my tent, I see that the scats are mere centimetres from where our small fire had been. They are elongated and stringy and it is not difficult to see the fur or hair of what were once living animals. Whatever it was that dropped the turds has walked right through our camp, obviously drawn by the prospect of an easy meal, perhaps straight from a tin.

When I crawled into the tent the night before, one thing I took with me was an empty orange juice bottle. Drinking between eight and ten litres of water each day, I unavoidably need to urinate through the night. Having grown increasingly frustrated by the need to hobble through the darkness to a place where Jane will not hear my pee splashing against the hard-baked earth, I decide being so delicate is no longer for me. From now on, I am sleeping with a piss bottle beside me. As such, one of my first chores for the day is to discretely empty the bottle. I am careful to return the emptied bottle to my bag. I do not want for it to be mistakenly filled with anything other than my pee.

I notice a rusty old tin lid that may have lain in the sand for years. I pick it up and place it in one of the bags we are using to store our rubbish. We are diligent about picking up all of our garbage and, with the tin lid in hand, I think that we can now say we have picked up 'everything, plus'—everything that was ours, plus someone else's garbage too. I want the place to be better for us being through it, rather than being worse for our presence.

Gee, I hate getting going in the mornings. As I take the first several hundred steps of the approximately 44,000 I will take today, I am as stiff as a board and, although it is my feet whose cries of pain are the loudest, there is not one part of me that does not hurt. My feet feel as if they constrict through the night and it takes a long while to get the blood pumping. In the meantime, to cover or hide my blistered areas from contact with the ground and the insides of my shoes, I involuntarily constrict my feet further, compounding my problems. I retrieve a couple of painkiller tablets from my pocket and swallow them with a big gulp of water.

As I hobble toward Hungerford, I wonder about how irritated I was by Moz's childish game of peekaboo. I know that they are unaware of my irritation and I am not sure that I was even irritated *by* them—I was just irritated. Perhaps subconsciously however, I am also irritated by the fact Moz has withdrawn and, as a result, he has the energy for such games when I do not. Moz refers to escaping inside a bubble when he is on the Track. When my wife and children drove in the support vehicle while we tackled the Track last time, he said they were intruding into that bubble—not through fault of their own, but just through their presence. He is right and I see that this is my problem this morning. Moz's repeated, 'peekaboos,' were an intrusion that rubbed away at me like poorly worn in shoes rubbing away at my feet, blistering the protective bubble around me. Like Henry, 'I was sulky, I was moody (I'm inclined to being broody).'

My thoughts turn from irritants when a gorgeous Red-capped Robin visits. It lands close by, left of the Track. It flits about briefly, keeping low to the ground while flicking its tail and wings. It might only be 10 or 12 cm from head to tail, but its vivid colours brighten my mood considerably. The red breast stands out vibrantly against its pure white body, just as the red forehead seems to shine from its jet-black head as the bird turns this way and that. I am mesmerised and stand transfixed. It is surely one of the most beautiful birds I have been lucky enough to behold. Alas, all too soon, the bird is gone, disappearing into the brush.

I notice a discarded Twisties packet on the side of the road. Oh, boy, Twisties would be so good right now. The sight of that golden packet induces an insatiable craving.

It takes more than half an hour for my feet to loosen although, even then, my calves still feel tightly knotted. That first 30 minutes consisted of slow, torturous walking—if that is what it can be called. Fortunately, I am moving a little better by the time Moz appears, having packed and put away the tents.

'How are yer poppin' up?' Moz drones in his mechanical, exaggerated imitation of a laid-back Bushy. Jane is not in the car with him but he tells me she is planning to walk the ten kilometres to where we will have breakfast.

'Have we got any Twisties?' I ask, hopefully.

'Nah.' Moz looks disappointed for me. Since taking on the role of support crew member he wants desperately to do all he can to make me feel better. 'Sorry.'

No Twisties? What I would do for a Twistie right now. Surely they will have Twisties in Hungerford, right? I guess I have just discovered what I would do for a Twistie: I will bloody well walk to Hungerford.

It has been a couple of days since Moz pulled the pin and so I venture to ask him when or if he is thinking about joining Baz and me back on the Track.

'No, I am going to let my feet rest and my blisters recover.' He says that he will reconsider once we get to Hungerford.

Before driving away, Moz tells me that I am looking good and that I appear to be moving well. Looks can be deceiving, I suppose.

Given the straight road, I can see Baz far in the distance ahead of me. I turn and see Jane behind me and, although I have a good head start, I am sure she is travelling at a better clip than am I, and so I expect she will catch up to me—probably within half an hour. My moods swing far and often out here but, mostly, I prefer just to be left alone. Very often, I do not even like seeing the support vehicle coming towards me, even if I expect that Moz or Jane are about to give me a nice cool drink, some tasty lollies, or an update on the cricket score. I figure I am best left alone, belying the notion that misery loves company. I want to have to think of nothing but just putting one foot in front of the other; not having to think about anything or anybody. 'Just keep going, just keep going,' I tell myself. Although I think not of a bubble like Moz—I prefer to think along the lines of something more substantial like a shield—the idea is the same. I endeavour to shut everything out—mostly shut out the pain. Intrusions bring pain to the surface. When the isolation inside my shield is broken, everything outside seeps in. This includes, perhaps foremost, the pain.

I turn to see how far Jane is behind me. What will she do when she catches me? Will she just extend a greeting and keep walking right on by? Or will she settle in beside me for a chat? I expect it to be the latter and that is not something I am interested in at this time. Thinking of Jane bearing down on me with every step, I set my jaw and clench my fists against the pain and work to increase my pace.

At about 8.30, my shield is penetrated. I become deeply melancholic as I think about Jennifer and my daughters. I said goodbye to the girls on Boxing Day and to Jennifer the following morning and so it has been 10 or 11 days since last I saw them. But I feel lonely and miserable knowing that this is just the beginning of more than six months away from them throughout the duration of my research study leave. I think particularly of my little mate, Bronwyn, just turned 11 years old. Back when I was last on the Track she was only nine, but I remember what a trooper she was. She left the comforts of the support vehicle and walked with me when I was at my lowest ebb, helping me to get through. It would be good to have the assistance of her company now. I blink my eyes against the tears.

I must be strong, I tell myself. Get the shield up. Get the shield up. Out here on the Track, there is no place for my wishful thinking. Thinking of Jennifer and the girls only adds to my woes.

The rough, scratchy calls of a flock of Apostlebirds draw my eyes their way. They fly from one tree to another, their grey bodies upheld by brown wings and long dark tails. Seemingly curious, they follow me for a short distance, like Biblical disciples following their Lord. Just as with some of those from ancient days, they soon lose interest and drift away, abandoning me when I could use their company. They leave me alone with the flies that are all over me this morning, constantly buzzing in my ears.

I approach the ghost town of Yantabulla. The place was originally called Yanda Bullen Bullen—an Aboriginal term for 'plenty to eat'. I wonder how many people have starved here through times of drought. I recall coming to this spot last time. It was one of our hottest and driest times. We had not seen Jennifer in the vehicle for some time and we were both out of water. From all appearances, people still lived in Yantabulla and so while Moz investigated the prospects from a water tank—which turned out to be dry—I knocked on the doors of a couple of houses. Alas, there was nobody and no water to be found. Today, the cloud cover has increased throughout the morning and, with the water I am carrying, I am neither dry nor thirsty so I do not stop.

As I walk past the town public phone, I recall that Moz called to speak with Jane in Sydney from there last time. Who would have expected a public phone still to work in a deserted ghost town? One would probably have to walk miles in a city to find a public phone that has not been destroyed by some dickhead with nothing better to do. Perhaps the ghost town absence of people is the explanation as to why this phone works.

Just as I leave Yantabulla, Moz passes, doubling back in the vehicle to check on Jane.

'How are you popping up, Grog?' he asks.

He notices the swarm of flies I carry with me. I ask him if he has any lollies. He offers me some jubes from a packet and I grab a big handful, they taste great. They do not satisfy my craving for Twisties, but they do get me to thinking about all of the calories I am burning on the Track. I figure I can eat anything and still lose weight out here.

Out from Yantabulla, I enter a plain where there are lots of birds. A Willie Wagtail lifts my spirits, waving its rump cheerfully in a 'Look at me. Aren't I sweet?' fashion. I watch as the bird flits from one spot to another, using its tail to stir up insects that will make a tasty mid-morning snack. A large flock of Galahs compete for my attention, calling raucously as they alight from the branches of a dead tree.

I walk another four kilometres past Yantabulla and I manage to stay in front of Jane. It occurs to me that she may well be as uninterested in my company as I have been in having her walk beside me this morning. I wonder if she has deliberately been staying behind me. Whatever the reason, just before 10.00, I make it to our breakfast stop well before she does.

Once I loosened up, I have been moving alright and I know that the breakfast stop is going to set me back but that is, I guess, just the way it goes. I sit down in the fold-up chair next to where Baz sits on a drink esky.

'Yantabulla was interesting,' says Baz.

'Yep.'

'There's an old cemetery there.'

'Yep.'

'I think I'll take a look through it on the way back—do a bit of exploring.'

I shake my head. 'Bloody rookie.'

Baz says that earlier in the day he had been talking with the bloke driving the water truck, spreading water to get the road ready for grading. The bloke is interested in our walk but when Baz told him that when we reach Hungerford, we will be turning around and walking back to Bourke, his reaction was, 'You're effing madder than I thought you were!' Word for word, it was the same thing Terry said the previous day. The weight of public opinion seems to be that we are not the most intelligent people in the Outback.

I open a small, individual-sized carton of Corn Flakes and tip some milk over the top. I watch the milk splash and trickle down the pile of cereal like lava flowing down the sides of a volcano. It makes me think of my volcanic crotch rot.

Baz tells me he used the public phone at Yantabulla to call his family in Adelaide. He tells me of their surprise and excitement at hearing his voice on the line. While I am happy for him, his joy merely adds to the depression I have experienced this morning about being away from my family.

Moz and Jane appear. Jane sits down with some breakfast. 'Gee, there are lots of flies around,' she says after a while.

'Yeah, I think Greg brought them in,' says Moz.

It is not a criticism or a knock. It is just a statement of fact and I take it as such. 'Share the love.'

Baz says that Moz had earlier made mention of the fact that I am craving Twisties.

'Yeah, we don't have any,' I say. 'I'll get some at the pub in Hungerford.'

'If they have any,' Baz says.

What are you trying to do to me, mate? 'They'll have Twisties.'

Baz and Moz look at one another a little uncertainly.

'Well...maybe,' one of them squeaks out.

'Hey? They'll have Twisties.' It is not a question. I look at their uncertainty. '...Won't they?' That is a question—one filled with hope, yet clouded by increasing despair. I need some Twisties.

≈

Although born in the Bush, after moving to Sydney to live with his mother when he was a teenager, Henry spent most of his life in the city. By 1892, he and Banjo Paterson had both established reputations as evocative poets. They decided they might draw some extra income if they sparred against one another. Paterson later remarked:

Henry Lawson was a man of remarkable insight in some things and of extraordinary simplicity in others. We were both looking for the same reef, if you get what I mean; but I had done my prospecting on horseback with my meals cooked for me, while Lawson has done his prospecting on foot and had had to cook for himself. Nobody realised this better than Lawson; and one day he suggested that we should write against each other, he putting the bush from his point of view, and I putting it from mine.

'We ought to do pretty well out of it,' he said, 'we ought to be able to get in three or four sets of verses before they stop us.'

This suited me all right, for we were working on space, and the pay was very small . . . so we slam-banged away at each other for weeks and weeks; not until they stopped us, but until we ran out of material.[1]

Early in July 1892, Henry fired the first salvo with 'Borderland'.

I am back from up the country—very sorry that I went—
Seeking for the Southern poets' land whereon to pitch my tent;
I have lost a lot of idols, which were broken on the track—
Burnt a lot of fancy verses, and I'm glad that I am back.
Further out may be the pleasant scenes of which our poets boast,
But I think the country's rather more inviting round the coast.

A fortnight later, Banjo's response appeared under the title, 'In Defence of the Bush'.

So you're back from up the country, Mister Lawson, where you went,
And you're cursing all the business in a bitter discontent.

Banjo labelled Henry a city swell. He tossed pointed barbs at Henry with only thinly veiled references to Henry's drinking problems. Banjo also made mocking references to Henry's poem, 'Faces in the Street'.

Other poets waded into the debate and so Henry's next piece was entitled, 'In Answer to "Banjo", and Otherwise'.

It was pleasant up the country, Mr. Banjo, where you went,
For you sought the greener patches and you travelled like a gent.

Henry claimed Banjo did not know the dry Outback heart of the country. In response to Banjo slanging at 'Faces in the Street,' Henry took a swipe at Banjo's famous 'Clancy of the Overflow'. Henry brought up the Great Shearer's Strike of 1891 and wondered aloud why the Bushmen would strike for better wages and improved conditions if things were as rosy as Banjo's portrayals suggested.

Ah! we read about the drovers and the shearers and the like
Till we wonder why such happy and romantic fellows 'strike'.

Banjo was a Sydney lawyer and Henry indicated the irony of Banjo enjoying the benefits of a city business, while supposedly simultaneously longing to get to the Bush. 'The city seems to suit you, while you rave about the bush,' Henry grumbled.

Others continued to insert their opinions in the 'Bush battle' but Banjo focussed primarily upon Henry and his 'dreadful, dismal stories' in the poem, 'In Answer to Various Bards'. Banjo claimed Henry was preoccupied with death. Banjo injected references to a number of Henry's characters who died.

Now, for instance, Mr Lawson—well, of course, we almost cried
At the sorrowful description how his 'little Arvie' died,
And we lachrymosed in silence when 'His Father's mate' was slain;
Then he went and killed the father, and we had to weep again.
Ben Duggan and Jack Denver, too, he caused them to expire,
After which he cooked the gander of Jack Dunn, of Nevertire;
And, no doubt, the bush is wretched if you judge it by the groan
Of the sad and soulful poet with a graveyard of his own.

≈

We are back on our feet by 10:20 am. As Baz and I walk along the Track after breakfast, I ask him about the scats at our camp in the morning. Although we both recognise the likelihood that it was a fox, we wonder at the possibility of a Dingo having come into camp.

'What animal would you most like to see out here, Gregga?'

'Oh, it'd definitely be a Dingo,' I say. 'I would also like to see an echidna too.'

'Would it be too dry for them out here though?' Baz wonders.

'Perhaps,' I concede, 'but I think they might get around just about everywhere.'

I ask Baz what he would most like to see.

'Dingoes, but then wild pigs,' Baz says without hesitation. I am surprised by that. I saw a wild pig out here last time and it was interesting, but I would have expected Baz only to pick animals indigenous to the country, rather than one that has been introduced. Neither of us said that we would like to see flies. We are seeing enough of them and more. We reach Cuttaburra Creek—the first of what we have been told will be five water crossings for the day. On 31 December, the day before we started our tramp, we drove this far and Baz and I got out of the car and waded through this crossing. The water had extended to about mid-thigh; a depth Moz decided was too deep for him to cross in the car. Yesterday, while Baz and I were on the Track, Moz drove across the Cuttaburra though. He felt the water level had dropped.

When I get to where the water has spilled across the road, I stop to roll up my pant legs. I roll them up over my knees. After all of my frustration at the first crossing—the Day Three crossing this side of Fords Bridge—I do not take off my shoes and socks. I decide it will be better to keep them on and, perhaps by so doing, to keep my bandaids and bandages in place. I will wait and see if I am able to continue on the Track with wet shoes and socks. If need be, I can always stop and get out my towel and spare shoes. The water over the road stretches for what looks likely to be 600-800 m up the road. Moz retrieves a couple of branches to use as walking sticks and then Baz and I step into the water. Baz has taken off his shoes and socks and replaced them with his thongs. I use my stick both as a support, but also as a depth finder, poking it ahead of me to detect potholes into which I might otherwise step and slip.

Beneath the clouded sky, the water is grey but much warmer than I expect.

'It feels like a bath,' I say to Baz as I take the first few steps.

Baz and I stay to the right-hand edge of the road. It tilts down to the left and so we have the shallowest path across. Although we proceed with caution, the road here is sealed and our footing is generally firm. A current runs from the right and so I focus my eyes on that side, wondering about the possibility of a snake riding the current. I know that a snake will not deliberately attack me in the water, but I worry that one caught in the flood and pushed along by the current might inadvertently become tangled up in me if our two paths should converge.

We go slow and steady, eventually crossing safely to the other side. Here, the road rises for 40 or 50 m to a bridge over what is labelled Cuttaburra Creek number one. The road dips then to where it is again buried beneath water for the better part of a kilometre. I walk on, completing another long crossing with caution but without incident.

Baz stops to dry his feet and to again smear Vaseline all though his shoes and socks before he pulls them on. It will take him a good half an hour to get ready to resume. I do not wait for him. Instead, with shoes still dripping water with every stride, I push on, crossing the bridge over what is labelled Cuttaburra Creek number two. I take my walking stick with me. With three crossings still to make today, I will need it again.

The sealed road soon gives way to red sand that sticks to my wet shoes. There are no trees to speak of, only woody shrubs and bushes like the Silver Cassia, Desert Cassia, Tangled Lignum and the soft Spinifexes lining a long, flat piece of road seemingly stretching into the eternities. As I wander along the road with only a swarm of flies for company, my spirits plummet so low that they disappear beneath my squelching wet shoes. Pain penetrates and then shatters the shield I have constructed. I become increasingly conscious of my blisters, my sore back, my deeply fatigued body and my overall discomfort. I feel overwhelmed by the challenge and, in that mindset; I am vulnerable to feelings of loneliness. I miss my wife.

I feel so alone out here.

A song from my youth starts repeatedly to run through my head. Talking Heads were an American rock band of the late-1970s and the 1980s and one of their songs—*Road to Nowhere*—starts playing again and again on the turntable in my mind. It does not take long for the road to nowhere to become the road to Hungerford—probably one and the same thing anyway.

I take the notebook and pencil from my pocket and, with this tune in mind, and my wife at the forefront of my thoughts, I scribble down some song lyrics, singing each new verse as I record it in my journal.

I'm on the road to Hungerford
Walk by my side
I'm on the road to Hungerford
I'll be your guide
Middle of nowhere
Open land
Middle of nowhere
Take my hand
We'll see it through together
Hot, blazing sun
We'll see it through together
One and one
I'm on the road to Hungerford
Across the plain
I'm on the road to Hungerford
Blisters of pain
Middle of nowhere
Swollen creek
Middle of nowhere
Feeling weak
We'll see it through together
Hot sky blue
We'll see it through together
Dreamtime with you
I'm on the road to Hungerford
Walk by my side
I'm on the road to Hungerford
I'll be your guide
Middle of nowhere
Open land

Middle of nowhere
Take my hand
We'll see it through together
Hot sky blue
We'll see it through together
I'm dreaming of you

I swing the walking stick onto and across my shoulders and then prop my arms up on it like the yoke of an ox. Some of the pressure is relieved from my back as I stretch my body in this manner. I walk along in this position, belting out my new song until I find that I am crying. I try to sing but I am unable to continue. I would kill to have Jennifer here with me now. Quickly, I wipe my eyes as I hear the approach of Moz's car. Before Moz has pulled alongside me, the salty tears evaporate from the back of my blistered right hand. I look skyward and notice that the clouds are starting to dissipate. It is approaching 40 degrees and it is going to be yet another long, hot afternoon.

≈

Henry was hurt by some of the things Banjo said in his 'Bush battle' poems. Henry felt Banjo's attacks were unnecessarily and inappropriately personal. Many years later, Banjo conceded, 'I think that Lawson put his case better than I did,' but Banjo felt he was right—the Bush was better than the city—and, therefore, 'honours (or dishonours) were fairly equal.' Banjo acknowledged the whole thing had been 'an undignified affair.'

≈

A little after midday, I come to the third water crossing for the day. It is the Back Creek crossing Terry Bates warned would be the most difficult to complete. I immediately see what he meant. I am near the eastern edge of the Cuttaburra Basin and the water is running fast from right to left, east to west. Most of the sealed road surface has washed away. White water cascades over ledges where gullies and pot holes remain on what had presumably once been a flat, smooth road. At the left edge of the road there is a sharp drop to the roadside verge. The white water and ledges are reminiscent of a mighty Colorado River, if on a modest scale.

Each water crossing builds upon the frustration of the one that preceded it. While the water offers an opportunity to soak my hat and bandana and to tip cooling water over my head, each crossing interrupts whatever rhythm I might have established. I am forced to stop and, although not taking off my shoes, roll up my pants and carefully navigate my way through the dark water. It leaves my shoes and socks as soaked magnets that pick up dust and grit and my feet bubble away in a cauldron of blisters.

With my walking stick in hand, I pick my way forward, testing the surface and the depth of the water as I go. A thin line of bitumen remains on the right edge of the road and I stick to that edge. I am, however, mindful that the edge is likely susceptible to crumbling and breaking under my weight, so I proceed with caution. It is not that my life is in danger at all—the water is not *that* deep—but to slip and fall would likely cause injury. At the very least, I can expect bumps and scrapes and will certainly be soaked and become more vulnerable to further rubbing and chafing. As I edge my way past the depth indicator sign, I note that it suggests the depth is less than 40 cm; however, that would only be true were the road still in place.

When I get to the other side of what is not a long crossing—maybe 40 m at the most—I call back to Moz, who has arrived behind me. 'How are you going to get the car across there?' I know that I would not like to be driving a car on that uncertain surface. I have doubts that he will make it, even though he is in a four-wheel drive.

'I just go slow and stick to the centre,' Moz calls back. He does not seem particularly worried, which is reassuring to me. If he is okay with it, I will not let it bother me. While Moz and Jane wait for Baz to catch up to them, I carry onward.

I am only a few hundred metres further up the road when a vehicle approaches. The vehicle rolls to a stop and, in what is the least friendly enquiry I have received since I have been on the Track, the driver leans across his female front seat passenger and asks, 'What are you up to?'

I am taken aback by the tone in which the question is posed but I respond, 'I'm walking to Hungerford with my brother.'

Paying my respects at Henry Lawson's grave in Sydney's Waverley Cemetery before driving to Bourke.

Baz and I contemplate John Longstaff's portrait of Henry Lawson at the Art Gallery of NSW. It seemed as if Henry wanted to warn me about the hell that awaited on the Track.

With the floodwaters coming down from Queensland, hundreds of trees stand in the blue water of the flooded Darling River.

One who has not carried a load in 47-degree heat cannot understand just how hot I feel as Moz and I steal what little shade we can find on the plain.

It is only Day One, but as we swelter under the sun Moz and I already realise we are in way over our heads.

The heat draws from them whatever spring might once have existed in Moz and Baz's steps.

A portrait of Henry Lawson taken later in the year that he completed his Hungerford tramp. This photo was taken in New Zealand by John Baillie. Image used with permission of the Alexander Turnbull Library, Wellington, New Zealand. (Photographer John Baillie, fl 1894. Ref. PAColl-8674.)

Sipping from the tube of my hydration pack as I fight the dangers of dehydration in the extreme Outback heat.

One can only guess at the thoughts going through his mind as Moz suffers in the dirt at the side of the Track.

After the frustration of the first of many water crossings, I tell the others that if a car comes along it can just run right over the top of me.

'You never told me that Baz is a freak,' Moz said to me. I know that it is not easy for him, but there are times when Baz makes it look easy.

Henry Lawson said that out on the Track you just 'get old and careless and dirty, and older, and more careless and dirtier.'

My favourite portrait of Henry Lawson taken by photographer William Johnson in 1915. Image used with the permission of the State Library of New South Wales. (Image no. A128724)

Holding the portrait of Henry Lawson and ruing the foolishness of ignoring his warnings about this undertaking.

Baz wades through another water crossing as he works his way ever onward towards Hungerford.

Baz and I push on through the heat.

Baz picks his way across more floodwaters.

Baz reads a birthday card from his son. Inside, the card reads, 'I hope you are having a good holiday with Uncle Greg.' This is anything but a holiday.

Whatever the obstacles, Santa can make it through. When I come across this sign on Day Six, I wonder if I can also overcome the obstacles blocking my path.

The Brindingabba Christmas decorations remind me that it is still the Festive Season. Despite the Christmas cheer, I do not feel like celebrating as I pass by.

Baz working hard to get more than a grunt of acknowledgement from me.

The support crew extraordinaire. Jane and Moz take a break from the sun in the back of the vehicle.

Wishing for respite from the heat as I begin to trudge across Whim Plain on Day Six.

Not far to Hungerford now. I just keep thinking of how good it is going to feel to have a shower and to get clean.

Baz rests his tired body against the Dingo fence before crossing into Queensland.

Having made it to Hungerford, it is time for congratulations and big smiles. But I cannot ignore the fact that we still have to walk all the way from Hungerford back to Bourke.

As we walk into Hungerford, I expect Baz is thinking that it has been a hell of a long walk to get to a place that looks like this.

'Oh, that's *you* blokes, is it?' The tone changes in an instant. 'We read about you in the newspaper.' The driver introduces himself as Paul Kaluder, with his wife Debbie beside him. They are a young couple and now that they know I am not up to anything sinister, they have friendly smiles across their attractive faces.

'Are you okay? How is it going?' Debbie asks me.

I tell them that things are going okay.

'Did you see any birds back at the crossing?' asks Paul. I look back to where Baz, Moz and Jane are still getting across the water. I tell Paul that I saw many birds as I approached. In addition to all of the egrets, herons, stilts, and other waders, I saw all three species of Australian ibises. There were many Straw-necked and Australian White Ibises, but I also noticed the distinct reddish sheen of a couple of Glossy Ibises as they took flight.

'When they are filming television documentaries on birdlife, we bring them here,' Paul tells me.

'It is a nesting site for lots of water birds,' Debbie says.

'So when we see someone out here, we like to know what they are up to,' says Paul.

Good for them. Fair enough. As a bird lover, I am happy that Paul and Debbie are keeping their eyes out for my feathered friends.

They tell me they live on Naree Station, up ahead, on the other side of the road from Tony Marsh's Kia-Ora. I tell them that I know Tony.

Paul tells me that he drives out to check the water at the crossing each day and that the level is down on what it was a few days back. He says that with floodwaters coming from Queensland, the level will rise again over the next several days. This news does not auger well for the return journey, but thoughts of the return journey are still only in the distant recesses at the back of my mind. 'I'll cross that bridge when I come to it,' I think, '...or, in the absence of bridges, walk through that water when I come to it.'

'That was our third water crossing for the day,' I say. 'Only two more to go and we'll be out of the water.'

Paul and Debbie look a little confused. 'You'll have a few more than that,' Paul suggests.

'No, our support crew drove ahead in the car yesterday and checked all the crossings to make sure we'd get through.'

'They must have missed a couple,' Paul says more definitely. 'You'll have more than two more crossings between here and Hungerford.

I am surprised by his certitude.

Paul and Debbie wish me well and then drive on to chat with Baz and, no doubt, to see if Moz is able to get his car through.

Moz does get through, because it is not long until he passes me with news from the car radio that Alastair Cook has posted another century for England. The opening batsman has had a phenomenal Ashes series. While Australia's best batsman for the series finishes with an average of 63, Cook's Fifth Test heroics push his average over 127 runs per innings. With three hundreds and two half centuries from just seven innings, he has slain the Australian bowling attack, posting scores of which Don Bradman would have been proud. It is an amazing performance and Australia has been put to the sword. As England plunder over 600 in Sydney, I realise that I am not the only Aussie taking an early January New Year hammering.

'How are your feet, Grog?'

'Good,' I lie. I do not want to talk about my feet.

Moz hands me some more lollies and then he and Jane drive on. For company, I soon have three Black Kites. I tilt my head back and watch the easy movements as the birds tip their wings to complete slow, full turns. They circle low overhead, obviously curious about this lone traveller in such an inhospitable habitat.

'Do you want to pick at my bones?' I ask aloud but, in response, I receive nothing more than a tilt of the wings as the birds circle less than twenty metres above me. 'I'm not dead yet,' I declare.

At 1.15, I complete another water crossing. There are dozens of Black-winged Stilts and a couple of Great Egrets wading in the water, picking at the aquatic insects that form much of their diets. The stilts move about mechanically on their long red legs, while the egrets stalk the shallows with poise and patience. Blue Strider dragonflies also dart about just above the surface, perhaps looking to take a mosquito in flight.

I lead Baz for the remainder of the pre-lunch session and at 1.30 pm I reach the spot Moz and Jane have chosen for lunch. They have set up my blankets so that they hang from a couple of shrubs, creating shade. Moz fusses around considerately, making sure I have everything I want.

'How hot is it?' I ask Moz.

'Forty-one degrees.'

Forty-one at lunchtime on what has been our coolest day so far? I am in hell. When Baz arrives, I ask him how he is going and then I tell him that I am cured of my yearnings for wandering in the Outback. I tell him I do not want to walk in it ever again. It is horrible out here and this will do me for the rest of my life.

I eat lunch heartily, dipping into my tin of tuna with the rice crackers that I have at hand. There is a lake nearby and we can see Australian Pelicans on the water and in the sky. I watch as they gracefully circle above, riding the air currents. They circle higher and higher, seemingly without effort.

'I wouldn't have expected Pelicans out here,' Baz says. 'I always thought they were coastal sea birds.'

I always think that pelicans look wise. When you look into their black eyes, they seem to know exactly what is going on in the world—and what is going wrong with it too. Yet, with all that wisdom, I wonder if they are more surprised to see us out here than Baz is to see them.

'The Australian Pelican has the biggest beak of any bird in the world.' It is just a useless piece of trivia that I offer, but I figure it buys me a ticket back to pained silence for a few more hours.

Maybe I do not need to purchase tickets to silence. Henry said:

There's no place in the world where a man's silence is respected so much (within reasonable bounds) as in the Australian bush, where every man has a past more or less sad, and every man a ghost.... They say in the bush, 'Oh, Jack's only thinking!' And they let him think. Generally you want to think as much as your mate; and when you've been together some time it's quite natural to travel all day without exchanging a word.

During the drought, this area would have been barren and devoid of vegetation. It is so different to see the land under the influence of heavy water. There is grass aplenty and I can only imagine the increase in wildlife numbers there must have been lately. The Emus and kangaroos must love all the food. Certainly, we have seen dozens and dozens of roos and Emus and, as we lunch, a flock of seven foraging Emus prod and probe their way across the plain, picking at the abundant plants under their feet and the hordes of grasshoppers that kick up as the Emus move.

≈

I have always considered that when they looked at the Bush, Banjo and Henry saw the same thing. The difference, however, was not what they saw, but the point of view from which they saw it. Banjo's view of the bush was that of a man of affluence whereas Henry was a tramp. One historian who penned authoritative biographies on both men identified Banjo as 'the bard of the men who ride' whereas Henry was 'the bard of the men who walk.'[2] I believe that simple distinction makes the world of difference.

≈

A few minutes after 4.00, Baz and I resume our tramp. I am as stiff as a board.

'Don't feel like you have to wait for me,' I say to Baz. 'It'll take me about half an hour to loosen up and get going again.'

I am wrong. An hour-and-a-half later, I am still hobbling badly, my feet feeling like I am walking on broken glass. My tight feet refuse to unwind and it occurs to me that I might be paying now for the generally good run I had this morning. The Track is littered with rocks and my blistered feet hurt so much that I decide that I would have been no worse off—if any—had I instead chosen to wear the same boots that caused me so much grief last time. This 90 minutes since lunch has been one of my toughest stretches since we left Bourke, full of energy and excitement, five days earlier. Were I to think back on that early energy and excitement, it would be as a worker in the sex trade thinking back on the naive innocence of their virginal youth. These days, I am hardened and embittered—and I have the sex trade crotch rot to go with it.

Despite (or perhaps magnifying) my horror stretch, Baz continues to travel well—as seems always to be the case. As I stop to pee, I see Baz disappearing into the distance in front of me. I eventually pass the gate that leads into Tony Marsh's Kia-Ora property and I am surprised I have not seen Tony on the road. Last time, he passed us a number of times. I was thinking that we might stop at his place; take him up on the invitation that he extended to us last time. However, we have decided to keep going so that we might get into a position whereby we do not get stuck on Whim Plain overnight—where there will be no shelter and where the ground will be hard and rocky. We want to position ourselves so that we camp before Whim Plain tonight and after it tomorrow night. We also want a relatively short run into Hungerford on Day Seven, thereby increasing our day of rest in Hungerford to perhaps a day-and-a-bit or even a day-and-a-quarter. Unlike my late arrival into Fords Bridge, I do not want to feel pushed or rushed—with Jane cheerleading—to try to catch a meal before the kitchen closes. I am hoping it will work out that I can have a leisurely shower and a rest before having to force down a meal. Tomorrow, however, will be critical. It will be necessary to do a big day tomorrow in order to have a smaller one on the seventh. Given the way I have been travelling since lunchtime, a big day tomorrow does not seem even remotely achievable.

My low spirits tumble further when I eventually catch up to where Baz is resting beside the road and he informs me that Jane and Moz have gone ahead and have asked for us to complete another ten kilometres to a spot where they will set up camp. I do not think I have another ten kilometres in my legs today.

'Ten will probably be twelve anyway,' I grumble.

I am growing increasingly frustrated by the water crossings and psychologically debilitated by the imprecision of the support crew. After lunch, Baz and I completed our fifth crossing of the day and I now know that Paul and Debbie were correct—there *are* more crossings today than the five Moz and Jane reported. I still have a couple of crossings ahead of me at Clarkes Creek—crossings that Paul said would not be pleasant because of the swamp-like conditions. I need the support crew to be precise. I think not just of the inaccurate number of crossings, but also of the imprecision of being told things like, 'You have done *about* three or four kilometres,' or 'You have *about* two kilometres to go.' With battered body and mind, I am becoming irritable and such imprecision and inaccuracies are wearing thin. When you are sitting in a car with an odometer staring you in the face, it should not be too hard to make accurate measurements. It is a kick in the teeth when I find out that things are actually worse than they were represented—when there are seven water crossings instead of five or when I have five kilometres to go instead of 'about four'.

It is 37 degrees now and with the heat, the water crossings, and the knowledge that I still have hours to go—at my present rate ten kilometres will take at least three and perhaps even four hours to complete—I reach a low ebb. I dread the next ten kilometres. I have to rest and get out of the sun. All of the clouds have disappeared, leaving behind a bright incinerator that burns my skin and scorches my eyes, mindless of the sunscreen and sunglasses that I wear. Although it is after 5.30 pm, the sun is still very high in the sky. Out on the Track, there is no escaping it. I have to get off to the side to shelter under some trees but to rest is only going to delay me further. Getting to camp late at night reduces the amount of time that I have to rest overnight. It adds to my ongoing sense of fatigue. I will be surprised to get to camp before 8.00 and resign myself to the miserable fact there is every possibility that I will not reach camp before 9.00 pm. Will this day never end?

I trudge on in misery; barely lifting my feet from the baking surface of the Track, yet trying to negotiate my way over the pathway littered with rocks the size of golf balls. My once-white hat and off-white shirt are now stained a grubby brownish-red from dirt, sand, and dust. The hat and shirt are soaked with the sweat of my body, carrying with me the stench of filth. I am sub-human—a beast: a miserable beast of burden toiling beneath a blazing furnace.

Baz walks beside me in silence, his presence somehow providing for me the support and encouragement that, at this time, his words would not. It is as if his will urges and cajoles me, coaxing from me one more step. One more step. One more, and then another.

For close to three hours I struggle in an agony of blisters, my feet never loosening up from the lunchtime break. The hard, stony road is a bad road to walk on and to have to do it at a bad time for me just adds to my discomfort and misery. The mixture of bad road and bad spirits knocks at me constantly.

It is after 6.30 when we approach Clarkes Creek. I dribble out a short pee. I remember the last time I was here. Moz and I camped here when the creek was dry. That night was the only time we got any rain, albeit just a sprinkling. Today, with the sun still shining brightly in what has become a cloudless sky, there is no sign of rain in the near future.

When we were last at Clarkes Creek it was the only time we had problems with mosquitoes. With all the water in the creek at present, I bet there will be millions of mosquitoes here in the evening.

The paddocks on either side of the road swim with water. Just before we enter the water, Baz sees movement in the paddock to the left. He retrieves his binoculars from their small carry case.

'I think there are pigs in there.' Baz is initially more hopeful than confident but as he trains the binoculars on the animals moving through the long grass and Tangled Lignum, his voice becomes more and more excited and more and more certain. 'There are pigs—wild pigs!'

I can see the movement too. Baz hands me the binoculars and I adjust the focus to accommodate my eyes. 'Yep, I see them. They're pigs alright.' We stand and watch for five minutes or so, straining to get a nice, clear view but our vision is mostly obscured by the grass and trees in the paddock. All the same, I am delighted we have seen pigs and especially pleased for Baz, seeing as he wanted to see them out here.

We walk the final several hundred paces to where the water from Clarkes Creek first spills over the road. Moz and Jane wait for us in the car on the other side of the crossing. Large gums cast dark shadows over big sections of the foetid, stinking water. The water reaches beyond my knees as I cross, working my way through the seventh water crossing of the day. Dragonflies and damselflies patrol the shady sections of the creek in their hundreds, feasting on the mosquitoes that move about in the shadowy swamp to the left of the road. That swamp is humming with mosquitoes. It must be alive with them. I am sure that, were I foolish enough to venture into the swamp, I would be eaten alive. What a horrible place it would be to camp anywhere near here. The more I look, though, the more dragonflies I see. They are everywhere, banqueting on the mozzies, doing their bit to help make my miserable world a better place. Unlike the other crossings, there is no current here. Rather than offering anything remotely pleasant or refreshing, this still muck is just something that one wants to rush through. A thick layer of green gunk pond scum is suspended across the water's surface. It stinks. It reeks. Perhaps, briefly, it even masks my own pong. After a hundred metres or so, I am glad to emerge from the water.

I move to Moz's car and, convinced that the last crossing is now, finally, behind me, I remove my stinking shoes and socks. I use my towel to dry my feet and then I pull on clean socks. I replace my dripping Nikes with my Merrells. While I am doing this, Baz is ready to resume and so continues on. He has not gone far from the car when some wild pigs with piglets cross the Track in front of him, affording him a good, uninterrupted view.

While I struggle on with the remaining distance I still must put behind me before camp, Moz captures an interesting photograph of a Bark-mimicking Grasshopper. The colour and texture make the insect nearly impossible to see but my gaze is momentarily drawn to grasshopper's double-banger unicorn-like antennae and the insect's ovoid eyes. It is a fascinating, almost otherworldly creature. Alas, I have not the energy to be gazing upon insects. If I am to make it to camp tonight, I must trudge on, wearily, seemingly endlessly.

I need to stop and rest as I am completely exhausted. Yet, I know that to stop does not get me any closer to the campsite. What's more, as badly as I am travelling at the moment, trying to get going again after another rest will be even worse so I just keep going, willing my weary legs to keep moving, one step in front of the other.

I clearly see where Baz has preceded me on the Track. Too tired to think for myself, I let Baz do the thinking for me and move across to walk in his footprints. Out here, we are always altering our paths, choosing the best line to walk. One picks a path and then constantly assesses it and adjusts his line, picking the flattest route, with the right level of firmness—not too hard and not too soft—and picking the clearest route and, on the bends, the shortest route. All of this takes thought and, for now, I am happy to abdicate that responsibility—that chore—and just walk where Baz walked.

A big Red Kangaroo bounds across the road and into the scrub to the right. The roo's long hops make movement appear so easy, and enviably so. The strong legs, perfect balance, and powerful tail are all designed for efficient movement. In contrast, my legs are buggered, the pack on my weary back throws my body forward uncomfortably and, if ever my ancestors had a tail, they left it in a deep, dark forest somewhere in the distant recesses of Time. The roo belongs out here. I do not.

Eventually, I top a rise and see the car parked at the side of the road a little more than a kilometre in front of me. I can see the spot chosen for our campsite tonight! I am so very happy to see that. It has been a rotten, tough session.

I have been on the Track for well over 12 hours today and, with a good morning and pre-lunch session, I have somehow put 33 km behind me. When I reach the camp, as buggered as I am by my horrible afternoon, I am still optimistic that I will feel okay tomorrow—that I can complete a big day tomorrow. I think positively—I will be able to bounce back. I am surprised that I was never able to bounce back this afternoon, but I tell myself that is a mere anomaly. I will bounce back tomorrow.

The spot they have chosen to camp is among some scribbly, scrawny trees. There is little undergrowth—mostly sparse tufts of grass—and the ground floor is mostly clear save for lots of fallen trees and branches. Baz points out some well-worn and fresh kangaroo tracks running right through our campsite. I notice one track seems almost to run right to the door of my tent.

'Imagine the mess a kangaroo's claws would make of you if it came barging through here at night and jumped right on top of your tent,' I say. It would get caught up in the fabric and ropes and would kick and scratch like mad. The tent would be torn to shreds, as would anyone unfortunate enough to be inside. Out here, if it's not one thing trying to kill you, it is another.

After, Moz prepares the food and passes mine to me. Tonight it is ravioli and meatballs. As good as it tastes, I am boiling in the heat and forking hot food into my body seems a strange and unhelpful thing to do.

'There is air-conditioning in the rooms at Hungerford,' Jane says. That seems almost as good a reason to get there as any other. 'The rooms in Hungerford are great.'

Baz replies, 'If we get to Hungerford and there is no air-conditioning, I am running back to Bourke.' I let that sink in but decide that he will be running without me. Baz continues, 'The air-conditioning in Bourke was the best I've ever felt.' In that, we agree.

As with every evening, I thank Moz and Jane for all of their help throughout the day. They really are doing a great job and I am mindful that, without them, I have no hope of making it to Hungerford. The others sit up; drinking billy tea and watching the sun go down. I notice the vibrancy of the sunset as I crawl into my tent but, for me, it is more important to try to cool down and rest. When I get into the tent and peel off my clothes, I notice that my skin is blotchy and red all over with heat rash. It has been a hell of a long day and I am done with it.

DAY SIX

Henry Lawson sent to Bourke

The poet's light was in his eye,
He aimed to be a man;
He bought a bluey and a fly,
A brand new billy-can.
I showed him how to roll his swag
And 'sling it' with the best;
I gave him my old water-bag,
And pointed to the west.

- The Bard of Furthest Out, *For Australia*, 1913

From the front verandah the scene was straight-cleared road, running right and left to Out-Back, and to Bourke (and ankle-deep in the red sand dust for perhaps a hundred miles); the rest blue-grey bush, dust, and the heat-wave blazing across every object.

- At Dead Dingo, *Joe Wilson and His Mates*, 1901

The first things I see when I awaken are swarms of thirsty mosquitoes, the females banging their proboscises against the mesh of the tent, desperate for my blood. Oh, how they would love to find a way inside—just a tiny tear in the mesh would do. I watch, fascinated by their persistence as they constantly test the defences of the tent. There are literally hundreds upon hundreds of them. As I start to get ready, I wish a ray of sunlight would find a way through the trees and chase away all the mozzies before I am ready to move outside.

I hear Baz moving in the tent beside me.

'Happy birthday, Baz,' I say.

'Thanks mate.'

He is 45 today.

Baz unzips his tent and crawls outside. As he pokes around, getting his water ready and other essentials loaded into his pack, I notice that the silly bugger is topless.

I ask, 'Aren't the mozzies eating you alive?'

'Um, no, I don't think so,' he responds. Maybe all the females have flocked to my side of the camp—there's a first time for everything, I guess.

It seems to take forever again to get ready in the tight confines of the tent but I know that to try to prepare my feet and to dress outside is a sentence to death by mozzie blood-letting. When I do eventually emerge, the mosquitoes descend on those few areas of uncovered skin—my face and hands. I thrust on my hat and drape my bandana over my face, leaving only my hands exposed. My fingers quickly disappear beneath a swarm of voracious mozzies.

I take to the Track at 7.30 while Jane and Moz pack the tents. Although Baz has started only 15 minutes before me, I cannot see him in the distance as I hobble those first several hundred painful steps. Unlike me, Baz does not seem to have trouble getting into his rhythm in the mornings.

I wonder at what point along the Track the Demons will spring their ambush today. I think of Baz out ahead and hope that he has a good birthday. My wish for him this day is that he enjoys good travels on a kind Track. Hopefully the Demons leave him alone. If I could influence these things, my birthday present would be to tackle the Track's Demons for him.

It is a very slow start for me. My blisters plague me and my constricted feet are slow to unwind. With each tortured step, I hope that it brings me closer to that point at which my movements will become freer but, after yesterday afternoon's ordeal, I wonder if that will ever happen. All the same, I must remain hopeful.

The support vehicle passes me after a short while. Whenever Moz drives up, he always asks if I want anything. On those occasions when the answer is yes, Moz invariably drives ahead, parks the car, gets out, and then retrieves whatever it is that I have asked for. I do not have to break my stride—let alone double back or wait for him to get whatever I want. It is not something I have asked him to do but, having been on the Track, he knows the things that are helpful. These little kindnesses make a difference. The little things can add up to a big thing—something as big as successfully walking from Bourke to Hungerford.

After they have checked that I am okay, they tell me they will drive on ten kilometres to pick out and set up a spot for our breakfast stop. Ten kilometres seems such a long way off.

In the evening yesterday, I passed the H60 road marker designating 60 km to Hungerford. It is good to know that I am inside sixty. We will do at least 30 km today and, therefore, have less than 30 tomorrow to get to Hungerford. Oh, how I am looking forward to our scheduled rest day. I cannot wait to put up my feet and to enjoy two nights in a bed. It will give me a chance to deal with my blisters and tidy up my feet. However, before Hungerford we must cross Whim Plain today. I know from last time that it is a hot, barren, shadeless place and it is going to be a difficult 20 km stretch that will test me in every way.

With yesterday morning's cloud cover, it would have been a good time to tackle Whim Plain. Alas, as I scan the sky, there is not a single cloud today. Above me is that beautiful, magical blue of the Australian sky. At present, it is a lovely azure, tapering down to a baby blue nearest to the horizon. I have travelled a lot of countries and no skies match the blue of the Australian sky and that is especially so out here in the Outback where the air is pure and clean.

Up until Yantabulla, we had been travelling constantly in a north-west direction. At Yantabulla, we turned due north and continued in this manner for 20 km. Now, we are heading west—almost due west—and so the sun rises behind me as I do my morning hobble.

The hatted head of my shadow extends 20 m or more directly in front of me. I turn to look back at the sun, still low in the sky. I hazard a guess that it is yet only about 20 degrees above the horizon. I gaze at the big fiery ball that is the sun—a shining star with bright white arms extending in every direction.

It really is a glorious morning. The birds are singing in full voice and I can hear insects off to my right, similarly enjoying the advent of a new day. Having cooled down overnight, I feel cool this morning. The low sun is pleasant and not yet hot, and a slight breeze caresses my face. After a hell of an afternoon and evening yesterday, 'my spirit revives in the morning breeze, though it died when the sun went down.' Buoyed by such thoughts, I start to feel better all over and my feet start to loosen up and allow me to stride out with a little less reluctance.

There are trees along this stretch that are mostly Mulga, but a few eucalypts rise above their neighbours. Unlike the blues of the sky, out here, the greens are not vibrant. Indeed, I note, there is a lot of blue to the greens within the acacias, casuarinas, and eucalypts that line the Track. As I walk along the roadside gutter reflecting on the vegetation, I dodge about the tufts of grass that grow here and there. I notice there are lots of dry prickly shrubs. Out here, the plants must be hardy to survive. Along with the adaptation of a tree's needle-like leaves to reduce water loss, many of the plants—including the trees—are prickly or thorny. This provides protection from animals. Because there is normally so little to eat, anything not protected in such a manner would be eaten, and eaten quickly, by the hungry creatures of the Outback. The result is that there are lots of prickles, burrs, and thorns to add discomfort to any human foolhardy enough to come walking. The prickles stick to my shoes and socks and so, as the grass increases in the gutter, I am forced again to move to the rocky centre of the road.

There is ample evidence of the harshness of the climate. Lots of fallen trees and branches litter the ground. This is a place of challenge—that, I know well. Thinking of the adaptations of the plants, I recognise it also as a place of survival. When the stink of death wafts along, riding the breeze, I know it is a place of death too. Not long after I catch the smell of death on the wind, I invariably come across the carcass of an animal decomposing in the sun. As Henry said, 'The skies are brass and the plains are bare, / Death and ruin are everywhere.'

Shortly after 8.00, something catches my eye in the distance ahead. With the sun glinting from it, it appears metallic, triangular, and manmade, but I don't know what it is. As I draw closer, I begin to make out a yellow sign suspended near the triangular 'thing'. Such small curiosities help to occupy the mind as I wander along. What is that in the distance? What is that thing up ahead? The diversions help remove pain from the forefront of my thinking.

The yellow sign indicates I have made it to the entranceway to Brindingabba Station, the home of Dave Fisher and his wife and children. I met Dave and his wife when I was last on the Track. The Fishers are a nice young couple and I am hoping to see them again. This morning, there is nobody at the turn off to their place though. My maps show me that it is a good seven kilometres from this entranceway to their homestead and so I am not walking in to see them. What I do see, however, is that the triangular thing is actually a metal Christmas tree the Fishers have erected. What I had thought from a distance were yellow flags waving in the breeze are actually half a dozen flashing yellow hazard lights. Some sheets of corrugated iron have been cut out, arranged, and then painted to depict Father Christmas in a small boat with his sack full of toys. A separate sheet of corrugated iron is attached to two metal stakes with loops of wire driven through nail holes in the iron. 'Drought or flood, Santa gets through,' the makeshift sign declares. In the middle of the flashing hazard lights and between the iron Santa and the sign proclaiming Santa's tenacity, sits the Christmas tree. It stands almost four metres tall and is a pyramid of wire to which has been attached a variety of decorations—a steering wheel, a shiny hubcap, ropes, wires, a green kettle, wheels from toys, a rubber-handled hammer, some plates, a thick metal chain, a red Tupperware box, the blade from an old circular saw, an old billy can, a red casserole dish, and various other sundry metal and plastic ornaments. Atop all else, where the star of Bethlehem might otherwise sit, is a four-bladed boat propeller.

I am grateful to the Fishers for bringing me a little Christmas cheer in the midst of my struggles. It was an otherwise short Christmas season for me. I said goodbye to my girls on Boxing Day and flew out from Winnipeg on 27 December. Seeing the Christmas tree makes me question what day it is. I do not know the day of the week. I only know the day of the walk. It is Day Six. That I know.

Let me see.

We started the walk on a Saturday. What does that make today? Is it Sunday?

I think some.

No, Sunday does not make sense. It must be Friday.

I figure some more, thinking as hard as I can. Thursday?

I can't figure it out. With my addled mind, thinking is work. I do not need more of that. Of that, I *am* sure. I have plenty of work just pushing forward.

I walk on and pass the entrance drive to Kilberoo Station to the left. I notice tinsel draped around the mailbox. It is nice to see these Christmas decorations. It is a reminder that it is still the Festive season.

I am less than 20 minutes past the Brindingabba decorations when my imaginary mouth-watering smells of Christmas baking are replaced by the reek of death on the still air. It does not take me long to discover the source of the stench. All that remains is fur and bones and, here and there, small flecks of flesh that the hungry scavengers were not able to tear from the bones. From the pulpy appearance of the underside of the skin, the animal seems not to have been dead as long as the smell would indicate. Although the skeleton is largely intact, it is twisted and contorted in such a manner that I cannot be sure what the stinking mess before me had once been. Perhaps it was a big goat, although it is hard to reconcile my image of a goat with the macabre, grinning pile of death before me. Whatever it was, the foxes and various raptors that have left their prints in the sand around the carcass have left little to be flyblown.

Out of the side of my eye, I catch a glimpse of the top of my right arm, almost black with flies. I shake my arm and brush at the flies. They give voice to being disturbed in this way. A loud hum ensues until they again settle.

A Country Energy truck rumbles towards me, heading west and the driver pulls over.

'Are you okay?' he asks me.

'Yep.'

'Did you breakdown or something?'

'No, my brother and I are walking to Hungerford.' I am surprised that he has not heard about us by now. 'My brother is somewhere up ahead of me.'

'Do you want a ride so that you can catch up?'

'Nah,' I laugh. 'I'm walking—as best as I can.'

'Hungerford? What do you want to do that for?'

'Well, Henry Lawson did it.' From the look on his face, my explanation does not seem sufficient. 'Henry's my favourite writer,' I continue, 'and so I wanted to come out here and do what he did.'

The County Energy driver scratches his head and then wipes his hand across his happy face. 'Good for you.' The prominent green strip and the golden lettering on the truck reflect the heart of Australia and I like this bloke—he represents the Nation's heart too. Another scratch. 'But, gee, Hungerford? How far is that?'

I tell him and he exhales involuntarily.

'Bloody hot for that,' he says.

As we talk, another Country Energy truck rumbles by in a cloud of dust. The two workers give hearty waves.

'Are they with you?' I ask.

'Yeah we are headed on towards Hungerford ourselves. We have some work to do at a property near there.'

I ask where they are headed.

'Warroo. You'll pass it. It is a big white place just off the road to the left.'

'I don't know Warroo, but someone else mentioned that place to me,' I say, thinking back to Terry Bates saying there were photos of Henry Lawson there.

'What have you been doing about the water over the road?'

'We just wade on through,' I say. 'It can be a bloody nuisance. I'm glad we have finished the last of the water crossings.'

The driver scratches his head again. 'You might have one to go, I reckon.'

What? More water? 'Oh, really? I was told we had finished.'

'You've still got Brindingabba Creek ahead of you. It is just a few kilometres up the road from here. That one often spills over the road.'

Bugger it, I think. More water.

'I'll find out for you,' the driver says, and he radios ahead to his work mates in the truck that passed by.

'Have you crossed Brindingabba Creek yet?'

I hear a crackled reply.

'Is there water over the road?'

'No water.'

'No water,' the driver tells me, and I can barely contain my joy.

You bloody beauty. 'Thanks for that.'

'I better let you keep going,' the driver says. 'I s'pose I've got work to do too.' He would probably be happy to chat until knock off time.

'Thanks for stopping.'

'I'll check on you on the way back,' he says as he pulls away slowly. 'You sure you don't want a lift?'

As with last time out here, I am impressed by the people who I meet along the Track. The people have always at least given a friendly wave. Most have stopped for a chat and others stop to express concern, to see if we are in any need, and to ask if they can be of assistance. These are Australia's finest.

I have been on the Track for over an hour-and-a-half. I have not seen Baz at all and have not seen Moz or Jane since they passed me in the vehicle at the start of the day. I wonder if they are up ahead, singing Happy Birthday to Baz. I am lonely this morning. I guess I have about an hour or so to go to get to our breakfast stop.

The sun is increasing in intensity. I turn to check its height and figure it must now be a touch less than 45 degrees above the horizon—perhaps 40 degrees. It is going to be bloody hot by the time I get to Whim Plain and the sun is high in the sky.

There are lots of flies on me today—more than any other day thus far. I know that I stink, but I am certain that, as yet, I do not smell as bad as I did last time, although Baz has told me that he is finding that hard to believe. It will be so nice to get to Hungerford and to have a scrub and a good soap and shampoo.

I pass a Desert Bloodwood, tall and strong, standing straighter in the heat than I can. It towers over the small greener shrubs at its feet, reaching to the brilliance of the blue sky above.

At 10.35, I reach our breakfast spot and learn that I am eleven-and-a-half kilometres down the Track from where we camped. Moz has set up my blankets under the shade of some small box trees. Baz has been waiting for me for almost an hour.

The flies are all over me—hundreds of them, burying me. Baz has taken to calling himself 'Two Shirts Bazza' because he has one more shirt than me. 'You're mistaken if you think you don't stink too,' I tell him.

'Did you get my note?' he asks.

'What note is that?'

'I left one for you.'

I think of how I have been walking with my head down, too weary to lift my eyes to look around. I think also of the arrow he scratched in the ground, pointing at the lizard on Day Four. 'I wouldn't be wasting too much time scratching notes to me. I am not seeing much out there.'

'No, it is on paper. You must have seen it. I left it back beside the 50 km marker.'

I had seen the marker. I remember being consciously aware of it as I passed. It was cause for momentary celebration—*only* 50 km to go to Hungerford. 'I saw the sign, but no notes.'

'You must have seen it.' Baz is insistent.

Do you think I am pulling your leg, Baz?

'There was nothing else around there,' Baz continues.

'Baz, when I get in a funk out here—which is not uncommon—I see nothing. I might as well have my eyes closed for all I see.'

He tells me he left one of his laminated red inspirational quotes beside the sign. 'You can't have missed the red.'

I guess I am pretty special, Baz. I missed the red.

I ask Baz what the note said. He will not tell me but he does say that he also left a handwritten note under a rock beside the sign.

'The red paper was on top of the rock and a white sheet of paper was under the rock. You *must* have seen them.'

I shake my head. Baz is disappointed. I am disappointed too. I could use a motivational pick-me-up.

'It is only two kilometres back, Moz. Maybe you can go back and pick it up,' Baz suggests.

Moz says that he will.

While waiting for me to arrive, Baz has opened birthday cards he brought with him from his family. He is keen to share them with me but I tell him I will look at them later. I do not have the energy right now.

Baz sets out again, disappointed that I do not want to look at his birthday cards and disappointed about the inspirational notes he left for me. I do not like to disappoint my brother, but I do not have the energy to care.

I brush my teeth. I swallow a couple of painkillers and then fill the small container in which I carry my tablets. I down a juice box, fill my water, grab my things and I am on my way.

Moz and Jane pack up our breakfast spot and a short while later, they pass me in the car. They do not stop to hand me Baz's inspirational note. That's strange, I think. Perhaps they are taking it up to Baz for him to give to me elsewhere.

≈

It is 11.45 am and, except for me, nothing moves on Whim Plain. The only sound is the constant buzzing of the flies that cover me, like the stippling of some great Renaissance artwork. It is hot—damn hot. It is well above one hundred in the old scale. The sun is not quite directly overhead. I have a small shadow in front of me—perhaps a foot in front. Whim Plain is a barren, featureless frying pan. I feel like lying down to die but I know that I cannot give up and so I ponder what Henry wrote:

When your head is hot and aching, and the shadeless plain is wide,
And it's fifteen miles to water in the scrub the other side—
Don't give up, don't be down-hearted, to a man's strong heart be true!
Take the air in through your nostrils, set your lips and see it through.

In the distance to my left I see puddles in the empty barrenness. As I stare, some disappear. I wonder if my mind is playing tricks. Are they mirages in the shimmering heat? As I stare into the heat haze, I am not convinced that they are all mirages because some have a permanency about their appearance that suggests Queensland floodwaters. It is very hard to tell out here. After all, 'it's the heat that makes us all a bit ratty at times' and it is hot enough for me to lose my mind. I stop to have a good look. I would not like to bet my life that any of them are real and, in different circumstances, that is what I might have to do. To wander out onto the claypan in search of a drink might be a death sentence. As warm as it is coming out of my sunbaked hydration pack, I am thankful for the water I carry. I stare into the distance until the water that I see comes from my eyes. Are they mirages out there or puddles? Who can tell? My mind and my eyes play tricks. Is it just an illusion? Maybe the whole big, barren plain is an illusion. The open vastness, devoid of vegetation, is otherworldly.

I resume walking—walking and looking into the distance. I do not see any animals or birds. It is not the time of day for things to be moving anywhere, but particularly so out here where there is no cover. There is no shade, no shelter, nowhere to hide from the sun.

I cannot see Baz in the distance. After breakfast, I started only a few minutes behind him but he has left me in his wake.

The sky is a magnificent blue. I stop to look around the awe-inspiring majesty of the enormous sky, searching for even a suggestion of cloud cover. I cannot see any sign of clouds except on my right, far, far in the distance to the north. Just peeking above the northern horizon I see a couple of very thin, small, wispy cirrus clouds.

Eventually, I see Baz in the middle of the road ahead. I am surprised to discover that he is less than a kilometre in front. Where he is, there are a couple of trees to the side of the road. I suspect he might have been sitting in the shade. He now appears to be facing back towards me, turned around to see where I am. He has his binoculars with him and I know that in addition to using them to get close-up views of birds and animals, he also uses them to look back to see where I am and to check that I am still coming along okay. He is his brother's keeper. He is always watching out for me.

≈

I approach the H40 sign—Hungerford 40 km—on the left. That's good. Just a few more steps and I will have only 40 km to go. A few more steps. I have only 40 km to go. A few more steps. I have less than 40 km to go. I am getting there. Slowly, but surely, I am getting there.

As I pass the road marker, I look around for another note from Baz. I thought Moz might have picked up the old note and given it back to Baz who, seeing as I missed it at 50 km, has put it here to try again at 40 km. I look hard, but no notes. I pass and turn and look back—still no notes.

Shortly after noon, I reach a spot where Moz has parked beneath a small stand of eucalypts and set up my blankets to create some shade. I take a quick drink but decide to keep going. I will not get across Whim Plain sitting on my bum. It is important to keep going while I can. Before I leave, I ask Moz if he went back to pick up Baz's motivational quote.

'No,' Moz says, to my disappointment. 'I figured that a motivational quote isn't of much use to someone if they are dying of exposure. We figured it was more important that we stuck close to you and Baz to provide shelter and shade.'

I walk on and Moz and Jane pass me in the car, providing a cricket update before they disappear. I found Moz's explanation for not retrieving Baz's note to be an interesting thing to say. He had not gone back to get Baz's note because of the perceived need to stay close. A dead man is hard to motivate. I appreciate the sentiment. I understand it, but I can no longer see Moz despite the fact that I can see a long, long way on the flat plain. It was two kilometres to drive back to retrieve the note, but I can guarantee that he is a lot further away from me than two kilometres right now.

≈

Whim Plain is incredible. It is so flat, so open, and so barren—absolutely incredible. I am out in the heart of Whim Plain where there is no shelter or shade of any kind. There are no trees whatsoever. There is nothing for miles to the left, to the right, or straight ahead. I must extend my vision into the far, far distance to see a tree.

Save for the scattered and sun bleached bones of deceased animals, I see little sign of animals having once existed out here. I approach the spot where, last time, a bull and his harem of cows put the wind up Moz and me. The bull stared us down and, after emptying its bladder in an impressive manner, it moved our way in a menacing fashion. Moz and I looked around at our barren, featureless surrounds. No trees to climb. Nothing to hide behind. I decided to go down fighting and instructed Moz to arm himself. Rocks are the one thing in ample supply out here. Fortunately, a white knight arrived in a dilapidated cattle truck, piercing the fast-closing gap between us and the approaching cattle. The white knight was Mick Fisher. I do not see Mick or his brother, Dave, around this time though.

Thankfully, there are no bulls or cows to worry about today either.

I see tree-covered rises in the distance far to my right. They are neither mountains nor tall hills, but merely more of Henry's 'ridges on the floors of hell.' I am walking on the floor of hell on Whim Plain all right. I expect Lucifer to pop up before me with a wicked smirk across his face, delighting in my suffering.

The sun is directly overhead. My shadow is straight underneath me. For the first time in my life, as rivers of sweat run down my back, I feel a preference for the cold of the Canadian winter over the heat of an Australian summer. I imagine the snow-filled backyard at my house in Winnipeg. Before I left, I dug out a snow cave in which my children could play. I imagine crawling into the cave to die now.

Here and there, tufts of prickled grass dot the claypans. Occasionally a grasshopper springs from one dry stalk of grass to another, somehow managing to cling—mid spring—before bouncing away once more. Plagues of locusts are eating swathes through country Victoria and the grasshoppers here just add to the foreboding sense of doom.

Although there is the ridge running in the distance off to the north, out in front of me everything is flat, flat, flat. I wonder what Baz thinks of Whim Plain. As horrible as it is, it certainly has an attraction that is all its own. I had wondered aloud whether or not Baz should accompany us when we went to bury our water and supplies before we started the walk.

'One of the great things about last time was that, like Henry, when we walked it, we were seeing it all for the first time,' I said. 'I wonder if Baz should stay behind so that he, too, can see things for the first time during the walk.'

Baz had decided that he wanted to assist us with burying the supplies. However, because of the water over the road at Cuttaburra Creek we had not driven past there. As such, everything that Baz is seeing today is a first time view. I am glad of that. He got to help like he wanted, but today, on his birthday; he is also seeing new country that he has never seen before. What interesting new country Whim Plain is to discover on a birthday. One would be hard pressed to argue for it being a beautiful place, but it is a fascinating place nonetheless. It is all I have to give to Baz as a birthday present. 'Happy birthday, mate,' I call out, but my words burn and disappear like smoke into the emptiness.

I hear footsteps racing up behind me. Baz, Moz and Jane are all far ahead of me. I spin around to see what is running my way. My heart racing, I see that the plain is as empty as ever.

'What the hell was that?'

I turn and continue walking along the plain but have only gone a few steps when I hear something running up behind me once again. Heart pounding even faster, I spin again, not knowing if I am about to be flattened by an enraged bull, pounced upon by a hungry Dingo, speared by the ghost of a long dead Aborigine, or even devoured by a bunyip. To a mix both of relief and consternation, I discover once again that there is nothing behind me. I stand in the emptiness of Whim Plain and slowly turn a full 360 degrees, surveying everything around me. The only thing out here is me—me and the sun. Slowly, I release the breath caught inside my throat. I see a small eddy of dust spiralling along the Track toward me. The fluky breeze ruffles and then rustles my bandana, forcing one side to flap against my shoulder. Flap, flap, flap. The breeze blows for just a few seconds. Flap, flap, flap—like footsteps running the Track.[1]

≈

I do not stop with Baz when I reach where Moz and Jane have decided we will have lunch. I merely refill my water containers and keep going. Knowing the difficulties I am having moving again after breaks, I decide that I am better to keep going, making cuts into the final session before I stop.

Moz tells me the cricketers have gone to lunch with England tied for their highest ever Test score in Australia. Although the Aussies are in an awful position, I appreciate the cricket updates. Henry would have said of me that I have 'a cramped mind devoted to sport' and that, as a lover of sport, my 'god is a two-legged brute with unnaturally developed muscles and no brains.' Elsewhere, Henry further expressed his distaste for Australia's unquenchable thirst for sport. He felt pressure to join the throng and sing praises to brawny gladiators.

In the land where sport is sacred, where the lab'rer is a god,
You must pander to the people, make a hero of a clod!

And then, fearing the fate that might await him as a writer in a land where sport was valued far above literature, he could see in his future

[Being] buried as a pauper; to be shoved beneath the sod —
While the brainless man of muscle has the burial of a god.

≈

At 1.30, I am only dimly conscious of three small Red Kangaroos as I pass the Wombah mailbox to my right. I am on automatic pilot. I try not to think—try not to feel. I recall Moz's description of me as a tractor. I try just to keep chugging along, not thinking and not feeling. It is a matter of putting one foot in front of the other.

I am so disengaged from my surroundings that, despite the emptiness around me, I pass a deserted old wrecked car in the paddock to the left of the Track and fail to see it. Moz and I stopped there last time and took photographs. Tomorrow evening, safely ensconced within Hungerford's Royal Mail Hotel, I will be looking through the digital photos on Baz's camera and see that he has photographed the car.

'I didn't see this,' I say.

'How could you miss it?' How, indeed? I am locked behind my shield.

With the sun beating down from the cloudless sky, the plain seems to be endless but, as the clock approaches 2.00, I see a big row of trees in the far distance in front of me and I know that those trees signal the end of the plain. I will get across this plain, eventually. The sun still seems to be exactly overhead, but it has seemed so for about an hour-and-a-half. Out here, the Earth turns very slowly. Looking down at my shadow, where the bigger part of that shadow is perhaps slightly behind me, the sun might have *just* passed across the top. It races to the sky's heights in the mornings, yet crawls ever so slowly throughout the remainder of the day and evening.

I notice that my sun blistered hands run with sweat. Throughout my tramp, I keep constant tabs on myself and my physical state, particularly with regards to hydration and the obvious danger of dehydration. I consistently check my hands to see if I am still sweating. I am always checking when I pee. I am mindful of how much I pee—in terms of how often, but also in terms of how much I pee each time—and the colour of that pee. An indicator of dehydration can be if one's pee is really yellow or tending to orange. Because of all the water that I am drinking, my pee should mostly be clear. Yesterday I noticed that I tended to pee less and less often. Despite having my piss bottle beside me in the tent, I did not pee through the night either, perhaps signifying that I did not drink as much as I might have—should have—and that I need to drink more.

I pass a couple of small waterholes. They each cover a distance of just a couple of metres square. Two stunted trees emerge from one of the waterholes. The sand around the water is trampled with the foot and hoof prints of hundreds of thirsty animals. The shallow water within is a runny baby poo brown. I could use some more water, but I am not yet so thirsty as to bend down and cup that into my mouth. A slight breeze caresses the bandana at the back of my neck. As much as I would love to stop and enjoy the breeze, I force myself to continue.

≈

At 2.30, I am still on Whim Plain. The only relief from the sun is a faint and occasional breeze from the east. The Track is rocky and hard and my battered feet are on fire. I have surely formed additional blisters today. I pass the H30—Hungerford 30 km—road marker. Our westerly change in direction from north of Yantabulla means that when the sun is at its hottest from early afternoon through to the early evening, it strikes right in our faces with blinding intensity.

I take my portraits of Henry from my pocket and gaze at them briefly. I do not allow my mind to wander or to wonder about what Jennifer and the girls might be doing. I expect some will find it peculiar that I do not carry with me a portrait of Jennifer, Bronwyn or Tegwen. Everything that is good and lovely has to be shut out. I am a base creature here that just keeps moving forward. I would not even say that I walk. I just move forward. I am a base creature—a machine. I am a dirty, stinking, gritty machine that does not think. It just moves forward. Everything else is shut out. I go through long periods where I do not look. I do not see. I do not think. I just move forward. It is all about progressing forward. Of necessity, there is hardness within me. That hardness is needed to keep me moving. The soft places—where portraits of Jennifer and the girls might exist—need to be denied. It is all about being hard—continuing always to be hard, and to move forward. I am a stinking, awful mess but, as such, I can get there. It is all about doing, and not thinking, not feeling.

≈

Despite—or because of—his success in the early 1890s, Henry was on a downhill path. The money he earned from publishing in the *Bulletin* and other periodicals would quickly be spent at the nearest hotel. The sight of him staggering drunkenly through the alleys of The Rocks—the rough, lower class area of 1890s Sydney—was becoming all too common. Among those who considered themselves of a higher class, Henry's reputation was taking a self-inflicted pounding. Early one morning, Henry and his mate Edwin Brady dropped in to visit Archibald at the *Bulletin* office. Henry was in a dark mood and, out of earshot, Archibald asked Brady, 'What's the matter with Lawson?'

'He's all right,' Brady replied defensively.

'No, he is not all right,' Archibald demanded. 'He is coming here in the morning with tobacco juice running down his jaw, smelling of stale beer, and he has begun to write about "The Rocks". The next thing he will be known as "the Poet of the Rocks".'

Brady knew that his friend needed to get away from some bad influences. 'If he got away to the bush he would be all right,' Brady suggested.

'Why doesn't he go back to the bush then?' Archibald asked.

'No money,' replied Brady.

Henry was a talent—one of the *Bulletin*'s stars. His talent was in danger of being wasted and lost to drink. Archibald knew also that there was sharp interest in the Union movement in northern New South Wales and Queensland, and with Bourke being a hotbed of Union activity Archibald thought this might entice him. The Amalgamated Shearers' Union and the General Labourers' Union both had a heavy presence in Bourke. At their 1890 conference in Bourke, the Shearers' Union refused to work alongside men who did not belong to the Union. At the same time, the aristocratic squatters would not be pushed around by their employees. Many of the wealthy squatters saw it as their God-given right to own land and to preserve their wealth and ownership as the natural order of things. They responded by forming what was essentially a union of their own, the Pastoralists' Federal Council.

Shearers were paid according to the numbers of sheep they shore. The pastoralists always wanted to lower the rate paid per sheep, while the shearers always wanted the rate increased. In 1891, the Queensland shearers went on strike. Emotions bubbled over and armed soldiers and policemen were needed to keep the peace. Henry weighed into the debate and landed firmly on the side of the workers. He saw the squatters as representing Old World tyranny. He saw the strike as a fight for freedom in the New World. The workers needed to 'knock the tyrants silly,' Henry proclaimed.

We'll make the tyrants feel the sting
O' those that they would throttle;
They needn't say the fault is ours
If blood should stain the wattle![2]

Archibald wanted Henry to get his drinking under control and he also wanted to help Henry do that. Having one of its best writers on the ground in Bourke could also serve the *Bulletin* well. Archibald, however, knew that Henry would be suspicious if he or his partner, William Macleod, handed him money and pushed Henry out of the city.

'Well,' Archibald said to Brady, 'you go to him and speak to him. If Macleod or I speaks to him he will think there is something behind it.'

Removal from bad influences was a tactic that had been tried before with Henry—and failed. Previously, when Henry had returned from his mother in Sydney to work with his father in the Bush, Peter was shocked at Henry's drinking. He rushed off a telegram to his estranged wife, urging her to find Henry a job in the city so that Henry might be removed from his hard drinking Bush friends. This time, Archibald and Brady were doing it in reverse—sending Henry from city to Bush rather than Bush to city.

In terms of his drinking, the plan of separation from bad influences was to work no better the second time than the first. Nonetheless, without the benefit of hindsight, of this Brady and Archibald were ignorant. They could only do their best. Brady spoke to Henry and an arrangement was made with Archibald. Henry packed together some things and, as he put it, 'towards the end of '92 I got £5 and a railway ticket from the *Bulletin* and went to Bourke.'

As much as I would like to go all the way to tonight's campsite, I am unbearably hot and need to stop and cool down. I continue up the road in the heat until about 2:50 pm and then begin to search for a shady spot to rest. The shade along the Track here is not good. I see a couple of Desert Bloodwoods 20 m or so to the left of the Track and make for them, hoping the shade will be sufficient. I pick my way over the rocky ground and through the prickly, albeit sparse, undergrowth, reach the trees, sling off my pack and take a pee. My fountain is only starting when I hear a vehicle approaching in the distance. In the openness out here though, one can hear vehicles well before they arrive. I finish my job before the car even gets to me. As it is, I am far enough off the road that the people within the vehicle do not see me as they pass.

I sink to the rough-barked trunk of one of the bloodwoods but, unfortunately, the trees are sparsely leaved and, in any case, they do not branch out very far from the trunk. What little mottled shade there is exists only in the undergrowth immediately beneath the trees. I take a fallen branch in my hand and, on the alert for snakes; I thump it on the ground several times.

I stay where I am for perhaps three or four minutes but, looking down at my legs, I see that I am as much out of shade as I am in shade. I will not get cool here. I see another pair of bloodwoods further in but I realise that if I continue to go from the Track, Baz and the support crew will pass me by.

I venture back onto the Track and wander another 15 minutes or so, searching for a clear and shady spot to rest. With every step, I feel my body temperature rising. With every step, I grow increasingly pessimistic about finding suitable shade. Spindly Mulga shrubs and Desert Cassia line the Track, but nothing offers much shade. Over the past 15 or 20 minutes, I have dismissed many insufficient spots that appeared shadier than anything I can see now. A bend forms in the road 20 or 30 m in front of me. I have no reason to believe that anything ahead of me will be better than anything beside me. From what I can see, things might even be worse around the bend. I decide that I should stop where I am, taking the Mulga Tree to my left as the Devil I know ahead of the Devil I do not know around the bend.

I unsling my pack, grab a stick from beneath the tree and use it to thump on the ground to warn off any nearby snakes. The shade is certainly not great, but it is the best that is currently available. At least I have a clear space where I can stretch out on the ground and also see whatever is sharing the ground around me.

I fall to my backside. I reach for my pack and remove a water bottle then detach my cup from my pack. I arrange the pack at my feet so that my shoes and socks are up out of the dust. I then stretch out with my head in the dirt, my hat over my face, my cup and water bottle within easy reach on one side and my snake-warning stick on the other. I give the stick a few taps on the ground and then close my eyes, listening to my rapid breathing. I feel like I might vomit.

I am not sure how long I lay there before Baz appears.

'How are you going, Gregga?'

'Not great. Bloody hot.' I do not bother to look up.

'Do you need water?'

'I'll get some when Moz appears.'

'He won't be too far behind, I don't think.'

Baz takes a seat on the ground beside me but does not lie down.

I look up at him. He seems to be comfortable and relaxed. 'I am glad to be off Whim Plain. It was effing hot out there.'

Baz nods his agreement.

After a time, we hear and then see Moz's car approaching.

'How are yer poppin' up?' Moz asks when he arrives with his big smile. That is usually the first thing he asks when he sees us. 'Do you blokes need anything?' That, or something similar, is usually the second.

'I'd kill for some water,' I say, but I do not have to. Moz brings it to me without anyone resorting to violence.

'Am I in still in the shade?' I ask after a while. I am melting and I know that the sun is moving.

'Well, most of you is,' Moz replies. 'You might want to move across a little further.'

Move? Me? That is what I was afraid of. It might be just as easy to move the tree. It takes me a long while, but I build up all of the energy that remains in my overheated body and force myself to the right, chasing the best of the spotty shade. I move 20 or 30 cm, but it is all I can manage.

'We were talking to a bloke back there who is the manager of Warroo Station,' Moz says. 'We told him about you two and he wants to talk to you. He is very interested in what you are doing. He says he has some things to do up the road and so he will be passing by here shortly.'

Warroo sounds familiar but I am too buggered to search my memory for where I might have heard of it. 'Where's Warroo?' I groan.

'Just back down the road there. It's a white house by the road.'

Lying here in the dirt, I do not recall seeing Warroo as I walked the Track. It may have passed before my eyes, but nothing was consciously recorded.

'He is a nice man,' Jane says.

'His name is John and he's really interested to talk with you,' Moz adds.

I exhale audibly. I do not have the faculty to engage in a conversation with anyone.

After a while, the manager of Warroo pulls up in his white one-tonner. I am flat on my back with my head in the dirt and, although I hear the words that pass between John and Baz, Moz, and Jane, I make few entries into that conversation.

'Is he okay?' I hear John ask at one point.

'I'm fine. I just need a rest, is all,' I manage, forcing the most reassuring and reassured voice that I can muster. I am not sure how convincing I am—I certainly fail to convince myself. As I lie here in the furnaces of hell, I know that I am close to death.

I do not know how long John visits. Fifteen minutes? Twenty? Thirty? Much more or much less? I have little idea of the passage of time. He seems like a nice man and I wish I had more to give but, hidden beneath the hat that shields my face from the sun, I am worried about my body temperature.

I hear the others talking about Cobb and Co., the famed coach company that holds such a significant place in Australia's historical record. John is saying something about coach tracks up ahead. I do not want to miss them, but my head is spinning and I cannot register my interest.

John says that he must get going. He says he needs to go up the road towards Hungerford but that he will pass us again on his return.

'I'm sorry to talk your ear off,' I say to John as he bids farewell. It is the spunkiest thing that I can manage. It is my way of saying that I am not beaten yet.

After John leaves, Baz decides that he will push on too. Having already left about 30 km behind us, together with Moz he decides we should do another three kilometres or so today, eating further into the distance remaining to Hungerford tomorrow.

Before they depart, I fill my water bottles again and check my hydration bladder. Baz asks if I want him to tip some water over my face and I tell him that I do.

'How hot is it, Moz?' I ask.

Moz goes to check the gauge in the car and reports that it is 38 degrees. 'That is the highest that I have seen it today,' he says.

Bullshit, I think. It was a lot hotter than that out on Whim Plain. I figure it must surely have been 45 degrees or more. Once again, I wonder about the accuracy of Moz's and Jane's reports.

≈

After the tramp is over and I am safely back on my hotel bed in Bourke, Moz tells me that he and Jane have been telling us falsehoods with regards to temperatures and distances. 'I've done a lot of reading about psychology and the workings of the mind,' Moz smiles to me. 'I knew that it'd always be better to believe it was cooler than it really was. During the day, we also said that you had travelled less than you had. That way, when you got to the end of each day, you would always get a nice surprise to find out that you had done more kilometres than you thought you had. It would be a nice boost to get you going for the following day.'

I look at him long and hard, not saying a word.

'How do you feel about that?' Moz asks. I think he is developing a good sense of what I am feeling. Nonetheless, he wants to hear me say the opposite. He is seeking reassurance that his plan was not as idiotic as it was.

'I'll have to think about it,' I say, remembering every mother's admonition. 'If you can't say anything nice, don't say anything at all.'

When we are all having a drink downstairs, Moz tells Baz about his amateur psychology games too. Baz is as unimpressed as me but perhaps does a better job of keeping his feelings from appearing on his face. 'I think it worked in reverse,' Baz says without equivocation though. 'It was a kick in the guts during the day to be told that you hadn't done as much as you thought. It made me feel more buggered.'

Underrepresentation of the difficulties of our situation was what Moz believed would help Baz and me to endure those difficulties. Given that I often did not believe what they were telling me, all it did to me was create an additional layer of difficulty through which I had to wade. I am afraid Moz got this one wrong. Not having accurate reports in these sorts of extreme conditions is, I believe, extremely dangerous. One makes decisions—whether to continue or to rest—based on all of the information that is available. If that information is deceiving, it creates dangerous circumstances. Not only were the reports not accurate, they were deliberately inaccurate. Unlike Jack Nicholson's *A Few Good Men* mates, Baz and I are both tough enough to handle the truth.

≈

It was a long rail journey from Sydney to Bourke but Henry 'took notes all the way up' and 'got a lot of very good points for copy.' Henry recorded the details of his train ride in the sketch, 'In A Dry Season':

Draw a wire fence and a few ragged gums, and add some scattered sheep running away from the train. Then you'll have the bush all along the New South Wales western line from Bathurst on....
Slop sac suits, red faces, and old-fashioned, flat-brimmed hats, with wire round the brims, begin to drop into the train on the other side of Bathurst; and here and there a hat with three inches of crape round the crown, which perhaps signifies death in the family at some remote date, and perhaps doesn't. Sometimes, I believe, it only means grease under the band. I notice that when a bushman puts crape round his hat he generally leaves it there till the hat wears out, or another friend dies. In the latter case, he buys a new piece of crape. This outward sign of bereavement usually has a jolly red face beneath it. Death is about the only cheerful thing in the bush....
Somebody told me that the country was very dry on the other side of Nevertire. It is. I wouldn't like to sit down on it anywhere. The least horrible spot in the bush, in a dry season, is where the bush isn't—where it has been cleared away and a green crop is trying to grow. They talk of settling people on the land! Better settle in it....

Henry was not impressed with one of the men with whom he shared the train:

About Byrock we met the bush liar in all his glory. He was dressed like—like a bush larrikin. His name was Jim. He had been to a ball where some blank had 'touched' his blanky overcoat. The overcoat had a cheque for ten 'quid' in the pocket. He didn't seem to feel the loss much. 'Wot's ten quid?' He'd been everywhere, including the Gulf country. He still had three or four sheds to go to. He had telegrams in his pocket from half a dozen squatters and supers offering him pens on any terms. He didn't give a blank whether he took them or no. He thought at first he had the telegrams on him but found that he had left them in the pocket of the overcoat aforesaid. He had learned butchering in a day. He was a bit of a scrapper himself and talked a lot about the ring. At the last station where he shore he gave the super the father of a hiding. The super was a big chap, about six-foot-three, and had knocked out Paddy Somebody in one round. He worked with a man who shore four hundred sheep in nine hours.

No doubt Henry enjoyed hearing 'a quiet-looking bushman...[take] the liar down in about three minutes.'

The train rattled on until, 'At 5.30 we saw a long line of camels moving out across the sunset. There's something snaky about camels. They remind me of turtles and goannas.

Somebody said, "Here's Bourke".'

Henry had arrived.

≈

My heart pounds in my chest. It feels as if it wants to escape the confines of my rib cavity. I drag myself across the dirt and reposition my body so that it more fully falls beneath the cover of the moving shade. I am almost cooked. I arrived hot and tired from Whim Plain and have not cooled down in the two hours I have been lying in the dirt. I want to get up and continue, but I cannot find the energy to rise.

My breathing is deep and rapid and my head throbs. I can feel my brain bubbling and boiling. As much as I drink, my mouth is still dry. I would trade all of my remaining water for one glass of ice though. As thirsty as I am, I am more concerned with my heat.

This is my hardest, toughest, most distressed rest stop yet—and the closest I have come to completely overheating and having my body shut down. I swish another mouthful of water. No sooner do I swallow than my mouth and the insides of my cheeks feel dry once again. I try to moisten my cracked lips but my tongue is dry. I remind myself to relax. It will not help to stress and worry about my predicament. I dribble some more water onto my face, letting it run in dirty rivers down from my eyes and into my ears.

I start to rouse. I must get down the Track. I hate the Track with a passion that burns, yet I know that is where I must be. Staying here, cooking, I am not moving closer to Hungerford. As much as I hate it, I realise that the Track is my only friend—my hate-filled merciless friend—because it is only on the Track that I can move towards my goal. As much as I would love to abandon the Track forever, I am bound to it.

When they left, Moz told me that he and Jane would drive about three kilometres down the road and look for a place to make camp. I know those three kilometres are going to put me through hell. I know that getting going again after such a long and miserable rest stop is going to be torturous. I know that I am going to look like a fool as I drag my broken body down the Track. I do not want an audience. I want only to get up and to complete the last stage without company and away from the eyes of other people. Then I can have all the water I need.

Gradually and painfully, I pull myself into a seated position. I hear the sound of a vehicle approaching from the direction of Hungerford. It is John on his way back to Warroo.

'Are you okay?' John is surprised that I have not moved from the spot where he last saw me.

'Um, yep.' I can tell he is worried. 'I was just thinking about getting going again.' I put on as brave and as cheerful a face as I can manage.

Hell awaits.

'The others have found a spot up the road a bit,' John tells me. 'They are camped beside the Cobb and Co. coach tracks.'

That's right, John had been talking earlier about wheel ruts in the ground. I want to see them.

'They were looking at them when I came along,' John continues. 'They didn't know what they were looking at. Sometimes things can be difficult to know until someone tells you.'

'Yep.' I do not know what that means. 'I'm keen to see them.'

'I told the others about some plans for Sunday. I won't bother to tell you now because it won't make a lot of sense.' He can obviously tell that my mind is not at its sharpest. He is very perceptive. 'The others will explain it all to you later.'

'Yep. Sounds good.' Sunday? When is Sunday?

'Are you sure you are alright?'

'Yep. Bloody hot is all. What was the temperature today, do you reckon?'

John says that he does not know what the temperature reached, but that the forecast for Bourke had been for 35 degrees. 'It is always a few degrees hotter than Bourke out here,' John says. 'They take the official temperature in the shade, so it is ten degrees hotter out on the road with what you have been doing.'

Thirty-five in Bourke. A few degrees hotter out here. Ten degrees hotter in the sun. It would put the day's top close to fifty. Moz tried to tell me the car thermometer had not gone above 38 degrees. Bullshit.

John bids me farewell and then leaves. I wait until I can no longer hear his vehicle noise. It allows me to listen for any other approaching cars. When I am convinced that no one is coming, I pull myself to my unsteady feet, feeling like a baby giraffe learning to walk. I have been off my feet for two-and-a-half hours. My painful, pain-filled feet cause my legs to wobble.

Slowly and uncertainly, I bend down and lift my pack onto my back. It is 5.45 pm when I take my first hobbled steps back on the Track. Camp may only be three kilometres up the road, but it is going to be a long three kilometres. I have a long, slow, painful stagger ahead of me and I suspect it will take me an hour to get to camp. At the rate of my first few hundred metres back on the Track, the remaining distance could well take me an hour-and-a-half.

The road is hard because of all of the rocks so I move to the side to try to find a softer line. I remove the tablet container from my pocket and jiggle two painkillers into the cup of my upturned palm. I swallow them one at a time, sipping some warm water from the tube of my hydration pack to help each one go down. My water bottles are empty, but I have about two litres of water left in my hydration bladder to get me through the next three kilometres.

I turn and look back at my footprints in the sand along the side of the road. I notice that the back edge of the lead foot is almost completely level with the front edge of the trailing foot. I am not walking, I am shuffling. I make deliberate and concerted efforts to extend my stride, but I cannot stretch my tortured legs and feet.

I am sure the others will be wondering where I am. They will expect me in much earlier than will be the case. For a start, they would not have expected me to rest for so long. They will also not expect it to take as long as it will to walk from my rest stop to camp. They should go ahead and eat their tea without me. I do not want them waiting for me. There is no reason to wait.

To distract the mind from my distress, I think about the Cobb and Co. ruts up ahead and the horses that dragged stagecoaches through the Outback and the Bush in Henry's day. I recite a stanza from one of my favourite of Henry's poems.

He had run with Cobb and Co.—'that grey leader, let him go!'
There were men 'as knowed the brand upon his hide,'
And 'as knowed it on the course'. Funeral service: 'Good old horse!'
When we burnt him in the gully where he died.

In my mind's eye, I see the 'old grey horse' carrying three or four children to Henry's Old Bark School.

My spirits buoyed by reciting poetry, I dig deeper and pull out another of Henry's references to Cobb and Co. I picture a lantern-lit coach racing through the Bush.

But on the bank to westward a broad, triumphant glow—
A hundred miles shall see to-night the lights of Cobb and Co.!

≈

When I reach camp, it is 7.10 in the evening. As I anticipated, it has taken me almost a full hour-and-a-half to walk the final three kilometres. Nonetheless, it has been a productive day. Baz says we have done 35 km. That does not tally with the 30 Moz said I had done before my big stop though. Either way, it means I have only 25 to 27 km to do to get to Hungerford tomorrow. I will be bloody glad when that is done.

The others tell me that John invited us to stay at his place at Warroo on Sunday, our first night of the return leg from Hungerford.

'How far off the road is it?' I do not want to go far from the Track, doing unnecessary kilometres.

'It is only about 50 metres off the road.' Moz knows what I am thinking.

'Yeah?'

'Yeah. You remember? It is the big white place right beside the road.'

I still do not recall seeing it when I passed.

'We told John about your interest in Australian history,' Moz says. 'He has the memoirs and some scrapbooks and things of someone who was in the Charge at Beersheba. John thought you might be interested to read the memoirs.'

'Yeah, good.' It sounds interesting. I just hope that I have the energy to get involved.

'How was your birthday, Baz?' I ask.

'Good.' He shows me some birthday cards from his two sons. They have both written, 'I hope you are having a good holiday with Uncle Greg.' We have to laugh. This is anything but a holiday.

≈

After tea, Jane goes to bed early. Knowing that tomorrow we will be in Hungerford, I feel sufficiently energised to sit up with Baz and Moz alongside the small, flickering campfire.

Baz looks at my hair, sticking up in sweaty clumps and badly in need of a wash.

'What do you reckon, Mozza?' says Baz. 'Greg looks like Grandpa Simpson.'

Baz and Moz have a good chuckle at my expense. There are no mirrors out here and so I do not join the hilarity.

Baz takes a stick and scratches the shape of a big television set into the sand. 'What do you fellas want to watch?' he asks.

'I'd like to see how this walk turns out,' I suggest.

Baz adds details inside the picture tube. Without narration, he draws a stick figure lying on the ground, his hair sticking up high. That must be me, dead in the dirt. Then Baz draws a dead little stick figure. That must be him. Then there is a great big giant of a stick figure, standing tall over the dead bodies, a big smile spread across his face. That is obviously Moz.

Bloody Bazza. I have always said he would sell his soul for a laugh.

Scratching the image of an aeroplane into his TV set, Baz says, 'And this is Jane flying out of here.'

DAY SEVEN

Henry Lawson in Bourke

And could I roll the summers back, or bring the dead time on again;
Or from the grave or world-wide track, call back to Bourke the vanished men,
With mind content I'd go to sleep, and leave those mates to judge me true,
And leave my name to Bourke to keep—the Bourke of Ninety-one and two.

- Bourke, *When I was King and Other Verses*, 1905

Bourke, the metropolis of the Great Scrubs, on the banks of the Darling River, about five hundred miles from Sydney, was suffering from a long drought when I was there in ninety-two.

- That Pretty Girl in the Army, *Children of the Bush*, 1902

From the interior of my tent, I see that there is a vibrant sunrise when I wake. I start to attend to my feet. Meanwhile, the violet sky gradually dims to periwinkle and then, as the orange glow at the horizon softens and is replaced by the yellow ball of the sun, the sky becomes increasingly bluer, adopting a powder blue shade that will grow brighter throughout the morning. As I watch the sunrise behind four Mulga Trees, I curse the fact that the second of my toe sleeves wore through yesterday, leaving me with a blister on the little toe of my left foot. It is not the only new blister from yesterday's hard slog so I try to use my bandages sparingly. Although my supplies were supposed to last me the entire journey, I am almost out. I began Day One thinking I had more than I would ever need but that was just one of my underestimations. This whole slog has proven many times more difficult than I expected.

Last time, reaching Hungerford was the cause of great celebration and even though I am proud of my achievement and of Baz's, I am also mindful that getting to Hungerford today is only the half-way mark.

≈

Henry's train rumbled into Bourke on 21 September 1892. His poem, 'Sweeney', suggests it was raining hard:

It was somewhere in September, and the sun was going down,
When I came, in search of 'copy', to a Darling-River town;
'Come-and-have-a-drink' we'll call it—'Tis a fitting name, I think—
And 'twas raining, for a wonder, up at Come-and-have-a-drink.

The poem also suggests that, as a destination to remove Henry from alcohol, Archibald and Brady had made a poor choice. It did not take Henry long to discover that 'Bourke has the name of being the most drunken town on the [Darling] River.' At the time, Bourke was home to somewhere between 1500 and 3000 residents.[1] While those people were served by only three Churches, there were 19 hotels at which to slake the thirst on a hot Outback day. Henry found drunkenness to be commonplace in Bourke. He claimed that one man's drunkenness 'excited no comment'. Even as the man remained drunk for weeks, nobody found it unusual.

He stayed drunk for three weeks, but the townspeople saw nothing unusual in that. In order to become an object of interest in their eyes, and in that line, he would have had to stay drunk for a year and fight three times a day—oftener, if possible—and lie in the road in the broiling heat between whiles, and be walked on by camels and Afghans and free-labourers, and be locked up every time he got sober enough to smash a policeman, and try to hang himself naked, and be finally squashed by a loaded wool team.

Henry booked into the Great Western Hotel and found there a copy of the *Bulletin* carrying one of his poems from the duel with Banjo Paterson. He sat down to scribble a letter to his favourite aunty and confidante, Mrs Emma Brooks. Henry had lived with her for much of the year prior to his journey to Bourke. As he wrote, Henry reflected on his battle against Banjo, writing, 'The bush between here and Bathurst is horrible. I was right, and Banjo wrong. Country very dry and dull... Had several interviews with Bushmen on the way up. Most of them hate the bush.' Henry was however, 'agreeably disappointed with Bourke. It is a much nicer town than I thought it would be.'

While in Sydney two years earlier, Henry read reports of flooding at Bourke[2] and of the townspeople being 'in great danger.' He asked around and his investigations led him to believe that, with the town cut off from the outside world, 'the chief danger was that the liquor would give out—the water having gotten into some of the pub cellars.' Given the reports that had reached Sydney, Henry was desirous to meet some of the life-saving heroes of the flood but he learned that 'most of the rescue work was done by a short man with his trousers tucked up and a bottle of whiskey in his pocket.'

≈

After the heat of Whim Plain yesterday, it was nice to cool off at night. I woke up during the night and was again amazed by the multitudes of stars in the sky. One of the constellations that I recognised was the hunter, Orion. On Christmas night, my wife and I saw Orion's belt shining brightly in the dark Canadian winter sky. It is comforting to know that even though we find ourselves at opposite ends of the world, the mighty hunter is watching over both Jennifer and me as we sleep.

One Aboriginal group recognises the stars of Orion as Julpan, the canoe in which a set of brothers made an eventful fishing trip. As members of the kingfish totem, the brothers were forbidden to eat kingfish as they were seen as family. Unfortunately for the hungry brothers, the only fish they could haul into their canoe were kingfish. As the brothers grew increasingly frustrated and angry, one declared that he would eat the next fish he caught, regardless of what type it was. As punishment, the brothers and their canoe were carried into the sky in a giant waterspout. The nearby Pleiades star cluster represents the fish that were still caught on the brothers' lines when they were taken up to spend the eternities paddling their canoe about the night sky.[3] With all of the Queensland floodwater we have had to cross, I wonder how Baz and I might fare with a canoe.

I fill my water containers and, with everything ready to go and a relatively short day ahead, I grant myself the liberty to wander over to where the Cobb and Co. wheel ruts are worn permanently into the ground. Those stagecoach tracks free my imagination. I think of the energy and excitement of the old Cobb and Co. days when:

Behind six foaming horses,
And lit by flashing lamps,
Old Cobb and Co., in royal state,
Went dashing past the camps.

Baz wanders over beside me. 'If those tracks could talk,' I say, thinking about all the people who would have ridden the coaches or seen them speeding by.

I break camp before Baz but lead only briefly. I had wished to start well ahead because he will walk a lot faster than me. I figured that if I got a good head start, he would catch me when we were close to Hungerford and we could walk in together.

Jane is alongside Baz as he passes less than five minutes after my 7.00 departure. He gives me the thumbs up. 'Morning Ralph. Morning Sam,' he says, borrowing from Warner Brothers' Ralph E. Wolf and Sam Sheepdog routine. With only 25 km to get to Hungerford, we can afford a little frivolity.

Although I have my usual pained start, the birds are chirping and the rising sun induces interesting patterns of light and shadow among the Mulga Trees. My own shadow stretches far in front of me, running perfectly parallel to the peripheries of the Track. My shadow will shorten as the sun moves higher through the morning but, for now, it suggests a long-limbed fellow perhaps more reminiscent of Henry.

Moz finishes packing the campsite and, with me only moving slowly, I have not gone far before he appears in the car at 7.30 am.

'How are yer poppin' up?' Moz drones and I wonder who the last of the laid-back Bushmen was to use that phrase before Moz reintroduced it to the Outback. For all I know, the phrase has not been used here since Henry asked Jim Gordon how he was getting along.

'You have done point nine of a kilometre,' Moz says with surprising precision. It is not far for half an hour of walking, but if I keep doing what I have just done, I will eventually be in Hungerford. It is just a matter of taking one step in front of the other and doing that over and over and over again. Just keep going. Just keep moving forward. Left, right, left, right, left, right, left.

'I'll just keep plugging away,' I tell Moz.

Moz wishes me well and drives on towards Baz and Jane who, by this time, are well ahead.

I hear a Galah in flight to my left and search the sky until I find it. The bird performs an acrobatic turn in the air and then alights on a tree. I see another Galah perched there as well. I watch as the birds greet one another. I hear other birds singing their morning greetings too. It is wonderful to see and hear all of the birds. I feel glad to be alive and I am excited and pleased that I will soon hit Hungerford. It has been a hell of a journey, but I will be there before the day is out. It will be great to have a shower. Since our departure from Fords Bridge, I have been without a wash for four days. I am also looking forward to doing my laundry and wearing clothes that do not reek of sweat and filth.

I notice that I am coughing and sniffling more today. At least I have ceased hacking up the big orange balls of phlegm with which I littered the Track over the first two or three days.

A fly chooses an inopportune time to descend from my hat and to buzz around my mouth and ends up stuck in my throat. This sets off another round of coughing as I try to extract the bugger. The fly does not make a tasty breakfast.

Eventually, I pass the H20 road sign. I am excited about having only 20 km to walk to Hungerford but this feeling is tempered by loneliness. I recall that the last time I passed this sign; I had Bronwyn as my companion. It was a long day and Bronwyn walked beside me for 12 km. It would be nice to have Bronwyn with me again this morning.

The brightness of the morning light is briefly dissipated by thin cloud cover passing before the sun. A breeze wafts by and, with the sun behind cloud, it feels suddenly cooler. I consciously increase the length of my stride and try to increase my pace. After 100 m or so, I stop to look back at my prints in the sand. Shuffling last evening, the back edge of my front foot was level with the front edge of the trailing foot. Now, consciously striding out, I see that there is a gap of a little over two feet between each footfall—a huge difference this morning compared to last evening. Maybe I can get to Hungerford sooner than I think.

I come across a dead kangaroo and see that it is crawling with maggots. There are literally millions of them—more than I have ever seen in one place. I think I now understand what Henry meant when he described this place as 'a land of living death.' The kangaroo's head is missing and the body is opened along the back and left side. The rib cavity is fully exposed, the flesh having been removed from the bones. Although the base of the tail is still fur-covered and intact, the fur and flesh have been removed from a point only a quarter of the way along the tail. The forearms, hands and feet all appear fine. I wonder how long it can be before the voracious maggots strip them too. Although it makes me squirm and sends a shiver down my spine, it is nonetheless fascinating to watch the maggots at work. I have been thinking that everything out here is trying to kill me but, as I watch the maggots crawling around inside the kangaroo's belly, I realise it is worse than that. Out here, nothing cares. It simply does not matter if I live or die on the Track. It is all a part of the cycle of life and death in this harsh environment.

I withstand the stomach-churning reek of death and this face-slapping reflection of my own mortality for as long as I am able, but then I must move on.

≈

The two rival newspapers in Bourke were the *Western Herald and Darling River Advocate* and the *Central Australian and Bourke Telegraph*. It was a coup of the highest order for the *Western Herald* to enlist Henry as a contributor. As was his want, Henry employed a pseudonym to conceal his authorship when he published a number of poems in the *Western Herald*. Henry said, 'The editor wanted to give me a notice, but I preferred to keep dark for a while.' Some of the poems appeared above the *nom de plume*, Tally—presumably a reference to a shearer's tally of sheep shorn in a day. On the train to Bourke he had, after all, 'had a great argument with a shearer about the number of sheep a man can shear in a day. I know nothing whatever about the business, but he did not know that.' Other pseudonym's employed by Henry over the years included Cervus Wright, Youth, Caliban, HAL, Joe Swallow, Rumfellow, and Jack Cornstalk.

The first poem appeared in the 27 September edition, just a week after Henry's arrival in Bourke. In Henry's *Western Herald* pieces he supported Unionism and passed comment on Bourke politics. Fresh from his 'Bush battle' with Banjo Paterson, Henry engaged in duelling verses with a poet publishing in the *Central Australian and Bourke Telegraph*. At this time, Henry was also employed to paint houses for a contractor named John Hawley. Almost 50 years later, and long after Henry's death, Hawley wrote a letter to the *Sydney Morning Herald* in which he stated, 'Henry Lawson told me he had written a poem each for the two local papers, the "Western Herald" and the "Central". Through them ran a competitive strain as one poet writing against another poet.' Despite the years that had passed, Hawley contended, 'the incidents and many more are fresh in my memory.' Having claimed victory in his battle with Banjo, was Henry now receiving payment for duelling with himself?

≈

Under the penname, Joe Swallow, Henry used the *Western Herald* to record his early impressions of Bourke.

When a fellow strikes the Darling, after coming from the East,
He will mostly take his bearing for a day or two at least;
And the roving rhymer always notes his first impressions down
For a life along the Darling isn't like the life in town.

Henry thought the men of Bourke were 'a trifle gone on sporting,' and, as one who should know, he found the local men 'over fond of beer'. All things considered however, life in Bourke was good because the 'girls [were] fairly pretty, and [the] liquor [was] pretty fair.'

After a week in Bourke, Henry wrote another letter to his aunty. He found some of the Outback women lacked the gentility one might expect in Sydney. 'This is a queer place,' Henry wrote. 'The ladies shout. A big jolly-looking woman—who, by-the-way, is the landlady of a bush pub—marched into the bar this morning, and asked me to have a drink. This is a fact; so help me, Moses!' A couple of barmaids might also have taken advantage of Henry's fondness for drink. 'The private barmaids sent me to bed boozed last night, but they won't do it again,—no. They are a pair of ex-actresses and cunning as the devil. I'm an awful fool.'

Henry acknowledged that the drinking culture in Bourke was bound to create problems for him. 'I won't do any good here. Everybody shouts. I must take to the bush as soon as I can.' After all, he had work to do for the *Bulletin* but, then again, Henry had no plans for working too hard. 'I am working up stuff for the *Bulletin* but—between you and me—I don't mean to sacrifice myself altogether.'

≈

I catch the others at the breakfast stop some time before 10.00 am. Seeing as I am moving okay, I decide just to fill up my water and take a bite of food with me to eat as I walk. Baz is stopped for a rest and a feed and so this provides me with the opportunity to move out in front. He will catch me soon enough and then we can walk in to Hungerford together.

I am filthy dirty. My clothes are a stinking, reeking mess stained with the red dust of the Outback. It will be lovely to shower and use soap and shampoo. It will be wonderful to emerge from the shower and to put on powder and deodorant and then freshly laundered clothes. I am so glad that I built the day of rest into the schedule for tomorrow. Psychologically, it has been good to have that carrot dangling in front of me. That said, for a brief moment I recognise what a bear it is going to be to get going again after a day of rest. For now however, I am not going to let thoughts of the future spoil my enjoyment of the present.

The sandy Track turns rocky for a period and, although the rocks of assorted sizes and shades of a burnt or rusty red are visually attractive, they make walking harder on my blistered feet. Never a slave to aesthetics, I curse the rocks.

At 10:20, I see a flock of more than 30 goats running through the bushes and trees to the right of the Track. The goats move in a hurry that, despite the Outback setting, is strangely reminiscent of waves of city folk rushing to work, 'flowing faster in the fear of being late'.

Five minutes later, a noisy flock of Grey Butcherbirds begin to accompany me along the Track. There are seven or eight birds in the flock and, while they remain within the cover of the branches of Mulga Trees, they fly from one tree to another and then to another, shadowing me. Their curiosity is undeniable. I am attracted by their unusual song. It is dimly reminiscent of a schoolchild's failed early attempts at playing a woodwind recorder, albeit more melodious and consistent. There is a somewhat carefree tone to the birds' songs. They bring to mind my Italian schoolmates' 'Devil-may-care attitudes' about music class—despite the imminent threat of a rap over the knuckles with the teacher's recorder. The birds perch only long enough for me to pass and then they fly ahead to another tree to wait for me. They peer at me from their dark faces, no doubt wondering what I am doing out here.

After accompanying me for almost ten minutes, the birds lose interest and trail off just as I pass the sign indicating I have only ten kilometres to go to make it to Hungerford. In the absence of the butcherbirds, it becomes almost eerily quiet, save for the occasional rustling of my bandana against my neck and ears when the fluky breeze blows.

Away in the emptiness to my right, I hear something that sounds like words spoken in a lost Aboriginal tongue. I turn and look, but see nothing but thirsty shrubs and tufts of grass on dry, sandy soil. Although it was most likely merely a product of my imagination, I cannot help but wonder about the Ancients who once roamed this land. I imagine stories of the Dreamtime were shared in voices sounding very similar to what I heard. More than a little unsettled, I continue towards Hungerford.

≈

Henry had not been in Bourke long when he met James (Jim) William Gordon. Henry was 25 at the time and Jim was just 18 years old. It was the beginning of a friendship that would profoundly influence the life and writing of Henry Lawson. In the rough surrounds of a bush town subsisting within a harsh environment, a mate was a necessity. Henry invited Jim to join him in the small house into which he had moved. Henry secured employment for the two of them and they set to work painting cottages. When Jim was sacked, he decided to tramp south to obtain rouseabout work in the shearing sheds. 'Eager for the bush and wanting a mate,' Henry decided to accompany him.

The two set off on 24 November 1892. Before departing, Henry scribbled a quick letter to a Sydney friend. The letter suggests Henry may not have carefully thought through what he and Jim were about to do. 'I haven't much time to write, also I am pretty drunk, so you'll excuse me.'

Henry and Jim obtained work at Fort Bourke Station and Toorale Station to the south. Henry was true to his ideals and Jim found him a good companion. Jim said, 'He was a stalwart mate, generous and unselfish, and ever ready and willing to take more than his share of hardships—and God knows there were plenty.'

They were back in Bourke in time for Christmas. 'It would take a very long letter to tell you all the news,' Henry wrote to a friend upon returning to Bourke. In walking to and from the shearing sheds, Henry and Jim had experienced their first tastes of tramping the Outback with swags on their backs. After Christmas, much more was to follow. Hungerford beckoned.

≈

I walk by another fly-blown animal carcass. Moz and Jane pass me in the car and drive on to Hungerford, creating a cloud of dust that rises from the dry red Track and encases me in its tomb. I struggle through the dust, ever pushing on toward my goal. A thick row of Mulga Trees line the Track on both sides. The camber of the Track slopes away towards the roadside, where it forms a gutter and then angles back up towards the trees. In the gutter and on the slope up to the trees, the vegetation is sparse and consists only of parched Buffel Grass and Spinifex growing amongst the dry leaf litter and fallen twigs.

The dusty bluish tinge to the Mulga Trees blurs the distinction between the trees and the sky in the shimmering heat-hazed distance. Rather than leaves, Mulgas have the much-needed needle-like, leathery phyllodes. The leaf litter and twigs below the trees are remnants of the so-called Drought of the Century, regarded in some quarters as the worst drought in Australia's history. For most of the first ten years of the new millennium, the country out here baked under a hot sun with barely a drop of water. While many farmers looked to the Heavens and prayed for rain while they went to ruin, these Mulga Trees dropped much of their foliage, further reducing transpiration, and also providing nutrients to the soil to sustain the life of the trees. Mulgas might not be the most beautiful of trees, but their ability to thrive in arguably the harshest conditions in the world is amazing. Aborigines used Mulga sap and seeds for food and the hard wood for a variety of implements, including a shield. The Mulga Tree derives its name from an Aboriginal name for the shield. Unfortunately, the Mulgas do little to shield me from the blazing sun.

I pass a fallen Mulga and stop to photograph the visually interesting pattern of the dry branches. The branches stretch towards me like the tentacles of an army of octopuses. If the tree were standing, the branches would be almost entirely vertical. Although this tree has not survived the drought, the orientation of the branches and the now missing phyllodes helps to reduce Mulgas Trees' exposure to the sun and also helps to funnel any rain that does fall so that the moisture runs down to the trunks of the trees and on to their deep tap roots.

When I reach the sign indicating Hungerford is now only another five kilometres, I stop and wait for Baz, who I can see walking in the shade alongside the Mulgas. Baz is wearing shorts and has his shirt unbuttoned save for the very top button. He catches up to me and we walk towards the Queensland border together. We are both bowed by all that is behind us, but also buoyed by how little remains in front of us. We are dirty and grubby and our patchy, bedraggled whiskers add years to our faces.

About ten minutes before noon, Moz drives out to meet us. He emerges from his car with two cans of cold soft drink. I could just about kiss the big fella.

'Oh, can I have the Fanta?' Although it was my favourite as a kid, I have not had a Fanta for years.

'This is great, Moz. Thanks,' I say.

'Thanks a lot, mate,' Baz adds.

'I have more good news, Grog. They have Twisties at the pub.' It might be January, but all my Christmases are coming at once.

Moz gets back in his vehicle, turns it around, and drives a short way ahead. He takes some photographs of us as we approach and then pass him. He drives a little further ahead and takes more photos. Shunting along in this manner, he gets lots of photos as we complete the final kilometres to the border.

We reach the turn off to Wanaaring, 99 km to the south. Baz and I haul our bodies over to the sign that says Hungerford is only 3 km away and that Bourke is 212 km in the distance. Moz takes photos of us with broad smiles across our dirty faces. He then disappears into Hungerford, leaving Baz and me to complete the last couple of kilometres with just one another's company. The rocks have again given way to nicely packed sand and, as we enjoy the good walking conditions, there is unspoken joy at our brotherhood achievement. Buoyant of spirit, we take several photographs of one another and of our surrounds, so our last few kilometres consists of a chopping and changing of the lead position depending on who happens to be the one who has stopped to capture a photo.

A little more than a kilometre from the Queensland border, the Track makes a sweeping bend to the north. It is not until we are around the bend that we can see the Dingo fence ahead. As Henry said, 'There is no distant prospect of Hungerford—you don't see the town till you are quite close to it.'

The Dingo fence was originally built in the 1880s as a rabbit-proof fence. It was intended to stop the invasion of rabbits after the initial introduction of 24 rabbits 13,000 km away in October 1859 by Thomas Austin on his Barwon Park property near Geelong. 'The introduction of a few rabbits could do little harm and might provide a touch of home, in addition to a spot of hunting,' Austin is reported to have said. Born in Somerset, England, Austin immigrated to Australia as a teenager. After originally settling in Tasmania, he crossed Bass Strait and became a member of the Acclimatisation Society of Victoria, introducing other birds and animals to Australia, such as hares, partridges, blackbirds, and sparrows.

When Henry visited Hungerford, he said the rabbit-proof fence was a 'standing joke with Australian rabbits'. After all, there were and are 'rabbits on both sides of it'. Henry was tickled by the whole idea of trying to stop rabbits with a fence. 'It is amusing to go a little way out of town, about sunset, and watch [rabbits] crack Noah's Ark rabbit jokes about that fence, and burrow under and play leap-frog over it till they get tired,' wrote Henry. 'One old buck rabbit sat up and nearly laughed his ears off at a joke of his own about that fence. He laughed so much that he couldn't get away when I reached for him. I could hardly eat him for laughing.'

In the 1940s, the various fences that had been erected to prevent the spread of rabbits were made higher to exclude Dingoes from the pastoral regions in the south. The fences were also joined to become one continuous fence—the longest fence in the world—stretching over 8500 km. The length of the fence has since been reduced to about 5600 km.

We soon can make out the large Queensland State Border sign to the immediate right of the fence gate. Looking through the wire fence, Baz captures his first glimpse of Hungerford on the other side. I expect that he cannot help but think that it has been a hell of a long walk to get to a place that looks like that.

Moz swings the gate open for us and at 12.36 pm we walk through the Dingo fence into Queensland.

> With blighted eyes and blistered feet,
> With stomachs out of order,
> Half mad with flies and dust and heat
> We'd crossed the Queensland Border.

There is an incredible sense of accomplishment, but I realise that we are only halfway. We still have to walk all the way from Hungerford back to Bourke. For now though, it is a time of celebration and congratulation. Baz and I shake hands heartily. I also place my left hand on Baz's shoulder in what, for the Bryan brothers, serves as a warm embrace.

'Well done, mate,' Baz says.

'Well done.'

'Well done guys,' Jane says.

'Congratulations,' says Moz.

We are happy and, at least for a moment or two, the weariness that we all feel seems to dissipate. It is time for photographs and big white toothy smiles all round.

Baz and I shuffle sheepishly towards the sign that, in large black lettering on a circular white backdrop proclaims proudly, The Pub. Built in 1873, the Royal Mail Hotel was originally a staging post for Cobb and Co. Stepping inside from the bright sunlight, our eyes take time to adjust to the dark interior of the barroom.

'You made it. Congratulations boys,' says a high-pitched voice coming from the darkness behind the bar.

'Thank you,' Baz and I say together. We have little time to say anything else before the woman turns to Jane, who has followed us inside, and says, 'Oh, you are right. They do stink!'

The lady gets no argument from me. She introduces herself as Sheree Parker. Together with her husband, Moc, she runs the Royal Mail Hotel.

'Can I get you a drink, boys?' Sheree asks.

'I'll have iced water, please,' I say. Baz orders the same. We do not even take off our packs while we drink.

Sheree is a cheerful, vibrant, talkative woman. Her smile is broad and easy. She is dressed in jeans and a lime green t-shirt. Her vivacious manner immediately gets one onside. Life in an Outback pub cannot always be pleasant for a woman but, as they say, you catch more flies with honey than with vinegar.

I loll the ice around in my mouth, enjoying the sensation. I notice that the pub's walls are still adorned with Christmas tinsel. Henry spent a blazing hot Christmas Day in Bourke in 1892. 'It had been a hot, close night, and it ended in a suffocating sunrise,' Henry wrote. 'I camped in a corner of the park that night, and the sun woke me.' Seeing as it was Christmas morning, 'there was peace in Bourke and goodwill towards all men. There hadn't been a fight since yesterday evening, and that had only been a friendly one.' Henry enjoyed 'a sensible cold' Christmas dinner that left the women:

> *fresh and cool-looking and jolly, instead of being hot and brown and cross like most Australian women who roast themselves over a blazing fire in a hot kitchen on a broiling day, all the morning, to cook scalding plum pudding and red hot roasts, for no other reason than that their grandmothers used to cook hot Christmas dinners in England.*

Henry's Union buddy, Billy Wood, wrote that he, Henry and Donald Macdonell passed the afternoon under the spray of a hose on Wood's veranda, sitting in their underpants drinking beer. 'The Bourke Christmas is a very beery and exciting one,' Henry wrote and, in veiled reference to the underwear, he added 'the hose going, and free-and-easy costumes'.

Sheree tells us that her husband will be back later in the afternoon but for now he is in Bourke attending to the mail run. From its early days, the Royal Mail Hotel publican has assumed the dual role of town postmaster and that tradition continues through to the present day.

'I might take my laundry over to the caravan park and put it in the washing machine and then have a shower,' I say.

'Jane is using the washing machine,' Moz informs me.

Bugger. 'A shower it is then. Is my bag in the car, Moz?'

'No, we've taken your things into your room. I'll come along and show you where everything is.'

Moz accompanies me to the room that Baz and I will share. It is tidy and spacious and, with an air-conditioner, a little fridge, and three beds, it has everything we need and more.

Once I have gathered up my shampoo, soap and other toiletries, Moz leads me out the back and shows me where I will find the shower and the toilet.

My first stop is the toilet. When I lift up the seat, I notice a number of small frogs enjoying the cool and wet of the bowl. I encourage them to evacuate. 'You're gonna want to be somewhere else,' I say, reaching an empty toilet paper roll down into the bowl to give them assistance. Rather than evacuate, the frogs choose to hide, crawling under the rim of the toilet. 'Your call,' I say, 'but don't say I didn't warn you.'

As I take my seat, I watch vigilantly for snakes. The cool ceramic of the toilet bowl in the shaded building seems like a great spot for a snake to escape the Outback heat, especially when there are so many frogs in here to provide an easy meal. I do not see any snakes, but a Green Tree Frog as big as my clenched fist sits on the corrugated iron dividing wall centimetres above the paper roll. I watch it move slowly but surely up the wall and across to the swinging door. Gee, it is a beautiful colour. It has horizontal black pupils in cloudy, creamy eyes. It has a few irregular white spots on its back and sides and a white stripe along the lower lip, but it is the dominant green of the skin that I find almost mesmerising. When I take my girls to the zoo in Canada, I like to see the Australian Green Tree Frogs. Here I am, with the best seat in the house, watching my own tree-frog display. Unfortunately, the frog seems not to enjoy its audience and I watch as it uses the suction pads on its toes to edge its way across the door and away toward the urinal.

With the job done, I hobble out of the toilet building and across to the shower. There are more frogs there—frogs by the dozens. I drag a metal chair from the small change room and place it under the shower. I turn on the water and adjust the temperature before settling down onto the chair. I will not be hurried. It feels unbelievably good to be wet and cool and to be able to use soap and shampoo on my filthy body. I can almost see the stench of my body dissipating as the water splashes onto my nakedness.

I see that the blister on one toe on my right foot is clearly infected. It has adopted a purplish colour and the swelling in that foot extends from the toe all the way up past my ankle.

I sit and soap and scrub for a long time before I decide that I should get out. If it were not for my painful, unsteady feet, when I emerge from the shower I would feel like a new man.

When I re-enter the pub, I feel clean and fresh in shorts and a t-shirt. It is also nice to have my feet out of my walking shoes and to be able to air them in a pair of thongs.

Sheree seems more than happy to do whatever she can to help us enjoy our stay in Hungerford. When she does not have raspberry for the pub raspberry that I request, she phones Moc in Bourke and asks him to pick up some. In the meantime, she mixes me a lime and soda and tells me it is on the house, as are the Twisties that I have been craving. When Moz, Jane, Baz, and I place our orders for lunch, Sheree kindly fixes a huge complimentary plate of yabbies for an entree. I follow the yabbies with a steak sandwich with the works and a meat pie. I am not in Hungerford to diet.

We eat our lunches in the room to the side of the bar. A couple of tables have been pushed together with ten chairs around the perimeter. There are no tablecloths, just plastic place mats. A small artificial Christmas tree is squeezed into one of the spaces between the tables and the wall. A strong fan blows cool air as we dine. The prime seats are those closest to the fan.

'Do you know Tony Marsh, Sheree?' I ask.

'Yeah, he is a wonderful bloke. Tony is a great man.'

'We met him last time we were out here. I am surprised we didn't see him on the road as we passed by Kia-Ora.'

'Tony is busy at the moment. Roby is out there mustering with him.'

'Oh. Who's Roby?'

'Roby is my son. He's giving Tony a hand. Tony had an accident a couple of days ago and got knocked around so I am not sure how he's feeling.'

'Oh, well I hope he is okay.'

'He's tough, Tony is. It would take and awful lot to hurt him.'

'We thought we might try to stop at his place on the way back. He made us the offer last time we were out here.'

'Yeah, well I'll call Roby and let him know your plans.'

'Thanks, Sheree.'

The walls of the dining room are adorned with photographs of successful race horses. There are also images and information about Roy Dunk, who is pictured aboard his mount in the Australian Light Horse Brigade during the First World War. 'That is the bloke from Warroo,' Moz says.

Dunk is a heroic figure, mounted on his horse. A swag sits on the saddle before him. He carries his rifle on his back, held in place by a strap slung across his chest. He looks the quintessential Digger, tall, proud and handsome, but perhaps also naive, and foolishly optimistic and virtuous. I wonder how well he was yet acquainted with Death at the time of the photograph.

'John Stephenson has his memoirs and things for you to look at,' says Moz. 'Remember? We are going to sleep there on Sunday night.'

'Great.' I am looking forward to it, but I am still worried I will not have the energy to look through old artefacts, no matter how interesting they might be.

After our hearty lunches, Baz and I wander across the road to the caravan park to put in our laundry. Baz then uses the phone box outside the pub to call his family. It would not be inaccurate to say that I am jealous. I retire to our room to rest and to read Henry Lawson, beginning with the Bourke-based story, 'Send Round the Hat'. I sprawl out on the Queen-sized bed. The other two beds are singles and, typical of Bazza's thoughtfulness, he has left the bigger bed for me. Resting on a chair in the room there are a couple of gossip magazines and a Bible. Although I will flick through the magazines later, like the third bed in this room, the Bible is surplus to my requirements.

By the time that I emerge from my room and return to the bar, the pub has started to fill with characters. Sheree's husband, Moc, has returned from Bourke. He is tall and wiry. Like his wife, he has curly brown hair. Save for his light skin, his sinewy frame suggests an Ethiopian distance runner. As with his wife, he is easy, enthusiastic company.

I order fish and chips for tea and then follow that with another packet of Twisties. When I pay for tea, I notice a small Australian flag hangs above the cash register. A place like this is the heart of the country. Moc and Sheree run a terrific pub. They keep the drinks coming and make everyone feel welcome. Despite the din of a bar full of thirsty Bushmen, Sheree's willing laughter can be heard over all.

I am surprised by some of the ways in which more than one person refers to Indigenous Australians. In some quarters, Henry has been decried for racist attitudes towards Aborigines.[4] 'The blacks may be low and degraded,' Henry wrote, albeit tongue-in-cheek, on one occasion, 'but we forgive...his dirt and laziness because of his cheerfulness and humour, which last is something akin to our own.' Henry's views, however, are clearly a product of his times. The same excuse is difficult to muster for some customers in the Royal Mail Hotel.

I watch Baz mixing free and easy with all of the locals. He has none of my reserve and he wins over people quickly and without bother. I suspect that Henry was more like me in this respect. One mate described Henry as 'intensely shy' while another wrote that he was 'reserved in company'. Another mate said that Henry was bothered when he learned of his mate boasting of friendship with the famous author. 'He was of a quiet and retiring disposition,' the friend wrote. Henry wished to avoid public attention as 'he had no desire for personal publicity.'[5] One time this friend was to recite one of Henry's poems at a lecture given in the poet's honour. Henry was so uncomfortable with public attention that he chose not to attend. The friend, Jack Moses, left the packed hall during the lecture to locate Henry and, upon finding him, Moses had almost to forcibly drag Henry to the hall. The poet and Moses entered to rapturous applause. The men took their seats but Henry was so discomforted by the adoration that he soon fled.

The people in the pub are kind, friendly and supportive. Jan, one of the women in the bar, looks around at all of the people inside and says to me, 'It is never like this in here—at least not any time I've been here.' Jane mentions that she heard some people have dropped into the pub tonight just to see us. This is really nice of them, but I rather suspect some might be using us as a good excuse for a night at the pub. Moc and Sheree are equal to the challenge of the larger than usual gathering and they soon have the place rocking with a steady flow of drinks and an equal measure of laughter and convivial chatter.

I strike up a conversation with Jess, a 22-year-old Canadian. She is here from Edmonton, experiencing the Australian Outback. With her blonde hair, pretty face, and figure-hugging black t-shirt and denim jeans, I am sure she is turning stockmen's heads.

Doug is the property owner where Jess is working. He is dressed in brown jeans and a loose-fitting khaki shirt. He hands Jess a drink and buys one for me as well. Perhaps in his fifties, Doug keeps a fatherly eye on Jess. Any drunken louts who might want to get to her will have to get through him first. Although protective, I notice that Doug's language is every bit as colourful as all of the other bar patrons. Swearing is common to the people out here and oftentimes every second or third word is an obscenity.

'What do your parents think of you being out here?' I ask Jess, wondering about the colourful language to which she is being exposed.

'They think it is okay, I guess.'

'What do your parents do back in Canada?'

'My parents are in the oil business.'

'Oh, that's interesting.' Is Jess a Lucy Ewing? 'So, I'm curious, Jess, what do you think of the language out here?'

She laughs. 'I'm fine with it. With my parents being in the oil business, I'm used to being around Rig Pigs. They speak as coarsely out there as they do out here.'

'What did you call them?' I ask, straining to hear above the growing bar noise.

'Rig Pigs.'

I laugh at that. It is not a term that I have heard before, but it seems pretty self-explanatory.

Later, Doug asks Jess what she wants for tea. 'What would you like to order, Dave?' Doug says, mistakenly calling Jess by another employees' name.

'What did you call me?'

Doug blushes.

'Dave?' Jess gives Doug a look of scorn and says, 'Oooh...' and then she says something that assures me that be it the influence of a Rig Pig or an Outback Bushman, she is fitting in fine.

Jess tells me about the assorted wildlife she has seen during her time in Australia. We get to the topic of snakes.

'I thought I killed one yesterday,' Jess tells me and Baz.

'What does that mean?—you *thought* you killed one?'

'Well,' Jess says, sheepishly, 'I killed it with a shovel...but I couldn't see the legs until I turned it over.'

'It had legs?' I say.

'It was a lizard?' says Baz.

Jess blushes. 'But it looked like a snake.' Kill first and check for legs later, I guess.

A drunken middle-aged woman edges into the conversation now that it has turned to snakes. 'I saw on a television programme the other day that you can live three days with a snake bite but then you have to get help or you're dead.'

What? If I get bitten within three days of Bourke, I am sure I will not be waiting to finish the tramp before I seek help.

Dark-haired Carol wears denim jeans and a white t-shirt. The Outback sun has shone long and hot on her brown skin. She is with Steve, who apparently has lots of what I am told is 'City Money'. Slender of build, blonde haired, and tanned, rumour has it that Steve is quite a wealthy man and I suspect many women would look upon him as a good catch.

Neil comes across from the caravan park to see us and to say g'day. Like everyone else in Hungerford, he is friendly and easygoing. Others come and go as the Royal Mail does a roaring Friday night trade.

Jan has come to the pub with Glenn. With his long face and prominent moustache, he bears more than just a passing resemblance to Henry Lawson. I can imagine Henry in this pub all those years ago. While downing his 'glass of sour yeast,'[6] he would have been busily taking mental notes of all the characters around him.

Jan tells me that she came here from Dubbo, where she retains her house 'just in case'. She helps Glenn maintain more than 125 km of Dingo fence, ensuring the Dingoes cannot cross from Queensland into the New South Wales sheep country.

I ask how many Dingoes she has seen in the several months she has lived in the Outback.

'Glenn shot one. He poisoned another, and a third one got away.'

'Why does Glenn kill them? Isn't that what the fence is for?'

Jan calls over Glenn, who leans his leathery face in close to hear what she has to say. 'Glenn, Greg wants to know why you kill the Dingoes.'

Glenn looks at me as if deciding whether to bother with explaining to this know-nothing-do-gooder-Outsider the ways of the Outback.

'Why don't we leave the Dingoes alone?' Jan asks.

'What, and leave them to get at the Queensland stock?' Glenn does not condescend to go any further than that but the smirk on his face says plenty.

'What do you get for a dead Dingo, Glenn?' I ask with a mixture both of fascination for the topic and abhorrence.

'There is a bounty of ten dollars a scalp.' I find the terminology interesting. For centuries, the practice of scalping has been associated with notions of barbarity and a lack of civilisation.

Jan's grandson, 14-year-old Tom, is on holiday from Dubbo. As the night moves on, mosquitoes enter the pub in search of some alcohol with their blood. The mozzies start to bother Tom and he retrieves from Glenn's truck a bracelet that works as a mosquito repellent.

I wonder what Tom makes of this night and this place, where the next youngest person is eight years his senior and the next gap after that is another 20 years and more.

I ask Tom what grade he will be going into with the start of the new school year.

'Grade Nine.'

'Do you like school, Tom?'

'It's alright...I guess.' Like me when I was his age, Tom seems to have what Henry described himself as also having as a child: 'The average healthy boy's aversion to school.'

'Have you read any good books lately?'

'Nah. I don't really like reading that much.' As a literacy educator, this is anything but music to my ears.

While Tom is the youngest in the bar, at 85 years of age, Mack is the oldest. He wears a battered old Bushman's hat. From the ends of his blue jeans extend a pair of long, thin shins and bony ankles. Other than my team and me, Mack seems to be the only one not wearing the Bushman's elastic-sided boots. On his feet are thongs and he wears his navy blue Jackie Howe singlet comfortably. He is obviously someone who has done a lot of shearing in his time.

Long after darkness has fallen, and after being in the pub all day, Mack decides it is time for him to venture home.

'Are you leaving already, Mack?' someone asks.

As quick as a whip, Mack replies drily, 'When you get to my age, you have got to start looking after your body.'

I cannot help but notice that, like hundreds before her, Carol has taken a shine to Baz. Although I am across the room, I can hear her saying, 'I am so proud of you. You should be in the *Guinness Book of Records*.' As the night goes on, Carol repeats the assertion over and over.

After some time, it finally dawns on Carol that Baz and I are related. 'You two are brothers?' She is surprised by that. She looks Baz and me over. 'But you're so skinny,' she tells Baz, 'and he's so...he's so... he's so *solid*.' Baz laughs aloud, but I will take 'solid' over some of the alternatives that Carol thoughtfully dismisses.

In looking me over, Carol becomes aware of the bloated state of my feet. Propped up on a barstool as I am, with my feet hanging below me, my feet continue to swell and I notice some of my blisters are leaking. In the space beside my infected toe, gooey pus sticks the toe to the one beside it. Carol insists that Sheree retrieves a bottle of Methylated Spirits for me to soak my feet. That is going to sting like hell, I think, remembering my father's cure for every external ailment is to apply Metho. Moz takes a photo of me sitting in the pub with a bottle of Methylated Spirits. Forget being a teetotaller, the prop makes me look like a hardcore alcoholic.

In place of where there might otherwise be tables, there are some big wine barrels. The wooden staves are bound by metal hoops, the barrels adding further charm to what is, indeed, a charming pub. Perched on a barstool, the barrels are the perfect height to prop an elbow and slouch. An accumulation of empty beer bottles and glasses soon occupy most of the spaces on the flat barrel tops.

Carol rightly describes what we have done as 'an incredible achievement'. She is amazed that we are heading back out onto the Track after a day's rest and insists this is why it is so important that my feet receive treatment. She is thinking only of my best interests and my welfare and I appreciate her concern, but her constant admonitions to Baz that he needs to take better care of me are starting to wear thin with my brother.

'He's going to die out there,' Carol tells Moc.

Carol decides that, if my brother will not do it, she will embrace the responsibility of looking after me.

'You need to elevate your feet,' she orders. 'My mother used to get swollen feet and her feet always needed to be elevated to get the swelling to go down.'

'I'm okay,' I insist. Actually, I am alarmed by the colour and size of my feet, but I do not want to bring attention to myself.

'Someone should go and get the policeman to take a look at you. He used to be a paramedic, you know?'

'I'll be okay.'

'Put your feet out the window,' Carol orders more forcefully. 'That'll be cool and it'll get them elevated.' As drunk as she is, she certainly has my best interests in mind.

I initially resist because I am trying to be discreet and to downplay all of this. Whatever else Carol might be; however, she is an irrepressible force. I decide it will be more discreet to comply and put my feet out the window than to try to resist her commands. As I move to swing my legs around toward the open window, Baz's body blocks my path.

'Get out of the way, you little monkey,' Carol snarls. Her affections have clearly switched from Baz's worthiness for a place within the *Guinness Book of Records* to my need for an Outback Florence Nightingale.

Baz moves aside while I laugh at him. It turns out that Carol is quite right. My feet immediately feel better once I have them elevated. Despite my protests, Carol starts to rub my calves to try to relieve some of the swelling. Now it is Baz's turn to laugh at me.

Not long after 10.00 pm, I limp back to my room. I am barely inside when I hear a soft knock at the door.

Oh, God. 'Is that you, Baz?'

Thankfully, it is.

I elevate my feet on pillows at the end of my bed. My blisters are weeping so I reach for the bottle of Methylated Spirits. This is going to sting like hell. Putting the Metho on my feet does not turn out too badly. My night might have ended much worse.

DAY EIGHT

Henry Lawson in Hungerford

He's somewhere up in Queensland,
The old folks used to say;
He's somewhere up in Queensland,
The people say to-day.
But somewhere (up in Queensland)
That uncle used to know—
That filled our hearts with wonder,
Seems vanished long ago.

- Somewhere Up in Queensland, *Skyline Riders and Other Verses*, 1910

As I sat up I caught sight of a swagman coming along the white, dusty road from the direction of the bridge, where the cleared road ran across west and on, a hundred and thirty miles, through the barren, broiling mulga scrubs, to Hungerford, on the border of Sheol.

- That Pretty Girl in the Army, *Children of the Bush*, 1902

I sleep until 10.00 am. I aim to make the most of our rest day and a good sleep-in is a pleasant way to start. Through the night, I dreamt of being pursued by an endless stream of people. In order to dissuade my pursuers, I constantly kept throwing my shoes at them. Dreamland is not always a place of logic, so I do not know where all of my shoes came from, but now that I am awake and look at my blistered feet, I have no difficulty understanding the subconscious desire to throw away my footwear. The space between the middle and next toe on my right foot oozes with pus.

I decide to forego breakfast and instead gather my towel and toiletries and drag myself toward the toilet and shower buildings. It is good to shower again this morning, particularly as I do not *need* to. Once I am dressed, I put toothpaste on my toothbrush and hold the toothbrush beneath the tap. When I turn on the tap, no water comes from the spout. I turn the tap further and, with a gush of water, a frog plops right onto my toothbrush.

Despite sharing my toothpaste with a frog, I am soon clean and refreshed and I make my way into the bar. 'Good morning,' I say as I enter.

'Hello.'

'What time did you finish up last night, Sheree?'

'Oh, about midnight.'

Despite her late night, Sheree is as cheerful as ever.

I look up, taking stock of all of the money pinned to the wooden ceiling. The collection is for the Royal Flying Doctor Service. The main denominations pinned there are five and ten dollar notes, but there are bigger notes too, including a number of fifties.

'I got a hold of Tony Marsh,' Sheree tells me. 'He said you can stay at his place on the way back.'

'Oh, great. Thanks for that, Sheree.'

'His house might be a bit messy,' Sheree says, almost apologetically.

'Okay.'

'You'd probably be best to tell him that you will stay in the shearers' quarters.'

≈

Two days after Christmas in 1892, Henry and Jim Gordon set out from Bourke for Hungerford. The two were accompanied by Ernest de Guinney but it is almost certain that Henry and Jim returned from Hungerford without de Guinney. Like Henry, de Guinney was a Unionist and writer who had published in several periodicals including the Union newspaper, the *Worker*. Another Bourke Unionist, Billy Wood, later wrote that the partnership with de Guinney did not last long because Henry soon tired of de Guinney's griping. That de Guinney was with Henry in Hungerford however, is confirmed in 'The Bush and the Ideal,' where Henry wrote:

I watched a mate of mine sit down in camp on the parched Warrego—which was a dusty gutter with a streak of water like dirty milk—and write about 'the broad, shining Darling'. The Darling, when we had last seen it, was a narrow streak of mud between ashen banks, with a barge bogged in it. Two weeks later this mate was sitting in a dusty depression in the surface, which he alleged was the channel of a river called the Paroo, writing an ode to 'the rippling Warrego'.... My Warrego bard was born in St Petersburgh.

De Guinney was apparently a Russian (hence the mate 'born in St Petersburgh') and the Warrego River is the one at Fords Bridge while the Paroo River is the one at Hungerford. De Guinney was obviously with Henry at the Darling River at Bourke, through Fords Bridge, and eventually at Hungerford.

When he reached Hungerford, Henry wrote another letter to his aunty.

Dear Aunt,

I found your letter in the Post Office of this God-Forgotten town. I carried my swag nearly two hundred miles since I last wrote to you, and I am now camped on the Queensland side of the border—a beaten man....No work and very little to eat; we lived mostly on Johnny cakes and cadged a bit of meat here and there at the miserable stations. Have been three days without sugar. Once in Bourke I'll find the means of getting back to Sydney—never to face the bush again....You can have no idea of the horrors of the country out here. Men tramp and beg and live like dogs. It is two months since I slept in what you can call a bed. We walk as far as we can—according to the water—and then lie down and roll ourselves in our blankets. The flies start at daylight and we fight them all day till dark—then the mosquitoes start. We carry water in bags. Got bushed on a lignum plain Sunday before last and found the track at four o'clock in the afternoon—then tramped for four hours without water and reached a government dam. My mate drank nearly all night. But it would take a year to tell you all about my wanderings in the wilderness.

....I'm writing on an old tin and my legs ache too much to let me sit any longer. I've always tried to write cheerful letters so you'll excuse this one. Will tell you all about it when I get down.

The letter is dated 16 January 1893. Given he left Bourke on 27 December, the tramp to Hungerford took 21 days. Baz and I covered the distance in seven days. Interestingly, in the letter Henry added, 'I start back tomorrow—140 miles by the direct road—and expect to reach Bourke in nine days.'

≈

For lunch, I order a steak sandwich and chips with gravy. As I wait for my meal, I peruse the information on the dining room walls and look closely at the various photographs pinned there.

As Moz, Jane, Baz, and I eat our lunch, Flo walks to and fro, beneath the table. Flo is Moc and Sheree's dog. She is a rather odd-looking thing. She is four years old and part Fox Terrier, with the colouring of a Blue Heeler. She is a long, low dog with big sailboat ears and seems a happy, careless character as she moves about the pub with ease. Her tail is curled in the shape of a smile that seems reflective of her sweet nature.

There is another dog lying at the doorway watching me eat. Rosie is a beautiful Kelpie bitch. Angular and small of stature, she cannot be much more than a pup and, 'Being a young dog, [s]he trie[s] to make friends with everybody.' She has obviously been trained to stay outside because she will not venture beyond the doorstep, even with my encouragement. Rosie wags her tail freely and the look on her face indicates that all she wants to do is to please.

I enjoy my meal but when I am finished I notice how swollen my feet have become since I got out of bed this morning.

After lunch, we move from the dining area back to the bar. Mack is in his customary spot on a stool in the corner. On the bar beside Mack's elbow, there is a framed photograph of a deceased hotel patron. Photographs of Don Bradman and Slim Dusty adorn the wall behind him, as does a portrait of Henry Lawson.

The portrait of Henry is a print of the John Longstaff painting that Baz, Moz, and I visited when we stopped at the Art Gallery of New South Wales on the day that I flew into Sydney. The painting was completed in 1900 when Henry stopped in Melbourne en route to London from Sydney. The painting had been commissioned by

Archibald. Archibald was apparently so happy with the final product that he decided to establish a prize for Australian portraiture artists. Archibald was not the only one happy with the portrait. Henry's wife considered it 'the truest and most living portrait of Harry ever made.'[1]

In the art gallery, I was amazed at how Henry's eyes followed me about the room. It seemed almost as if Henry knew what hell I was about to get myself into and wanted to warn me. Longstaff did a remarkable job portraying those eyes. Those who knew Henry were captivated by his eyes. One mate said of Henry that, 'wherever he was, his beautiful dark brown eyes were keenly observant, and quick in perceiving the incongruities of life.' Another wrote of 'those great dark eyes—woman's eyes, dog's eyes—full of sympathy and emotion.' In a book entitled *Henry Lawson by His Mates*, published in tribute after Henry's death, four themes emerge repeatedly among his friends' recollections: his stunning dark eyes and, additionally, his strong handshake, his sense of humour and his unfailing generosity.

In addition to the Longstaff portrait, there is another tribute to Henry at the Hungerford pub. At the main entrance, a wooden carving of him has been screwed to the door. The carving is a depiction of Henry when he haunted the streets of Sydney in later life, on the lookout for someone from whom he might sponge money for the price of a beer. He is in his characteristic pose, walking stick in his left hand while he salutes with his right. Although it is supposed to be a humorous portrayal, knowing the depths to which he had sunk by that stage in his life, I cannot help but feel that the face is that of a sad, shattered shell of a man.

With my feet seemingly swelling by the minute, I decide to retire to my room to rest and do some more reading. As I exit the pub, a young couple arrives. Although I barely notice *him*, *she* has blonde hair and makes me think of what I told Terry I would do in Hungerford.

When I get to my room, I read some more of Henry's Outback and Bourke-based work. Despite its harshness, I believe Henry was at his happiest during his time in the Outback. He developed a fondness for his Bourke mates that stayed with him for the remainder of his days.

I close my book and think about Henry and his time at Bourke and Hungerford and those points in between. I feel that I am in a position from which to make the immodest claim that I know Henry Lawson better than any person alive today. My claim is not based on knowing facts, figures, dates, or statistics. I make the claim based on having been out there on Henry's Track, having felt the blazing sun, having fought with bubbling blisters, having struggled with the long, often tedious miles and, like Henry, having come out the other side at Hungerford. Henry said, 'If you know Bourke you know Australia.' I say, 'If you know the Track, you know Henry.'

No doubt, this period of Outback swagman life was an incredible challenge for Henry. He was to face many more challenges throughout his life but they tended to be more mental than this physical one (although the mental aspect out here is enormous as well). It is staggering to think that, not a dozen years after doing this tramp and being inspired to create some of the greatest writing in Australia's literary history, Henry had a failed marriage, had been in a mental hospital and had attempted to take his own life. They are the mental challenges that lay ahead, but this physical one in the heat and the dust, with insects biting and buzzing, being filthy dirty, covered in sweat, grime and grit—this, I believe, was physically Henry's greatest struggle and, therefore, Henry's greatest physical accomplishment.

I consider my present tramp to be much harder than the winter journey of 18 months earlier. At the time it seemed to be the hardest thing I had ever done physically. However, I consider this to be harder—much harder—with the added burden of the extreme heat and, of course, also the weight—the load—of having to carry water all the way. And my blisters are as bad as last time, if not worse.

I put my head on my pillow and lie on the bed with my feet elevated. It is good to be off my feet and to relax in the air-conditioned room. With thoughts of Henry foremost in my mind, I drift off to sleep.

≈

Henry said Hungerford was found at the end of 'one of the hungriest' roads in New South Wales. He had learned that Bushmen rarely were slaves to the facts and so when he heard a story after his arrival in Hungerford, Henry reserved doubts about the story's accuracy. After all, as Henry wrote, 'the man who told me might have been a liar. Another man said he was a liar, but then he might have been a liar himself—a third person said he was one. I heard that there was a fight over it, but the man who told me about the fight might not have been telling the truth.'

Henry said that he and Jim and, presumably, Ernest de Guinney camped 'on the Queensland side of the fence.' They got to yarning with a certain Clancy, who was 'an old man who was minding a mixed flock of goats and sheep.' Henry asked the old man whether Queensland or New South Wales was the best state.

[Clancy] scratched the back of his head, and thought a while, and hesitated like a stranger who is going to do you a favour at some personal inconvenience.
At last, with the bored air of a man who has gone through the same performance too often before, he stepped deliberately up to the fence and spat over it into New South Wales. After which he got leisurely through and spat back on Queensland.
'That's what I think of the blanky colonies!' he said.
He gave us time to become sufficiently impressed; then he said:
'And if I was at the Victorian and South Australian border I'd do the same thing.'
He let that soak into our minds, and added: 'And the same with West Australia—and—and Tasmania.' Then he went away.
The last would have been a long spit.

Henry and Jim sat down against the rabbit fence 'to talk things over'. Because of Henry's deafness, Jim had to shout from time to time. According to Henry, 'a trooper in Queensland uniform came along and asked us what the trouble was about.' Henry told him that he and his mate were 'discussing private business'. The policeman explained that, with all the shouting, 'he thought it was a row, and came over to see.' Satisfied that nothing was amiss, the policeman wandered off. 'Later on,' Henry wrote, 'we saw him sitting with the rest of the population on a bench under the hotel veranda.'

≈

'Come out of that, Rosie!'

I awaken when I hear Moc and Sheree shouting.

Although I am rested, I am still in pain. Some of my blisters are leaking and the infection in my toe is crawling up my foot, leaving it bloated and red. I hobble around like a cripple and I have no confidence in my ability to get back to Bourke. Still, I keep my reservations to myself.

Baz enters the room. He has a selection of t-shirts of different colours. Adorning the chest pocket is a picture of the hotel and some lettering promoting the pub.

'Which colour do you want?' Baz asks me.

I look through the selection and decide on a white shirt. I try it on for size.

'It's yours,' Baz says. He also buys one for himself and one for Moz. 'The Hungerford policeman was in the pub earlier, but he has gone now, Gregga.'

'Is the vision of loveliness still in there?' I ask.

'Yeah,' Baz says, 'are you heading in?'

'I might wander in.'

'I'm going to have a rest,' Baz tells me.

I hobble to the bar with difficulty.

I ask Moc what trouble Rosie had been into earlier.

'Did you hear me yelling at her, did you?'

Henry was hard of hearing but even he would have had no trouble hearing the shouting.

'She chases cars. She is going to get run over one day,' Moc explains.

Fortunately, there are not a lot of cars out here but, then again, it only takes one car to get squashed.

I take a seat at the bar beside the young couple and introduce myself.

'G'day, I'm Will,' he says.

'I'm Narelle,' she says.

Narelle's button-up shirt is hot pink and she wears snug blue jeans. She has a warm smile and sparkling eyes that brighten the dim light inside the pub. Were the show still running, I fancy Narelle could wander straight onto the set of *McLeod's Daughters* and adopt a role.

Will tells me they have been busy fencing, walking long miles in the heat. I admire Narelle's soft, clear complexion. Her smooth, pretty face has not yet been ravaged by the harsh Outback sun but if she stays out here, she is doomed to become one of Henry's 'leathery women, with complexions like dried apples' and, unfortunately, it will not be long in coming 'for women dry quickly in the bush.' If Will wants his companion to retain her youthful beauty, he needs to know that trudging along a fence line in the broiling Outback is 'no place for a woman.'

Like Narelle, Will is dressed in blue jeans and a long-sleeved shirt. They both wear the obligatory elastic-sided boots. Will is tall and angular, but robust. His hair is close cropped. He is reserved, but he smiles freely.

They look at my thonged feet and ask how I am feeling.

'Why do you have so many blisters?' asks Will after a time.

He knows about the walk from Bourke.

'Don't tell me that you have new shoes,' Narelle says, 'because that'd be the stupidest thing *ever*.'

Well, they are not *new*—newish, maybe.

I ask Will and Narelle if they have their own property. They tell me they live at Will's parents'.

'We live at Ningaling,' says Narelle.

'Where's that?'

'You'll pass the turn off on the way back to Bourke. It is just before John Stephenson's at Warroo.'

The more Will drinks, the more talkative he becomes. He tells me he would enjoy walking the Track with me.

'I'd love to be doing what you're doing.'

Carol and Steve arrive at the pub. They both look seedy. Steve orders a 'hair of the dog', but Carol is on the soft stuff today. She asks me if the policeman has been in to see my feet.

'He's gone fishing,' Moc says.

Carol looks disgusted. 'I gave him a call and told him to come and see you,' Carol informs me.

'Well, he was apparently in earlier but I missed him,' I say in the policeman's defence.

Carol leaves the pub and goes outside to sit on the veranda—perhaps she can use the fresh air today.

I wonder who lives at 1 Hansom Street in Hungerford. Above the bar, there is a hand-lettered sign proclaiming:

To Let
½ Double bed
Reasonable rates
FEMALES ONLY!!
Apply: no. 1 Hansom St., Hungerford

Sure to be a winner, that one.

Mack is sitting on the veranda as I hobble back to my room.

'Oh, geez,' Mack says when he sees me, 'you'll make about a mile a day.'

Yes, it is going to be a long walk back to Bourke. When an 85-year-old suggests he could walk faster, I am in a bad way. Mentally, I feel fresher today; however, I am also mindful of the pain that lies before me. This remains as a black cloud above. Because of my feet, I know that the next seven days are not going to be pleasant. Only time will tell, but I do not think that I have any cause for optimism. I think that, as tough as it was to get here to Hungerford, it is going to be even tougher on the return.

Alone in my room, I ask myself if I am confident of getting back to Bourke. There is no thought of me quitting but I am not sure that I can go so far as to say that I am confident of making it back.

'Do not fail,' I tell myself. 'Remember, you will be a long-time failed.'

I realise that I am extremely afraid of the pain that I will feel over these next seven days. I know that I am going to have to fight hard to keep myself together. There is not going to be any place out there on the Track for vulnerability, softness or weakness. On the way back, I am going to have to be strong. I am going to have to shut out everything that is not just about me taking the next step. It is going to involve locking myself behind my shield and working like mad to keep the pain out.

When I return to the bar, a man that I have not seen before says hello.

'How are your feet?' he asks me.

'Yeah, they're good, thanks.' Unless he is blind, he can see for himself that they are not good.

'You're not moving too well. Do you mind if I take a look at your feet?'

It is a strange request but, the way my feet are, unless he takes at them with a knife, they are not likely to get any worse.

'Sure.'

Although it is a Saturday night, it is a quieter, emptier bar than the previous night. I suspect there might be one or two sore heads in the district.

We move into the dining area and each take seats beside the table.

'I used to be a paramedic in a former life,' he says.

'Oh, you're the copper.' It suddenly dawns on me who this bloke is. He is not just some weird Outback foot fetish murderer.

He laughs. 'Yeah, I'm the copper.' He introduces himself as Dean Hutchinson.

As the sole Hungerford policeman, Officer Hutchinson serves a wide area. Moc tells me later that Dean's medical training has saved more than one Outback life that would otherwise have been lost during the wait for the Flying Doctor.

Dean is a handsome bloke and I would bet that some lonely Bushwomen have invented an emergency so that Dean would drop by to investigate. He is dressed in blue jeans and an R.M. Williams t-shirt. I hope I have not spoiled what might otherwise have been a rare day off.

'What have you been doing today?' Dean asks.

'I've been resting—mainly just taking it easy.'

When he looks at my bloated feet, Dean says, 'If this is what they look like after a day of rest, I'd hate to see what they'd be like without the rest.' He keeps his head down when he talks to me.

'The thing is, I usually have thin feet and long thin toes,' I tell him.

'Well, I can see that they are badly swollen. There is no definition to your feet at all.' He pauses. 'Do you mind if I touch it?'

He has a gentle way about him and his soft voice has probably prevented many potentially volatile Outback situations from escalating. He pokes at my puffy right foot, checking my circulation by performing the capillary refill test, watching to see how slowly my skin colour returns.

He thinks my left foot is 'not too bad'. 'Put it this way,' Dean says, 'the left foot looks like I would expect it to look after walking all the way from Bourke.' Fair enough. 'But this right foot is the worry.'

He is concerned about the risks posed to my safety if the infection continues to spread and he talks to me about what is happening with the swelling and the infection in my right foot. He explains the manner in which veins work and talks of things like venous sinuses and valves and refluxes. He is worried about the valves collapsing and suggests the possibility of varicose veins as a result of what is happening with my right foot and leg. He also expresses concern about other, more sinister, possibilities and mentions thrombosis, which can lead to a stroke. The things that he says are sobering and disconcerting, but I remain committed to my task. Maybe that commitment is evidence enough that I am insane. Maybe, like Henry, I should be hoping for the policeman to take me off the Track and 'to lock [me] up, for [my] own protection, as a person unfit to be at large.'

'Do you have a support vehicle with you all the time?'

'Well, a lot of the time they are nearby. It might be two or three hours between seeing them though. If they head in to Bourke, it might be several hours.'

This response does nothing to allay Dean's fears. 'If things turn for the worse, you might not have "several hours" that you can afford.' He advises me that I should pull out because of the potential for a dangerous situation to escalate rapidly.

I make it clear that I am not willing to pull out. 'I am here to walk from Bourke to Hungerford and back and I am only halfway done.'

Dean thinks about my response in silence. After a few moments, he breaks that silence with, 'Seeing as stopping is not an option, we have to figure out the best ways to manage the situation.' Dean is not here to argue. He does not resist my determination in any way. Once I have indicated that I am not stopping, he talks only of ways to minimise the risks.

The pus between my toes is gooey and sticky. 'We need to get those toes dry and get air circulating. We'll need to put gauze in there to keep them separate.'

I nod.

'One of the problems you face is all the water on the road. Much of the water out here is dirty and full of all sorts of messy things. Where it is stagnant and dirty, you'll get things in your toes that could cause new infections and magnify the one you already have.'

He talks to me also about the warning signs that I need to watch for—indicators that my situation has reached a critical point.

'If you see dark tracks creeping up your calves, you need to get to medical assistance—fast.'

'Yep.'

'If your toes or feet turn dark, you need to get medical assistance—fast.'

'Yep.'

He asks how I am for bandages. I tell him I do not have any left.

'I'll go to the clinic and bring you back some things,' he says.

Dean leaves and I join the others. Baz asks how I got on talking with the policeman.

I don't want to talk about it. 'He just gave me a prognosis,' I say.

'And not a good one?' Baz asks.

'That's right.' I am done talking about it.

≈

Sheree notices that all four of us have placed exactly the same orders as we did for tea last night. 'That is evidence of how good it was,' I assure her. 'We have come back for seconds.'

'Well, I hope it is as good today as yesterday.'

It is. My fish and chips are wonderful. It is nice not to be living on food from a can.

The policeman returns with bandages, scissors, gauze, bandaids, and other supplies for me. He has lots of advice to help me manage my feet and to minimise the risks under which I am operating. I take it all in and I will ever be grateful for his concern and his help.

Dean talks of the need to elevate my feet at every opportunity. 'Keep your feet above your heart whenever you can.'

He wonders about my shoes. 'Maybe you are tying your shoelaces too tight. Loosen them off as much as you can.'

'I am wearing two pairs of socks,' I say.

'Maybe try wearing just one pair. You need to get that blood circulating and moving away from your foot.'

'Okay.'

'Use the gauze to keep your infected toes apart and to help them dry.'

'Okay.'

'Don't put Vaseline in there because it is making it oily and interfering with the body's function.'

'Okay.'

'Sprinkle the antiseptic powder on your feet.'

'Okay.'

My head is full of things to do and to watch for.

'Thanks so much,' I say. 'I really appreciate everything you have done.'

'Good luck,' Dean says.

Unbeknown to me, Jane has overheard the policeman's earlier warnings of the gravity of my situation. After I return to my room with the medical supplies, she tells Baz what she heard. Jane and Moz are concerned for my safety and think I should stop. I do not want to die, but I am willing to face Death. In any case, in many ways, having survived the ordeal of earning my Doctorate, I consider every day to be a bonus.

In the months leading up to this endeavour, I have consistently told Baz that one of his jobs is to make sure no one stands in the way of me realising my dream. Baz comes to the room to talk to me one-on-one, brother-to-brother, and he tells me what Jane overheard.

'What do you think, Gregga?'

'I am okay to continue.'

That is good enough for him.

'I won't let the others get to you.'

'Thanks, mate.'

DAY NINE

Henry Lawson's Marriage

They say that I never have written of love,
As a writer of songs should do;
They say that I never could touch the strings
With a touch that is firm and true;
They say I know nothing of women and men
In the fields where Love's roses grow,
And they say I must write with a halting pen—
Do you think that I do not know?

- Do They Think That I Do Not Know?, *Skyline Riders and Other Verses*, 1910

'What sort of a husband did you make, Joe?'

'I might have made a better one than I did,' said Joe seriously, and rather bitterly, 'but I know one thing, I'm going to try and make up for it when I go back this time.'

'We all say that,' said Mitchell reflectively, filling his pipe.

'She loves you, Joe.'

'I know she does,' said Joe.

- Mitchell on Matrimony, *On the Track*, 1900

When I awaken in the morning, I see that my feet are still blistered but the gauze I used to keep my infected toes separated overnight seems to have done its job. The antiseptic powder also seems to have worked well. My toes and feet do not look as bad today as they might. They are not the angry red colour of the previous several days.

I shower and brush my teeth. When I turn on the tap, the water again fails to emerge. This time, I keep my toothbrush back from the spout when I turn the tap further. As expected, a frog pops out, followed by a torrent of water.

I swallow a couple of painkillers and then Baz and I retrieve our water bottles and hydration bladders from the fridge. It will be nice to have cold water for the first few hours of the day. We extend our thanks to Moc and Sheree for their many kindnesses during our time in Hungerford. They have been lovely hosts and I am sorry to say goodbye. I do not expect ever to see them again.

Just walking around, getting ready for my departure, I see that the swelling in my feet is already returning. This is a big worry. Although it is not yet 7.00 am in Queensland, it is almost 8.00 in New South Wales. We will be away much later than I hoped. It will be a long day in the heat walking 30 km to get to Warroo to accept John's invitation to stop at his place for the night.

Despite the late start, we still need to line up for photographs aplenty. The cameras click manically. 'It is like the first day of school,' I whisper to Baz.

Moz sets up his camera with the timer to take a photo of us all together—Moc and Sheree included. In the photograph, we all wear smiles although, in the bright sunshine, mine appears forced. I am not looking forward to returning to the Track today but, given that I must, I would prefer to have been out there for two hours by now.

Moc and Sheree wander down the road with us to the Dingo fence—the gateway back to hell. 'Welcome to New South Wales,' a sign proclaims. Baz and I pass through and begin our return journey.

The sky is a vibrant blue, devoid of even a hint of clouds.

With backward glances and waves, Baz and I leave Hungerford in our past.

'Look at the footprints,' Baz says. We have passed this way before. In the barren stillness, our footprints remain in the sand like the evidence of Man's visit to the moon.

'One small step,' I say, taking my next step. I know it brings me closer to Bourke.

At about 8.50, we reach a detour sign at the Wanaaring turnoff and discover that the road to Bourke is closed.

'Things have obviously gotten worse at the water crossings, Gregga.'

'Yeah. I wonder how bad it is.'

Four steel fence droppers, some twisted lengths of wire and a couple of yellow hazard signs are not about to stop us. Regardless of what lies ahead, we will push on. Baz echoes my thoughts when he says, 'We'll keep going and see what it is like when we get up ahead.'

We will not be able to 'cross that bridge when we come to it,' but we will presumably be able to wade those creeks when we come to them.

'We can walk through anything that is lower than, say, our waists, right Baz?' It is hard to imagine the water could be that high so Baz and I continue.

'In the car, Moz and Jane might eventually have to take the detour though,' says Baz.

'They should drive forward today and then see what the depths are like when we get to the water in a couple of days,' I say. 'At that stage, perhaps they will have to turn back and take the long way around.'

Baz stops to write a note to Moz explaining our thinking and pins the note to the detour sign. I am travelling quite well and so I dare not stop to wait while Baz attends to his correspondence. In my mind, I am toying with the idea of trying to go the whole day without stopping. That way, I might avoid the problems that I have in trying to get going again after morning or midday stops.

Whereas we previously had the sun behind us as we walked to the west, we now have the sun in our faces. In the heat it is wonderful to have cold water in the hydration bladder. It seems a luxury. The simple things can mean so much.

Although I am travelling well this morning, I cannot eradicate from my brain Dean's warnings about my feet. 'If this is what they look like after a day of rest, I'd hate to see what they'd be like without the rest,' he had said. Well, tonight is when I will see what they are going to look like without the rest. It is a worrying prospect.

Despite my worries, I am in good spirits. My feet are tender, but they feel okay.

Sometime later, the support vehicle appears.

'How are yer poppin' up?' Moz chirps, his smile spreads across his face in the manner that it has ever since his withdrawal from the tramp. He settles his car into a slow roll beside me. He obviously has no intentions of any more walking. This tramp is, after all, *my* dream. He is here to support me in achieving that dream. Initially, his intention was to do that by walking. Now, he is doing that by driving. Either way, he is supporting me and, for that, I am grateful.

'I'm good, thanks.' I feel optimistic.

≈

Although it is not possible to know when Henry and Jim made it back to Bourke, he was certainly there by early-February when he wrote another letter to his aunty.

> *I got back again all right, and am at work painting....It's hot as hell here—too hot to think or write.* Bulletin *hunting me up for copy, but they must wait....I find that I've tramped more than 300 miles since I left here last. That's all I ever intend to do with a swag. It's too hot to write more.*

After his return from Hungerford, Henry stayed in Bourke for another four months before catching the train in June 1893 back to Sydney. However, he then felt disillusioned and disappointed for he had had enough of Australia and on 18 November he sailed for New Zealand.

> *An' if you meet a friend of mine who wants to find my track,*
> *Say you, 'he's gone to Maoriland, and isn't coming back.'*

Henry noted that on the journey to New Zealand there were 'about fifty male passengers, including half a dozen New Zealand shearers, two of whom came on board drunk.' The other passengers 'were chiefly tradesmen, labourers, clerks and bagmen, driven out of Australia by the hard times there, and glad, no doubt, to get away.'

Henry was pleased with what he found in New Zealand or, more to the point, *who* he found: 'Meet old chums at every corner. Lot of Bourke people on the boat yesterday,' Henry wrote. Although Bob Brothers—Henry's good friend and Bourke-based character, the Giraffe—had been in New Zealand, unfortunately, he was gone before

Henry arrived. As Henry wrote, Brothers 'got back to Sydney about the time I left. [Brothers] got a letter to say that his baby and wife were ill. Baby dead and buried when he got back.'

Henry and a mate secured two weeks' work felling trees for a sawmill, but 'at the end of the fortnight,' Henry wrote, 'the boss said we weren't bushmen.' Henry's pride was stung. He said this criticism 'hurt me more than any adverse criticism on my literary work could have done.' Shortly after, Henry returned to Australia.

≈

The uniform Mulga Trees lining the Track are four metres in height. Here and there, a Desert Bloodwood interrupts the uniformity, perhaps stretching eight or ten metres towards the sun. The orangey-reddish tinge to the flakes of their scaly trunks adds interesting colour and texture. It is the Mulgas that dominate though. They are dusty, dirty-looking trees. As much as they represent the heart of the country, they do little to stir the soul and to inspire me in the way of a regal gum. They are said to live for 200 or 300 years. I wonder how many of these trees stood when Henry passed this way. As I kick at things on the rocky Track, I wonder what physical evidence exists—if only we knew—of Henry's passing. Did he displace a particular stone? Did he carve his name in a tree? If we knew he was responsible, could any of the roadside rubbish be identified as evidence of Henry's passing?

The sun climbs higher and higher so that by 10.00, Baz and I are trailed by only short shadows. We move to the left to try to avail ourselves of some of the sparse shade that the Mulgas offer.

≈

A good friend of Henry's introduced him to Bertha Marie Louise Bredt in 1895. While on holidays from her nursing duties in Victoria, Bertha was visiting her mother, who owned a small bookstore in Sydney. Henry and his friend came to the bookstore and the introductions were made. 'At once we became very great friends,' Bertha recalled. The two enjoyed discussing books and they quickly discovered a shared passion for the work of Mark Twain. 'Harry came every day to see me,' Bertha wrote, 'and at the end of a week begged me to marry him.' Henry moved fast—perhaps too fast for Bertha as she refused the proposal! Not to be deterred, Henry tried again and, six months after meeting, the two became engaged.

≈

Just before noon, Baz and I pass a road sign indicating we have completed 15 km. Moz has set things up for us to stop and rest and Baz accepts the offer. I do not. As I pass, I grab some Arnott's Choc Ripple Biscuits from Moz. I will eat as I walk. I used to love these biscuits when I was a kid. When I take a bite, the taste brings to mind younger, carefree days when, although there were times when I might have thought I was 'very wicked,' as Henry said, 'you never realize how innocent you were until you've grown up and knocked about the world.'

The balloon of my reminiscences bursts when a burr digs into my left foot. I lean against a small white roadside marker and raise my foot so that I can remove the burr. I am on a long straight stretch with lots of big and loose rocks, each one of them with a special commission to cause me grief: 'Burst his blisters and make him squeal.'

I work my way down into the gutter in the hopes of finding a better line to take. A breeze blows. Insects hum in the Bush to my right. A fly buzzes around my head. Wearing freshly laundered clothes, I do not yet have the same number of flies that accompanied me to Hungerford but I am sure that will come.

At 12.15, I pass two dry white skulls, bleaching in the sun. Henry said that, out here, it is not just that 'the stars are brightest,' but also that 'the bones of the dead gleam whitest.' This is certainly true of the skulls before me, which are perhaps 20 m apart. Each of them is missing the lower mandible of the jaw. The upper jaw remains intact—the teeth made strong for chewing the dry, woody grasses out here. The empty round eye sockets stare into the eternities, perhaps seeing places where things are neither so dry nor so harsh.

Although the sun beats upon my head, I am happy with the way things seem to be going with my feet. With my comfortable Nike shoes, Dean's bandages, plus the day-and-a-half of rest, my feet are coping well.

Baz closes the gap between us. He walks in the shade right beside the Mulgas. When he catches me, he has a small dead snake in his hand. On this hot day, dead snake in hand, he is an otherworldly creature of the shadows.

≈

As much as George Robertson was a supporter of Henry's career, he feared for the innocent young girl to whom Henry was engaged. Robertson tried to dissuade Bertha, saying that, with three daughters of his own, he would 'rather see them dead than to marry a temperamental genius, who was a drunkard as well.' His admonitions fell on deaf ears though, for Robertson could see that Bertha looked upon Henry 'with large and deep brown eyes filled with worship of him and his works.'

On 15 April 1896, at the age of 28, Henry married Bertha. She was 19 years of age and, for all Henry's good qualities, she had just made the worst decision of her life. Henry and Bertha found 'a cosy little place' to begin their married life. When the landlord demanded the rent in advance, Henry asked if the landlord was able to provide change for a £10 note. Possession of such a fortune suggested wealth and prosperity and, unable to provide the change, the landlord waived the requirement. The waiver was fortunate for, in reality, there was no £10 note. Henry had no money at all with which to start married life.

≈

At 12.50, we get to where Moz is set up for another rest.

'You've done 20 kilometres,' Moz announces.

Baz takes a seat, but I do not. I am mindful—nay, afraid—that if I stop and sit down, I will not be able to get back up to get going again. I am fearful of the after-break crash. To keep going while I am feeling able to do so, the next ten kilometre stretch might take a little over two or two-and-a-half hours. If I stop and rest, the next ten kilometres might become a four-hour ordeal.

I quickly spread some Vegemite over a couple of pieces of stale bread.

'That's gone mouldy,' Jane says.

'It won't kill me.' If I can handle Henry's Track, I can handle a bit of mouldy bread.

I grab an orange juice and refill my water containers. I stuff a handful of Clinkers into my pocket.

'I'm gonna keep going, Baz,' I say.

'Are you okay, Gregga? How are your legs?' It is not the first time Baz has asked about my legs today. He is reminding me to keep an eye on things and to watch for the signs Dean warned me about.

'I'm fine. See ya.'

I toss the Clinkers in my mouth before the chocolate melts. Moving away from the others, I notice the ground beneath my feet is cracked in the form of some bizarre M. C. Escher tessellation. The ground is an intricate spider web of cracks and crevices, torn apart to surrender whatever moisture it held.

The pattern of the cracks reminds me that despite the floods, this land is still recovering from the Great Drought. It was like this when Henry was out here too. 'In many places the surface of the ground was cracked in squares where it had shrunk in the drought,' Henry wrote. In this unchanging place, droughts have been going on out here for centuries.

I approach a mailbox and see that it is the Ningaling mailbox. Somewhere up the dusty road to the north, Will and Narelle live. Finding an old probationary driver's P-plate on the ground, I take my pencil from my pocket and write a greeting on the back of it.

'G'day Will and Narelle. Just walking by. 1.15 Sunday. Best wishes, Greg.' I open the mailbox lid, ready to retreat if a snake happens to be inside. Nothing is in there but an old leather mailbag. I toss the P-plate into the mailbox and carry on.

At 1.40 pm, Moz and Jane catch me. 'How are you popping up?'

They tell me Baz is about two kilometres behind. I cannot see him, but I am on a long sweeping bend and he might be anywhere back around the bend.

Moz restocks and refuels me while I continue walking.

'I left a note back there in Will and Narelle's mailbox.'

'Yay! Grog's back,' says Moz. 'You must be feeling okay if you are getting back into your famous Grog correspondences.'

'It was more a note than an epic.' I smile. In our youth, Moz and I regularly used to pen long letters to one another. 'Another Epic,' we would proclaim. Moz still has my letters from 30 years earlier. The last time I was home, he brought some to share with me. We laughed until it hurt.

After keeping me company for a while, Moz says, 'We'll drive ahead and check things out with John at Warroo.'

'Yep. I suppose you better check that the offer of a mattress still stands.'

A few minutes after Moz leaves, I see something wandering down the Track, coming towards me. From its general shape and the way that it is moving, I make it out to be an animal, but I cannot tell what kind of animal. Eventually, to my surprise, I discern that it is a sheep. As I draw closer, I see it is a big old thing—maybe a thousand years old. It sees me and wanders off the side of the road into the scrub. I see it in the scrub though, walking parallel with the road to get around me. Sure enough, I look back after a time and see it re-entering the roadway about 100 m back. It stops on the Track and looks at me, as if to assess the situation. The sheep is shaggy and haggard. Loose clumps of wool hang from its belly and breast. The wool is stained coral pink from the red dust of the Track. I wonder how long it has been since it last was in the grips of a shearer and how long it has been on its own. It would be fascinating to know its history—to know how long it has been wandering around out here; how long it has been since it saw another sheep; how far it has travelled since its owner last saw it; how long it has been since it saw another person.

I watch the sheep cross the road and enter the scrub on the other side and I guess that it sees Baz approaching. On Day Three, Baz took a photo of a shaggy old sheep. Perhaps this is the very same one. Baz's photo would have been taken somewhere out from Ford's Bridge. Has it wandered this far along the Track over the past six days? Maybe it has spent a thousand years wandering this Track, back and forth.

I think of one of my favourite novels, Natalie Babbitt's *Tuck Everlasting*. Winnie Foster sprinkles a toad with water from a spring of eternal life. I wonder where the Ancient sheep gets its water out here in the desert. Would I want to drink from the spring and live forever? I feel as if I have been walking forever. Maybe the Ancient sheep saw

Henry pass this way. Would Henry have been in better or worse shape than me at this point in the journey?

Despite its haggard appearance, there is wisdom and nobility about the animal. It is a survivor in a harsh land. Everything out here—everything—has to be tough to survive. In an email after we have finished and recovered, Baz comments on how, when I was struggling he 'really admired [my] toughness'.

I am committed to plugging away and I figure I have now probably done about 23 of my 30 km for today. If I can be feeling this good at this stage of each day of the return journey, I will be over the moon with joy, and relief.

'Bourke, here I come,' I say with confidence.

However, I am mindful that Dean warned me that things can get worse in a hurry. I constantly monitor my situation, taking stock of how I feel and making occasional visual checks, rolling my pant legs up or pulling my pants down to see my legs and ankles.

As I roll up my pants for another visual check, I see that my legs are red and splotchy with heat rash, but there is no sign of darker tracks going up my legs. I wiggle my toes within my shoes and they feel sore, but okay. Dean suggested the pain could become so unbearable that I could not put any weight on my feet at all. At present, it is painful but bearable.

≈

Thick cloud is building up on the horizon everywhere except to the rear.

At 2.00, a Little Crow calls to me. 'Hey, hey. Hello, hello.'

I approach the spot where I rested for about two-and-a-half hours on Day Six. I veer off the Track a little to walk briefly into the shade where I stopped in hell and suffered. The place now brings back unpleasant memories of overheating and distress. The crow calls again. Does it wonder how I made it out of here alive? Does it wonder if I have returned to die? 'Where the white man lies dead / The crows will fly over,' Henry wrote. The place reminds me of my vulnerability and the tenuous nature of my progress. I move on, eager to put memories of that low spot behind me and to leave the crow to search for other pickings.

My right foot is starting to hurt; however, I notice that the worst pain is not around the infected toe. The pain is primarily on my instep. It seems to be a new injury. It is possible that I am creating new problems in trying to compensate for the infection in my right foot—trying to shield and shelter one sore spot and creating additional problems elsewhere. I will have a better idea of how things stand when I get to Warroo Station and take off my shoes and socks.

Moz and Jane have ventured back out in the car after watching some tennis on the telly at Warroo. 'We thought you might be along soon,' Moz says. 'You have been going really well.'

Yep, I have. With 30 km of the return journey completed without a stop, I feel good about the day.

'How's John?' I ask.

'Yeah, he's good. He's expecting you. He wants us to make ourselves comfortable.'

Moz says he and Jane will drive back to check on Baz.

'I won't miss Warroo will I?' I do not want to walk past and have to double back.

'No, it is a big white house just off the road on the right.'

'You can't miss it,' says Jane.

'Yeah?' When I am out here, if I am behind my shield, I can miss a lot.

'There's a sign on the gate, Greg,' Jane reassures me. 'You won't miss it.' They drive off, but return soon, reappearing almost at the very moment I reach Warroo.

Beside the 'Warroo' sign on the gate, there is another sign: 'Beware of the dog.' I open and pass through the gate at 3.07 pm. Seven hours and seven minutes of walking without a stop. With the road closed, I have not seen a single vehicle on the road today with the exception of Moz's car. I hold the gate open for Moz to drive through. He gives me a cheery wave and drives ahead, eventually disappearing behind the house. As I walk the 100 m or so to the homestead, I remain on the watch for dogs. The house is a big one—it's huge. It looks like the veranda might run all the way around. The roof is corrugated iron. Although the house could use a fresh coat of paint, I love the look. The heart of the Outback has long beat in homes like this one.

A couple of dogs move to my right, but I notice they are on chains.

I walk around the back to where Moz parked his car. He and Jane are already inside, back in front of the tennis. I venture to the back door and am greeted by John who is very welcoming.

'How are you feeling today?'

'Better than last time I saw you.'

The last time I saw him, I did not really *see* him. I was too buggered for that. His smiling, inviting face narrows from his slanted ears towards his jaw, lending an interesting V-shape to the happy face. He still has a good crop of hair but it is thinning. It would once have been red but, save for his eyebrows, is now more blonde than anything. He has a ruddy complexion—sun damaged skin. We white fellas were not made to be out here.

John invites me to fill my hydration bladder and water bottles and to leave them in the fridge and freezer overnight. It will be fantastic to have cold water on the Track again tomorrow morning.

John guides me through the house to the veranda, where there is a long line of beds.

'Take your pick.'

Each bed has a mattress, but no blankets. Blankets are not needed out here. The veranda has mosquito netting all around. It will be cool here and, although there are a few small holes in the netting, largely devoid of mozzies.

I choose my bed and throw my pack down.

'The shower and a toilet are back here,' says John and he leads me to the 'john'.

'This is great, John. Thanks so much for your kindness.'

'Just make yourself comfortable. When you have rested and feel up to it, I'll show you around the rest of the house.'

When I strip down to shower, I notice I am covered head to foot with a bright red heat rash. Although I have the rash at the end of each day, this is the worst it has been so far. Fortunately, unlike with Baz, my heat rash is not unbearably itchy.

After a visit to the toilet and shower, I lie down and use the foot of the bed frame to elevate my feet. It feels so good to be off them. They are swollen, but there is no sign that they are worse than the day before, and possibly a good deal better. Still, as I lie here, I begin to realise how buggered I feel. It has been a long day. To knock over 30 km without stopping has been tough on me physically and mentally. Having lain down, I do not feel like getting back up any time soon.

Baz arrives 20 minutes after me. He collapses onto a bed, saying, 'I've had a tough afternoon. I've got a bastard of a headache.' His hands are swollen and he cannot make a fist. 'A bull ant crawled into the water line of my water bladder. When I took a sip, I swallowed the ant and the bugger bit me in the throat.'

Baz looks exhausted. It is strange to see him obviously feeling worse than me.

'Rookie,' I say, resurrecting a label I had discarded after spending so much time watching Baz in the distance far in front of me.

I look over to where Baz lays with his eyes closed on the bed across from me. He has his bare feet elevated too. His arms are raised above his head in the manner of a newborn infant. As with a baby, his arms look short and stunted. His cold necktie rests on his forehead and temples. The poor bugger looks rooted. I notice how white his feet and ankles are compared to the red-stained colour of his legs from the calves upward. He has been ploughing through the red dust. The demarcation between clean and dirty shows clearly how long his socks were.

Moz comes in from where he was watching TV to check on Baz and to see if we need anything.

'Yeah. Can you get my bag please, mate?' Baz says.

'Can you get me a bottle of water, please?' I ask simultaneously.

'Yeah, no worries.'

In a few moments, Moz returns with the bag for Baz and then leaves.

'Hey, Moz,' I call. I am not certain he heard me when I asked for the water and so I mistakenly believe he is going back to watch the television.

'Yeah?'

'You didn't get me the water, mate?'

'There's about six loads of stuff to bring in and I am getting to it!' Moz snaps.

Moz only rarely gets snappy. It does not suit the big friendly giant with the enormous smile. Maybe Henry would say, 'The truth seem[s] that we [are] getting on each other's nerves—we [have] been too long together alone in the Bush.' I understand that Moz and Jane are tired and hot too, so I do not tell him where to shove the water bottle once he gets it. After all, 'Old mates seldom quarrel, because they understand each other's moods.'

Moz brings me the requested water.

'Thanks, mate.' Did you spit in it?

I gulp down the refreshing cold water.

I close my eyes and rest for two hours.

≈

Baz stirs and works his weary way to the shower. I wander out to John and tell him I am ready to accept his offer of a tour of the house. We enter a room with bookcases overflowing like volcanoes spewing lava. This is my kind of room.

'These are some photos of the Dunks.' John draws my attention from the books. The Dunks are the station owners.

This reminds me of what Terry said about photos at Warroo. 'John, I was told there are some photographs here of Henry Lawson when he was out this way. Do you know anything about that?' I would love to see those photos.

'I'm afraid I can't help you there.' He can see my disappointment. 'I guess if there were photos, they would have gone with Mrs Dunk when she moved.'

There is a big piano to one side of the room. John tells me a little of its history. I imagine the old time sing-alongs that might once have occurred here.

We enter the next room where dozens of framed photographs plaster two of the walls. In each photo is a different woman. They all wear bright smiles. They all wear wedding dresses.

'Who are all of these people, John?'

'Oh, they are all the brides from the district over the years.'

It is a fascinating collection—so much history.

'How old are these photographs, John?'

'Oh, the oldest would be a hundred years or so.' He takes one down and turns it over. On the back is written the wedding date. 'This one is from 1890.'

The blushing bride would still have been a newlywed when Henry passed through. She would not yet have gotten 'past carin'' about love and romance.

Past wearyin' or carin',
Past feelin' and despairin';

I peer into the photographs. Some of the dresses are pretty. Some are strikingly beautiful. More than one of them is a 'bride of frivolous fashion.' One or two of the dresses are ugly. Same with the brides, I notice. I can see in the eyes of each bride that 'She has her dreams; she has her dreams.' The time is yet to come when, despite having 'built the usual castles in the air...all her girlish hopes and aspirations have long been dead.'

How many of the women in these photographs have been brought out here by their husbands, to battle in loneliness with the Bush? The Bushwoman's life is one ceaseless battle against the never-idle threat of ruination to drought and fire or, in times such as these to be buried beneath floodwaters: 'She stood for hours in the drenching downpour, and dug an overflow gutter to save the dam across the creek. But she could not save it. There are things that a bushwoman cannot do. Next morning the dam was broken, and her heart was nearly broken too.'

Many of the brides possess a face still flushed with the vitality of youth. When Henry's character, Joe Wilson, brings his young bride to the Bush, his concerned neighbour, Mrs Spicer, asks, 'What-did-you-bring-her-here-for? She's only a girl.'

More often than not, no groom appears in the photographs. In those where he does, his appearance is secondary to the bride yet he looks happy nonetheless. 'There are times when a man is happy. When he finds out that the girl loves him. When he's just married....some men make fools of themselves then—I know I did,' wrote Henry.

I hope there were not too many of the happy women who later regretted their wedding day or their choice of partner. Observation and experience compels me to agree with Henry where he writes, 'She picked the wrong man—girls mostly do.'

'Make the most of your courting days, you young chaps,' Henry advised. 'They're the days that the wife will look back to, anyway, in the brightest of times as well as in the blackest, and there shouldn't be anything in those days that might hurt her when she looks back.' Things change after the wedding day. Henry felt the 'courting days' were 'about the only days when there's a chance of poetry and beauty coming into this life.' I remember my courting days—how I could not bear to part from Jennifer. With almost 18 years of marriage under my belt, here I am, voluntarily on the other side of the world—and for six months apart. Is it just the memory of my courting days that make the prospect of those six months now seem interminable? 'Make the most of your courting days, you young chaps, for they will never come again.'

I wonder how Jennifer is coping with my absence. Does she miss me? Is she lonely? Has she become another Mrs Spicer?

'Don't you feel lonely, Mrs Spicer, when your husband goes away?'

'Well—no, Mrs Wilson,' she said in the groping sort of voice. 'I uster, once. I remember, when we lived on the Cudgeegong river— we lived in a brick house then—the first time Spicer had to go away from home I nearly fretted my eyes out. And he was only goin' shearin' for a month. I muster bin a fool; but then we were only jist married a little while. He's been away drovin' in Queenslan' as long as eighteen months at a time since then. But' (her voice seemed to grope in the dark more than ever) 'I don't mind,— I somehow seem to have got past carin'. Besides—besides, Spicer was a very different man then to what he is now. He's got so moody and gloomy at home, he hardly ever speaks.'

Have I become another *Mister* Spicer?

John hands me a large pile of typewritten pages, pulling me from my reverie. 'These are some of Roy Dunks' memoirs. When you feel up to it, you might be interested to have a read.'

We complete the tour of the house and I settle down to read the memoirs of R. J. Dunk, who was born in 1892. 'When the First World

War started I was working as Oversear [sic] on Currawinya Station on the Paroo in Queensland near Hungerford. In May 1915 I left there to go to Adelaide to enlist,' Dunk wrote.

At the age of 25, Dunk was involved in the Charge of the Light Horse Brigade at the Battle of Beersheba. 'Perhaps never since the far off days of Abraham had the water in the old wells of the patriarchs been so needed by parched men riding in from the southern desert,' I read. I know something of being parched. 'All rode for victory and Australia,' Dunk continued. It is a sentiment Henry would have cherished. Such thoughts would have inspired Henry to wax lyrical. 'Australia! My country! Her very name is music to me. God bless Australia!...And heaven send that, if ever in my time her sons are called upon to fight for her young life and honour, I die with the first rank of them and be buried in Australian ground.'

Henry was lots of things, of that there is no doubt, but there is equally no doubt that he was a patriot. He loved his country and would happily have surrendered his life for her. He offered his services during the First World War but, with his deafness and alcoholism, Henry was never any chance of being enlisted.

Henry would dearly have loved to have been in the ranks of the Light Horse, galloping under the Ottoman sun alongside Dunk in 1917 on that historic and heroic October day. More than 20 years earlier, Henry had dreamed of just such a charge.

But, oh! if the cavalry charge again as they did when the world was wide,
'Twill be grand in the ranks of a thousand men in that glorious race to ride.

At least briefly, however, my thoughts are not so much with Henry as they are with Roy Dunk. I read that he consistently attended the annual Adelaide Test match and, as such, he was among the record crowd at the Adelaide Oval in January 1933 during the infamous Bodyline game. He describes the game as having 'very nearly wrecked the Ashes series'. He leaves it unsaid that Harold Larwood's thunderbolts also very nearly wrecked Bill Woodfull's heart or Bertie Oldfield's skull.

≈

Henry's marriage was not a happy one. He found the restrictions and responsibilities of marriage galling. As one biographer wrote, Henry was a 'vagabond on the face of the earth.' Restless of spirit, 'there was...no place [and]...no person who could slake the thirst of Henry Lawson.'

Henry was not suited to what he saw as a 'dull domestic life' and he grew quickly to resent his 'bitter nagging Missis.' Henry saw alcohol as his release—his 'secret whisky cure' from his marital problems. The pub became 'Heaven' while 'his own home was a hell' of what, having lost taste for the institution of marriage, he considered 'the everlasting friction that most husbands must endure.'

On the day of his wedding, Henry borrowed money to pay the marriage celebrant. Ever after, Henry was heard to say that the only thing he had against that particular friend was that he had lent Henry the money to get married. Henry so loved his mates that, almost inevitably, it created problems between he and his wife. Friends who were 'eager to celebrate his success' crowded around. According to Bertha, 'this was an ever present temptation' and an influence that often led Henry astray. After seven years of watching her husband led astray and of Henry spending most of his money before she ever saw any of it, Bertha had had enough.

'Twas drink and nag—or nag and drink—whichever you prefer—
Till at last she couldn't stand him any more than he could her.
Friends and relatives assisted, telling her (with motives pure)
That a legal separation was the only earthly cure.

In April 1903, Bertha formally applied for a judicial separation. Two months later, the separation was granted. A Sydney newspaper reported: 'DIVORCE COURT. (Before Mr. Justice Simpson, Judge in Divorce.) LAWSON v. LAWSON: By consent a decree for judicial separation was granted in the case of Bertha Marie Louise Lawson, formerly Bredt, versus Henry Lawson, author and journalist.'[1]

The marriage may have been over, but Henry's troubles were not.

DAY TEN

Children in Henry Lawson's Life

Mrs Douglas had a hard time, with her two little girls, who were still better and more prettily dressed than any other children in Bourke.

- 'Lord Douglas', *Children of the Bush*, 1902

I don't believe in parents talking about their own children everlastingly—you get sick of hearing them; and their kids are generally little devils, and turn out larrikins as likely as not.

- Brighten's Sister-In-Law, *Joe Wilson and His Mates*, 1901

We have a big day planned, as our aim is to cover 40 km to Tony Marsh's Kia-Ora. John tells me to help myself to breakfast cereal but I am happiest with toast and Vegemite.

'Here's the butter, Greg.'

'Oh, I always just have my Vegemite without butter. I like it best this way.' I am still a Vegemite kid. It is the way I have always eaten it, and I will continue to do so throughout the eternities. If there is a Heaven, there will be Vegemite there.

When I am ready to leave, John walks with me to show me where an old Aborigine used to live near the homestead. The sunrise away to the east is stunning. Flames of gold slice through the grey clouds to bring light to the day, bathing the plains in golden warmth. All of the Aboriginal tribes have their own Dreamtime legends to explain the advent of the sun but a common denominator amongst many of them is the notion of a great campfire burning in the sky.

'Tommy Spider's humpy used to stand over here,' John says, directing me towards a small pile of iron and rubble amongst some Mulga Trees.

'Did Tommy Spider work for the Dunks, did he?'

'Well, yes, he'd help out a bit here and there and do whatever he could.'

Who knows? Perhaps Warroo's Tommy Spider could trace his ancestry back to one of the Aborigines Henry encountered and upon whom some of Henry's characters are based. 'Old Black Jimmie lived in a gunyah on the rise at the back of the sheepyards, and shepherded for my uncle,' Henry wrote. 'He was a gentle, good-humoured, easy-going old fellow with a pleasant smile; which description applies, I think, to most old blackfellows in civilisation. I was very partial to the old man, and chummy with him.'

'Every year, they'd give Tommy Spider a new suit and then he'd take a bath and put on that new suit and walk around as proud as a peacock,' John tells me.

I laugh a little uneasily at the image. 'Yeah?'

'He wouldn't take that suit off until he got his new suit the following year.'

'Yeah, really?'

'When they had to start paying the Aborigines the same wages as the other stockmen, Tommy said to me, "This'll be the end of it" and he was right.'

'Why's that?' I ask, not knowing why equality would spell 'the end of it,' whatever it might have been.

'Well, the station owners were just helping them out and giving them food and a place to stay. Tommy knew that when they had to be paid too, there would be no more handouts because there was not the work to employ them.'

I stop and wonder about the old ways—both of the white landowners and the traditional Aboriginal landowners. There is so much history out here.

As I prepare to leave, Baz joins John and me. I take a photo of Baz and John together. They are both thin, wiry types. I know that, in many ways, Baz would love to be living John's life out here in the Outback, working the Land.

'Will we see you out on the road at all, John?' I ask, wondering as I leave if this is the final farewell. 'Will you be heading into Bourke at all in the next few days?'

'No, I don't think so. I have plenty to do around here.'

'Well, thank you so much for your hospitality, John. It was very interesting to stay here and to read all the things about Roy Dunk.'

We shake hands and I depart, leaving Baz to chat further with John.

There are two black sheep grazing on the dry grass in the driveway. They seem friendly and I think for a while that they are going to follow me like puppies.

I walk through the gate and I am away and back on Whim Plain by 7.00 am. Within the first 15 minutes, I see a kangaroo, five Emus, a large flock of Crested Pigeons and various other birds including crows, parrots and a small falcon hovering on rapidly beating wings. I have not gone far before I see more kangaroos and Emus in the emptiness and a number of bearded dragons on the Track. It is a good way to start what will be a long day.

Although there is good cloud cover this morning and a breeze blows in my face, it does not take long until I am hot and dry, like the plain. Fortunately, I have ice in my water bottles.

After a while, Moz drives up. 'How are yer poppin' up, Grog?'

'Good, Moz. Is Jane walking, is she?'

'Yeah.' He tells me that John walked out a kilometre or so with her. 'His two dogs and those two black sheep went for a walk with them too.' I thought those sheep looked like they wanted to go for a walk.

'The sheep just jumped across the stock grate. John was surprised to see they could do that.'

'I guess it kind of defeats the purpose of the grate, eh?'

Moz tells me that the group encountered a snake that reared up at the sheep but, fortunately, did not strike. Jane tells me later that it is the first snake that she has ever seen.

'It is 26 degrees, Grog,' Moz informs me. 'That's good. This is the sort of temperature that we had when you and I did the walk last time. We were able to do 40 km days back then so it should be okay today.' I think about that. 'Of course, we hadn't already walked all the way to Hungerford beforehand,' he says, giving voice to the very thing that I am thinking.

'Yeah, good point.'

I look up at the layer of grey stratus clouds to the east and wonder how long they will hold out the rising sun. Behind me, the puffy cirrocumulus clouds resemble the fleece of a lamb at its first shearing. Once the sun gets higher, those clouds will do nothing to restrict the heat.

'John said he will give Tony Marsh a call to let him know that we have left Warroo and are on the way.'

'Great.'

'I'm going to drive ahead to check out Clarkes Creek and see how that is,' Moz tells me before he disappears.

I see the deserted old car wreck that I missed as I blindly walked by on the way to Hungerford. I realise how fortified behind my shield I must have been to miss the car because it is out in the open and, frankly, difficult to miss. I wander into the field to take a couple of photographs. There are thousands of grasshoppers or locusts hopping about. I wonder how many snakes have called the shell of the car their home over the years. I decide to move on.

I see a herd of cattle away to my right, in the distance. I hope they will stay over there because if there is a bull amongst the herd and it decides to play some games, there are not a lot of places for me to run to and hide out here.

At 9.00, I pass a small waterhole. The grey puddle appears to consist less of water than mud and shit from where thirsty cattle have churned up everything. A wizened tree reaches from the water, its leafless branches arming this way and that. A pair of Galahs rest upon a couple of those branches, the feathers of their breasts catch the morning light in such a manner that they are the pinkest Galahs I have ever seen. With a backdrop of dry grass on the otherwise treeless plain, the Galahs look out of place—too pretty and colourful to be out here.

A farmer drives up in his Toyota Hilux and stops for a chat. Despite his dark sunglasses, I recognise him immediately. In his dry, inimitable fashion, he does not bother to say hello. 'How is Henry treating you?' he asks.

'Like a dirty old drunk,' I respond. 'Mick Fisher, it is good to see you again.'

'Are you getting a good taste of what Henry went through?'

'You bet I am.'

'How has it been?'

'It's been tough.'

'You have a pretty good day for it—nice and cool. I expect Henry probably had it quite a bit warmer.'

'Not every day has been like this one though, Mick.'

He smiles. 'Your mate is cheating.'

'How's that?'

'Well, he's in a car.'

Mick has encountered Moz down the road. Moz tells me later that, as with me, Mick did not bother with a 'hello'. 'You're back,' he said simply.

'Sean had to pull out on the third day. It was very hot then,' I say, reinforcing my point that not every day has been a pleasant 26 degrees. 'My brother is back behind me. He and I are continuing with the walk.'

Mick is relaxed and laid-back. It is the way that I remember him, even in the grips of the Great Drought. He has a big esky beside him in the vehicle. No doubt it is full of cold drinks and sandwiches for the day's work.

'Mick, last time I was out here talking to you, you said your herd was down to about 20 cows. How are your numbers now?'

'Oh, they are building up again nicely at the moment. Plenty of feed for them. Things are on the improve.'

Last time, he told me his herd was down from 400 head. 'It's a bit of a sore spot,' he said, before proceeding to walk alongside us, reciting a Bush poem.

'The place is certainly different to when I went through last time, Mick.'

'I'm not sure what it is that draws people back here. They reckon it's the red dirt.'

'Well, it's certainly a unique part of Australia. That's the reason we come back—to have a second look.'

'Did you see the young blokes earlier?'

'You're the first person I've seen today.'

'My boys are supposed to be back behind you somewhere. They should be catching goats.' He tells me that he will be driving past Hungerford with the goats and so he asks me about the road.

'Well, there is a detour at the Wanaaring turn-off, but from here to Hungerford there has been no water.'

He nods. 'You're stopping at Marshy's tonight?'

'Yeah.' I am not sure if Moz told him that or if he has just heard it on the 'Bush Telegraph'.

'Be careful.'

'Why's that?'

'Just watch he doesn't give you any of that poisonous home-brew of his. He is a pretty wild man, Marshy.'

I laugh.

'I better let you keep going, I s'pose.'

I am pleased to have seen Mick again. He is the salt of the earth. In another time, he would have enlisted with Roy Dunk. I can hear him making dry quips for his mates as they charged the Turkish guns.

Mick drives on and stops to chat with Baz. Later, Baz tells me that Mick said, 'You're holding them up.'

'Who said that?' Baz asked.

'Your brother.'

'Well, I'll have to catch up and give him a hiding then.'

No doubt Mick is the type who is tickled to stir the pot a little, sprinkling nuggets of contention here and there. It is a source of humour and entertainment. It is the Australian way.

≈

I am off Whim Plain and back amongst the trees where tall gums stretch into the increasingly blue sky. The yellow of the grass and the green of the tree foliage reflect the colours of my country—as with Henry, 'The land I love above all others.'

I find myself thinking back to Day Two and my lunchtime stop where, filled with pain and self-doubt, I was almost at breaking point. I had done so little and had so very much still before me. My butchered feet cried for relief. It was a low, low point. It was a period of serious doubt and of fear. I doubted my ability to be able to continue to withstand the pain and I feared the additional pain that still lay ahead. Now here I am, several days and many, many, many kilometres later, still moving forward. Indeed, I feel optimistic, hopeful and confident. I must be getting close to 20 km already today. I have not stopped. My body has toughened. My feet have hardened. My mind will not break.

I look, but cannot see Baz back behind me. After his struggles yesterday afternoon, I wonder how he is holding up. Yesterday and today, so far, have been easier days for me—*less difficult* is probably a better, more accurate description. Baz might be struggling though.

I approach Brindingabba Creek where there is a lot of vegetation. I cross the bridge and see Moz's car a short way up the road—maybe 100 m in front. I expect he has stopped at the 20 km mark. The water in the creek is a dirty grey colour and I notice a lot of movement as insects dart this way and that and bounce about on the surface. I hear lots of birds singing, accompanied by insects and, occasionally, the croak of a frog. The Bush symphony continues despite the fact that the sun is high.

I reach Moz's vehicle a few minutes after 11.00 am. I see him running through the trees towards me. He stops where he has his camera set up for some artistic shots across the water and through the Tangled Lignum shrubs, River Red Gums and Black Box eucalypt trees.

'I am okay to continue walking,' I say when he joins me on the Track, 'but I could use some more water if you have it.' With Moz's help, I fill my water bottles and hydration pack before I leave. The water is still cold from being in the freezer at Warroo.

Moz also gives me a ham sandwich John knocked together for me after I left. 'John thought you might like a bit of a change.'

'It'll be as good as a holiday.'

I walk on and, although I would not describe myself as having a spring in my step, the cold water and ham sandwich are invigorating.

When I get to the Brindingabba mailbox, I take out my pencil. I might yet see Dave or Kylie Fisher further on but, in case I do not, I figure I will pen a quick cheerio. I find a piece of paper on the ground and leave a note inside their mailbox.

'G'day Fishers. Just walking past on the way back to Bourke. 11.20 am Monday now. I ran into Mick earlier today. Best wishes for 2011. Greg.'

Shortly after, I approach the Hungerford 50 km marker. This is the marker where Baz left the motivational note that I missed on the way to Hungerford. I keep my eyes peeled and wonder if it still might be possible to find it. Will it still be here, five days later? Sure enough, not ten metres on the Hungerford side of the marker, I detect the red note, face down in the dirt. It has not travelled far. Beside the marker, under a big, heavy rock, I discover another note, scribbled in pencil on white paper. It says, '7 km down. 4 km to breakfast. We did 33 yesterday. We are inside 50, mate.' Going in, at that stage we were inside 50 km. Going back, it will be nice when we reach the point when we are inside the final 50 to *Bourke*. With detours here and there, we figure our whole trek will be about 450 km. I have now completed 275 and have about 175 left. I turn my attention to the red, laminated note. It says:

The credit belongs to the man who is actually in the arena:
The man whose face is marred by dust and sweat and blood;
The man who strives valiantly...

It is not a quote with which I am familiar but I learn from Baz later that it is Theodore Roosevelt. They are inspirational sentiments. I am certainly out here in the arena—in the cauldron. My face is covered in dust and sweat. Although not streaked with blood, I have all sorts of sore spots that are equivalent to the blood referred to in the quote. I know that I am—and I am sure that Baz is too—striving valiantly to get this incredible undertaking behind us and to achieve success. Inspired, I continue plugging away—striving valiantly.

I have only gone another 100 m or so when I hear a vehicle approaching from behind. I mistakenly assume it must be Moz. I am on a narrow part of the road, where the laneway is raised to allow for a culvert to pass underneath. The one-tonner and trailer pass me, close, grinding to a halt about ten metres in front. I feel lucky they did not clip me. Given they went past before they stopped; I wonder how late they saw me.

I catch up and peer into the vehicle's window. The driver and his mate are kids.

'G'day,' they say.

'G'day.' I cannot get over how young they look. 'How old are you blokes?'

'I'm 14 and Mikey is 12,' the driver says. He looks 11 and his mate looks like he is 9 years old.

'How long have you been driving?'

'Since I was eight.'

I laugh. It is a different world out here.

'I'm Greg.'

'I'm Joe,' says the driver.

'I'm Mikey,' says his companion.

'What is your last name, Joe and Mikey?' I can see a family resemblance.

'Fisher.'

That's what I thought. These are Mick Fisher's boys. With his dark hair and blue eyes, Mikey looks a lot like his father.

'I was talking to your dad earlier. He was wondering where you two had gotten to.' It will not hurt to put a bit of wind up them.

'Yeah, he found us. He's gone to sell some goats.'

Although Mikey seems quieter than his brother, they both project the same cheerful, happy, relaxed persona of their father.

'Do you want some water?' Joe asks, wiping his brow.

'No, I'm okay thank you. I just filled up my water bottles a short way back.'

'How about an orange?'

That sounds good. They have an esky in the tray of the one-tonner and both boys alight from the vehicle to offer me my choice.

'I'll take this one. Thanks boys.'

'Take an apple too.'

I ask where the boys live.

'Kilberoo.' I will keep my eyes open for it.

After an enjoyable chat, I tell them that I should continue.

'Thanks for stopping,' I say.

Joe turns the vehicle and drives back in the direction from which he and his brother arrived. He leaves me in a cloud of dust.

≈

Henry and Bertha's first child was born in New Zealand in 1898, less than two years after they were married. The baby was named Joseph Henry but was known as Jim. Two years later, a daughter was born and named after her mother, but the girl was given the pet name, Barta. In 1903, another child died at birth.

Although eventually estranged from their mother, Henry was always pleased to see his children. Friends who ran a dry-cleaning business recalled that 'there were days when [Henry] came in, dressed in his very best, and would ask us to rub any small spot or crease from his coat and hat.' When asked for the reason behind Henry's determination to look his best, he would reply, 'I'm meeting Jim.' His daughter recalled that it brought 'special happiness' to her father to spend time with his children.[1] Separated from them, Henry treasured any letters or cards that he received. He kept them carefully, at one time saying 'I value them very highly.'

Henry attributed much of his monetary trouble to spending too much on his children. 'I got into the hole I'm in now by spending too much on my children,' Henry wrote, 'and I got into debt on their account.'

Henry's separation from his children brought great anguish to the troubled poet who could see 'the man he used to be' and could see what he had lost.

The father sees, with laughing eyes,
His little son and daughter wait:
They race to meet him as he comes—
And—Oh! this memory is worst—
Her dimpled arms go round his neck,
She pants, 'I dot my daddy first!'

In his 1905 book, *When I was King and Other Verses*, Henry included two poems written especially for his children. One was entitled 'To Jim,' while the other was simply called 'Barta'.

In the poem written to his son, Henry worried that 'a life of storm and stress' lay before his young boy with the 'dreamy eyes'. Henry encouraged his son to 'be generous' and to do good things for people. Well aware of his own failings, Henry wrote sadly but optimistically:

These lines I write with bitter tears
And failing heart and hand,
But you will read in after years,
And you will understand:
You'll hear the slander of the crowd,
They'll whisper tales of shame,
But days will come when you'll be proud
To bear your father's name.

In the poem to his daughter, Henry worried again what people were saying about his 'disgraced' name. Henry took comfort from his daughter's 'Wide solemn eyes that question me' and the 'Wee hand that pats my head.'

Let friends and kinsfolk work their worst,
And the world say what it will,
Your baby arms go round my neck—
I'm your own Daddy still!
And you kiss me and I kiss you,
Fresh kisses frank and free—
Ah, baby girl, you don't know how
You break the heart in me!

≈

Moz and Jane pass me a short time after the Fisher boys leave. They check with me and, for a change, I am honest when I tell them that I am feeling good and travelling well. The weak breeze is refreshing and my spirits are strong.

Twenty-five kilometres from Warroo Station, I pass some old stockyards. The history in those yards must be incredible. I wonder at the stockmen who have worked there, black and white. I am only 10 or 20 steps past the yards when I am surprised by John pulling up in his vehicle. He pulls a bowl and spoon from the cabin and then proceeds to serve me ice-cream and frozen peaches! Is this bloke an angel?

Still scared to sit for fear of not getting up again, I remain on my feet for my ice-cream treat. It goes down well. It must now be well over 30 degrees and to have ice-cream is a joy without limits.

'Thanks so much, John. You can't know how good this tastes.'

'I thought you might like it.' John is chuffed with himself, and rightly so. His thoughtfulness and kindness is a wonder.

'Those stockyards must have some history,' I suggest, directing John's attention across the road. 'I wonder how long they have been there.'

'Oh, they would have been built more than 50 years ago now. Mack built those yards.'

'Mack? Old Mack from the Hungerford pub?'

'Yes, old Mack. He built them when he was a young man.'

I ask John about the trees and he points out some Beefwoods and Box Eucalypts amongst all the Mulgas.

Once my stomach is full and will not hold any more ice-cream and peaches, John drives back to feed Baz too. I will never see John Stephenson again, but I will never forget him.

It is only a few more steps from the Outback Ice-creamery to the Kilberoo driveway. I scribble my second mailbox note for the day and leave it with the leather mailbag.

'Dear Fishers, It was nice to see you on the road today, Mick. Your boys Joe and Mikey were also kind enough to offer me a drink and give me an orange. They are nice boys. I'm sure you and your wife are very proud. Best wishes for 2011. Greg. Monday, 12.40.'

The Brindingabba turn off is only a few more steps along the Track, on the opposite side of the road. I notice that they have removed their Christmas decorations. Like me, Santa won't be back this way any time soon.

≈

Henry loved children. His daughter believed Henry had a special ability to know what children were most in need of during different periods of their lives.[2] One friend felt privileged to accept an invitation to accompany Henry on a walk through the slums of the Rocks district of Sydney. Upon spotting Henry, a poor little girl called excitedly,

'Here's Mr Lawson! Here's Mr Lawson!' The call brought forth a rush of children from the lanes and alleyways of the neighbourhood. Rather than being set upon by pickpockets and thieves, the friend watched in admiration as Henry passed money to all of the children and then ushered them to a nearby shop to feast upon unexpected treats. Another friend noted the joy on Henry's face as he played with a publican's children. Henry wrote in the children's autograph books and answered questions about his characters including which characters were real people and which ones were products of Henry's imagination.

Late in his life, when there was little that Henry desired more than beer, he and a mate camped beside a river. They had brought with them the mate's young son, Frank, who, like all boys, was excited to have the opportunity to sleep outside and to enjoy the Bush. The campers were joined briefly by some of Henry's friends who, knowing his tastes, presented the author with a case of beer. After the visitors had gone, Henry looked at the case and said, 'We won't open it. If we did, it would spoil everything for Frank.' In recalling the incident, the mate wrote, 'I was always fond of Henry, but I loved him, then, for that generous consideration of the boy.'

≈

At 1.45 pm, I arrive at Clarkes Creek. Initially, as I approach the crossing, I have water on my right and then, another 30 or so metres up the road I also have water on my left. It is still another several hundred metres along the Track until I get to the spot where the water is across the road. I see many ibises in the swampland away to my right. The downward curve of their long beaks makes them easy to identify.

Moz and Jane have gone back behind me to see how Baz is travelling. I have not seen him for hours. My most recent report is that he is doing it tough today, just as he did yesterday afternoon. The humidity is a challenge. I feel sticky and sweaty and I am sure Baz is the same. Is it possible that he won't make it? Surely not.

There is a lot of grey cloud overhead and it is darker and heavier on the horizon in front of me. Behind me, the sky is blue. Because he is struggling, Baz is taking regular breaks. Because I have taken the opposite approach—avoiding breaks for fear that, if I get down, I might not get back up—a sizeable gap has opened between us.

I encounter another Eastern Bearded Dragon. I have lost track of the number that I have seen today but because I am looking off into the swamp to my right, I do not see the lizard until I am almost on top of it. Its tail is curled upwards like a crescent moon. It remains so perfectly still that it has ants crawling over the top of it. I expect that it probably eats ants and, if so, I wonder how soon the ants will realise that when they crawl on a dragon, they are playing with fire. The dragon's tail is darker than any other part of its body, almost as if it might have scorched itself at one time. The spiny sides of the lizard and the beard beneath its chin make it easy to see why it is called a dragon. It is a primitive and ferocious-looking creature but it seems not to be bothered by my presence and is certainly of no bother to me.

The frogs are croaking loudly in the swamp. In one Dreamtime story, Tiddalik the Frog awoke one day with a huge thirst and drank the water from all of the rivers, creeks and waterways. This left a parched, desolate landscape. When an eel named Nabunum wiggled and wormed his way from his dried up creek bed, Tiddalik watched the eel get tied in knots and the frog began to laugh. Water spilled from Tiddalik's mouth, covering the Earth with a great flood—the equivalent to that flood over which Noah floated his ark.

Moz mentioned that, according to his calculations, it is about seven kilometres from the end of the Clarkes Creek crossing to the turn off to Tony Marsh's place. I figure it is about two-and-a-half kilometres along the driveway to Tony's place. Once I complete this water crossing, I will have less than ten kilometres left for the day. My feet are moving well, and although I have chafing under my right armpit, my spirits are good. My optimism is high.

I see a couple of Straw-necked Ibises in the trees, their black wings, backs, and necks failing to conceal them in the shadows because of the white of their breasts and undersides. I also see a White-necked Heron perched on a low branch. An ibis passes across the road and then doubles back over the top of me. A White-necked Heron flies by with its neck tucked in, flying with slow, easy wing beats. It croaks audibly, the harsh sound not at all in keeping with the gracefulness of its appearance.

The walking stick that I used the other day is still propped against the Clarkes Creek road sign. I leave my shoes on and just walk on through. In the breeze, a ripple creases the water surface. With the ripple, Clarkes Creek looks cleaner than the dirty, stinking crossing I completed on Day Five. I cross without problems and take the walking stick with me to the next crossing, which is not far ahead.

A Straw-necked Ibis stands on the road about 30 metres in front of me. I suspect it is probably wondering what the heck I am doing out here. It flies off, heading down the Track before it veers to the left and gathers height as it makes an arc before disappearing over the trees.

≈

Late in his hard life, Henry was invited to dine with the family of an old friend. After the meal, Henry playfully read the children's palms, delighting them by identifying one child as a future musician and another as a future author. One of the children asked, 'What about your hand, Uncle Harry? What does it say?'

'My hand?' Henry replied with a mix both of melancholy and humour. 'Oh, it says I oughter be dead.'

≈

At the third and fourth water crossing for the afternoon, I take my hat off and fill it with water, repeatedly tipping the contents over my head. It is overcast, but the sticky, humid conditions are tough. To have water trickling down my whiskered face is both refreshing and reinvigorating. It has been a long day.

Many birds are in the swamps and fields to my right. They take flight when they see me. An Australian Magpie flies into a small eucalypt to my left, where it proceeds to sway with the breeze. A Willy Wagtail darts across my line of vision.

While I am in the middle of the third crossing, I see a Sacred Kingfisher. Like all of the kingfishers, it is a beautiful bird; however, the strong, thick beak looks almost too heavy to hold up with its head. If I had to lug that thing around, it would give me a sore neck. In talking of the various ways in which he carried a swag, Henry said, 'I tried carrying a load on my head,' but unfortunately 'got a crick in my neck and spine for days.' I rub the back of my neck. I know about back and neck pains.

As I approach and complete the fourth and final crossing, two Emus splash through the swamp to the right. It seems odd to see Emus in water. In their efforts to avoid me, the Emus do not seem as sure-footed, and are certainly not as fleet-footed, as they usually are on dry land. I see Moz's car driving up behind me. I point out the Emus. 'Look at that, you two. It is unusual to see Emus in water.'

'I thought you might want a cold drink,' Moz says, handing me a Staminade from the esky. It is very good—still cold from being in John's freezer. 'You only have another 10 or 15 minutes of walking until you get to Tony's gate.'

Moz turns back to Baz and I continue to Kia-Ora Station. Eventually, I pass through the gate and start wandering down the roadway towards where the house must be. Every step in to Tony's place is a step that I must take to get out tomorrow, so I hope it is not far.

Along the two-wheel rutted Track, I notice a number of newborn lambs with their mothers, including one set of very small twins. A Wedge-tailed Eagle drifts overhead, the long fingers of its wings fully extended. It banks this way and that. I suggest to the bird that it might want to make itself scarce for I do not think Tony will take kindly to having an eagle looking over his flock of lambs. I wonder how many lambs Tony loses to predation from eagles.

A few minutes later, a flock of ten Australian Pelicans flies overhead. I watch them as they slowly descend and then splash into the Cuttaburra Basin lake formed beside the driveway. I pass by some old buildings to the left and shortly afterwards, I run into Tony Marsh driving out to meet me.

'Tony, it is good to see you again,' I say as I shake his hand. Tony has a welcoming gapped smile across his bearded face. A couple of friendly brown Kelpies wag their tails in the tray of Tony's truck. 'I heard you had a bit of an accident.'

'Yeah, a calf cradle fell on me.' The iron cradles are used to hold cattle for branding and dehorning. Tony shows me the huge bruise all the way up his side. Little wonder that he is moving stiffly—he is moving like he has walked to Hungerford.

'How are you travelling, Greg?'

After seeing Tony's black bruise, I say simply, 'Oh, I'm doing alright.'

I tell Tony about how things have been so far. As I do, we hear Moz and Jane approaching in the car.

'I reckon you'll be best off in the shearers' huts,' Tony says, pointing back to the buildings I had passed on the left. 'There's beds, lights, showers and toilets back there for you. It's nothing fancy though.'

'We don't need fancy.'

Tony leads the way in his truck, with Moz and Jane behind him in the car while I walk. When we get there, Tony points out the shower building, the toilet building, and the various sleeping quarters. 'Make yourselves at home.'

The main building has a big bed and a kitchen. I figure that will be where Moz and Jane will want to be, so I throw my pack into a small room in an adjacent shed.

Tony says, 'There is a big Carpet Snake about here somewhere. Just give him a wide berth and he won't bother you.' I would like to see a Carpet Snake, as long as it doesn't choose to share my bed.

'I can fill up the bore for you to soak in. It'll only take an hour,' says Tony.

I do not want him to go to trouble on our account. 'That's okay, Tony. We can just take showers.'

He leads me to the shower building, which is made of corrugated iron. A water tank sits on a stand above the roof. I notice the holes under the building and wonder what lives in there. There are two showers and a bath inside, and a lot of frogs. Wooden framework and more corrugated iron partition the interior into separate washing areas.

'You should have a bath. It'll feel good. Fill it right up to the top and soak in there.'

The bath is covered in possum excrement. I imagine lying back with the water to the top and the shit floating around in front of my face.

'Oh, I reckon a shower will be good enough for me, Tony. Maybe my brother will want a bath when he gets in.'

Baz gets to the shearers' quarters a short while after I get out of the shower. After introductions, Tony disappears briefly. He returns with cold beers for Baz and Moz and soft drinks for Jane and me.

'There's some steak and hamburgers and eggs for you here too,' Tony says, handing over the booty. With his bushy white beard and moustache, if we dressed him in a red suit Tony would pass for Father Christmas. Given Tony's generosity and John's ice-cream treat from earlier in the day, I cannot help but think that while they do not look like it, these men of the Outback really are angels.

Tony enjoys a good chat while he has a beer with us. I am now in thongs and shorts and he notices my blistered feet. 'Have you tried going without socks? Maybe that'd be better.'

The very thought gives me blisters. No socks? Out there on the Track? No, thanks!

'Did you get the postcards that I sent, Tony?' I thought he would enjoy getting postcards from across the world. I sent him one that featured a photograph of a wolf. I thought that might make struggles with Dingoes seem slightly less troublesome.

'Yes, thanks.'

'Do you run anything other than sheep on your land?' Baz asks.

'I have a couple of hundred head of cattle.'

'How many sheep do you have now, Tony?' I ask.

'Oh, about 800 or so.'

'I notice you've got some tiny twin lambs there. How old would they be?'

He tells me they were born yesterday.

'Do you lose any lambs to Dingoes?'

'No, not really. The real killers are the wild dogs. Doggers come out pig hunting and lose their dogs. They kill everything they see. The dogs get lost and get hungry, see? They can bring down anything they come across. I shoot any dogs I see.'

I am surprised when Tony leaves. I thought he would stay and chat with us all night but I guess he has things to do. He has only gone a short while when two others arrive.

'I'm James Clark.'

'I'm Josephine Birch.'

'We are with the *Warrego Watchman* newspaper in Cunnamulla.'

'G'day.'

James is a good-looking man with long dark hair and a set of sideburns that would have pleased Elvis. Josephine is a tall woman with long curly hair the colour of the Track when the going is good—when there is red sand and no rocks. They both wear jeans and dirty wide-brimmed hats.

'We have been following your tracks, trying to catch up with you,' James tells us.

'It's a small world,' says Josephine. 'I used to go to school with Tony Marsh's daughter.'

The reporters have a number of questions for the four of us.

When they ask about the likelihood of me not being able to finish, Baz tells them, 'He'll make it. He's a stubborn bastard.'

Josephine asks about our motivations for being out here. I have difficulty explaining why it is that this has always been something I wanted to do.

'Why are you here, Barrie?' Josephine asks.

'Loyalty,' Baz says. 'I'm here to support Greg.'

'And you Sean?'

'Mateship.'

These blokes are here for me. A bloke can be bloody lucky in life.

'These people are here to help me realise my dream,' I say. 'Jane too. I hardly know her but, in supporting Sean, she is out here to support me in something I have always dreamt of doing. It is impossible to find the words to express how much I appreciate that.'

James and Josephine stay for almost an hour-and-a-half. Josephine asks most of the questions and scribbles down notes, while James takes some photographs. We enjoy their visit but at about 6.45, they say that it is time for them to leave.

'Good luck,' Josephine and James say together. They know that we still have a long way to go, despite how well the return journey has gone so far.

'What was the highest temperature today, Moz?' I ask.

'It got to 38 degrees.' The cooler morning temperatures were squashed once the sun got up.

I paint my infected toes with Betadine before I lay down on my dusty bed. The ramshackle shearer's quarters are neither elaborate nor extravagant, but they are all that we need. Anywhere up off the dirt will do. It gives me a good place to get my feet elevated while I rest. I feel good about the return journey thus far. We have 70 km behind us already. While I am confident, I remind myself that I must not get overconfident. A lot can go wrong in a short time out here.

DAY ELEVEN

Life Gets Too Much for Henry Lawson

When you've been too much and too long alone in a lonely place, you begin to do queer things and think queer thoughts—provided you have any imagination at all.

- 'Water Them Geraniums', *Joe Wilson and His Mates*, 1901

'Brother,' he said, 'do you not think we should offer up a prayer?'

'What for?' asked Peter, standing in his shirt sleeves, a rope in his hands and mud from head to foot.

'For? Why, for rain, brother,' replied the parson, a bit surprised.

Peter held up his finger and said 'Listen!'

Now, with a big mob of travelling stock camped on the plain at night, there is always a lowing, soughing or moaning sound, a sound like that of the sea on the shore at a little distance; and, altogether, it might be called the sigh or yawn of a big mob in camp. But the long, low moaning of cattle dying of hunger and thirst on the hot barren plain in a drought is altogether different, and, at night, there is something awful about it—you couldn't describe it. This is what Peter M'Laughlan heard.

'Do you hear that?' he asked the other preacher.

The little parson said he did. Perhaps he only heard the weak lowing of cattle.

'Do you think that God will hear us when He does not hear that?' asked Peter.

- 'Shall We Gather at the River?', *Children of the Bush*, 1902

When I wake, I watch frogs hopping across in front of my door. A mouse crawls cautiously across too and I wonder where the Carpet Snake is hunting. I kept my feet elevated as much as I could through the night as they were puffy after my arrival yesterday, but they look a lot better this morning.

The day is slowly stirring when I emerge from the shower at 6.30 am. The sun is just peeking through the trees and cloud cover to the east. The purple clouds promise to block some of the heat from the day ahead.

I sprinkle my feet with antiseptic powder and then I redress my feet. I also remember to slop some Vaseline under my armpit where it has been rubbing and bothering me for a few days.

I have a hamburger patty and a couple of fried eggs for breakfast, compliments of Tony Marsh's generosity and Moz's cooking. Moz also tosses me an orange juice to take with me. I bid farewell and leave with my breakfast in hand, knowing the others will be ready to follow shortly.

The walk out to the main road is a nice one. The birds are up and about early. A beautiful Eastern Ringneck takes flight, the blues and greens of its feathers sparkling in the radiance of the morning light. Several pelicans pass overhead with graceful ease, their long wings extended in an effortless glide. Flocks of Galahs alight from the shrubs lining the Track, the soft pink of their crowns in contrast to the bold vibrancy of their breasts. There is so much elegance and beauty out here that I briefly find myself wishing this experience would never end.

A rabbit crosses the Track in front of me. I think of the Wedge-tailed Eagle I saw here yesterday and, borrowing a sentiment from one of my favourite books, Richard Adams' *Watership Down*, I say out loud, 'Run. Run for your life you Prince of a Thousand Enemies.'

I pause to take some photographs of Baz as he walks behind me. When he catches up, we walk together on the soft red sand of the driveway, our shadows stretching far to our left. We see many sheep and envy the exuberant movement of the lambs.

We pass through the gate out of Kia-Ora at 8.03 am and take photographs of one another beside the big drum that serves as Tony's mailbox.

'While I live, I grow,' says Baz, reading from a sign attached to the mailbox. 'Very profound, Tony.'

Behind us, the sky has developed into a dreamland blue sprinkled with strands of fairy floss clouds. Baz and I see a kangaroo with a joey as we pass the mailbox for Naree Station, whilst more pelicans pass overhead. Just before 8.30, we approach our first water crossing for the day. At this spot, the water had not been across the road on the outward leg. The water must be higher than when we last passed, which does not auger well for the day ahead. We see dozens of pelicans, spoonbills, ibises, egrets, herons, stilts, and a Brolga. The Brolga takes flight and we lose it amongst the long grass along the creek. With the aid of Baz's binoculars, we eventually relocate it and watch the easy, graceful beauty of the bird. In the Dreamtime stories, Brolga was a beautiful Aboriginal girl who loved to dance. A wicked spirit became so enamoured with her that he captured the dancing beauty. When Brolga's People tried to free here, the evil spirit decided that if he could not possess her, no one would. He turned the woman into a bird.

The first crossing is a short and easy one and we make it through just before we are joined by Moz and Jane in the car. Tony Marsh is in his one-tonner behind them.

'Good morning, Tony. Where are you off to?'

He tells me he is heading into Bourke to have a doctor look at his calf cradle injuries. He thinks he might have broken ribs.

'I wonder if we will get some rain.' The clouds are starting to look threatening.

'I don't think it'll rain today,' Tony says, looking at the clouds. 'There's nothing much in that.' I do not know how well Tony knows Henry's writing,[1] but he certainly knows the Bush. Tony scratches his head for a moment. 'Maybe we'll get a bit of rain tonight. There won't be anything much in it though.'

The sky looks ominously grey to me. I think we will get rain well before nightfall.

'I hope things go well at the doctor's,' I say.

'I'll follow the car through to Back Creek in case they get stuck,' he tells me. 'I'll be there to pull 'em out if they need it.'

The two vehicles roll away and Baz and I continue our journey.

We see a fox ahead and several feral pigs in amongst the trees and Tangled Lignum shrubs in the swampland off the sides of the Track.

At 9.45, we get to Back Creek where everything is lush and green. It is an oasis in the desert. In the shallows, waterlogged stalks of grasses and shrubs periscope above the surface, defiantly refusing to buckle beneath the inundation of water. Moz's car is on the other side and there is no sign of Tony so he has obviously gone on to Bourke. Because Back Creek is running fast, I find a suitable branch for use as a walking stick.

The water is gushing and, as I venture in, I am glad I don't have to drive the vehicle through. Almost the entire bitumen surface is washed away. Only a thin lip of bitumen remains on the left-hand side of the road and so Baz and I stick to this lip. The water cascades over the lip and bubbles away below us. Baz carries his shoes and socks in his left hand and we both carry our walking sticks in our right. I follow close behind Baz, walking where he walks in the belief that unless he falls, I am safest to stick with the path he chooses.

Despite the rushing water, we get through without trouble but only with a lot of caution.

'Do you blokes want a lemon cordial now?' Moz asks, emerging from where he has been snapping photographs through the bushes. Baz and I both accept the offer.

'Moz, it still has ice in it,' I say when he pours the drink into our cups. 'Bloody hell, you look after us, mate!' Moz accepts the thanks with a big, satisfied smile. The drink was in the freezer at Kia-Ora overnight and it goes down nicely.

The stretch after Back Creek is the same stretch where I suffered the blues on Day Five, missing Jennifer and composing my *Road to Hungerford* song. I have Baz for company this time and we also have Moz and Jane in the vehicle nearby. Moz takes a series of photos and then darts ahead to park the car and take more photos as we approach and pass. He climbs on the car, lies on the ground and climbs through fences to get the best possible shots. He then jumps back in the car and darts ahead again to park and take more photos. Baz and I egg him on with comparisons to Australia's renowned Outback photographer, Steve Parish.

'Eat your heart out Steve.'

'Look out, Steve. There's a new kid in town.'

Laughing with Baz helps the time to pass. I am in vastly different spirits to when I walked this stretch a few days ago.

'Gee, you stink,' I tell Baz. I cannot have him thinking I am having too much fun out here. Besides, he has it in his head that it is only me who stinks.

Despite the build up of clouds, I notice that Baz and I still seem to be in the sun though. The shade of cloud shadows beckon ahead, but as we move, so does the shadow. It is still hot enough that I would welcome shade but we never seem to be able to catch it. The shadow rolls ahead of us like a wave and Baz and I follow like novice surfers, never quite able to catch the crest of that wave. The Lignum shrubs on the sides of the Track offer no shade and so we must continue to chase the shadows down the road.

'Do you want a bit of hamburger patty?' Baz asks.

'That'd be great. Have you got one.'

He asks me to retrieve from his backpack what he calls his 'wet ones' container. I retrieve the container and, sure enough, he has a hamburger patty stuffed inside.

We come to Cuttaburra Creek number two before Cuttaburra Creek number one. There is a good 700 m or more of roadway under water. Baz and I cross together. With my walking stick in hand, I feel like Moses ordering the waters to part. They do not. There is deep water to both sides of the road as well as over the road. Moz waits in the car with Jane to see how deep the water is before they follow. The walking stick gives me something to lean upon for a bit of extra support where my footing is uncertain. The current seems considerably stronger towards the end of the crossing than was the case at the beginning. Unlike Baz, I do not wait for Moz to cross in the car. I want to be in the water at the next crossing before Jane arrives because I am planning to strip down before I tackle it.

Just after 11.00 am, I prepare to cross Cuttaburra Creek number one. It is the fourth and final crossing for the day and, like its predecessor, the water extends over the road for 700 m or so. I strip down to my underwear because this crossing was the deepest on the outward leg and looks to be higher than before. When I am about 50 m into the crossing, I see Baz behind me, crossing the bridge between the two Cuttaburra overflows.

The sky above is a beautiful, deep blue. However, there is a wall of solid cloud to a point of about 30 degrees above the horizon in every direction. Although it is not hot today, it is sticky and so I find the water refreshing. With all the water on both sides of the road, this has the appearance of a lake, rather than a mere creek. There is water as far as I can see through the numerous trees. Some of the trees are thick-trunked, solid specimens. Below the towers of the Black Boxes, Coolabahs, and River Red Gums, there are lots of Tangled Lignum shrubs. Dozens of birds rest in these numerous trees.

I use the walking stick to prod my way forward, carefully testing the ground for potholes before I move ahead with all of my weight. In the same hand as my walking stick, I carry my pants. In my left hand, I hold my voice recorder. I have my pack on my back with my camera and watch dangling down from straps. Mostly, the crossing feels firm underfoot, but there are spots where the bitumen has been washed away. In these spots, the water extends to my thigh and I do not have a firm base under my feet, making my path more treacherous.

The current runs from north to south, left to right, in the same direction as a small yellow *Coliadinae* butterfly that flutters past close by. There is a lot of whitish-yellow blossom in the Black Box trees and yellow blossoms in the River Red Gums. The butterfly heads to them.

Where the road is washed away and I have only one foot on bitumen, the other leg might be as much as a foot further under the water. Moz will not see the unevenness of the road when he is driving through and so it will be a tricky crossing for him. The water is a soft, dirty grey and there is no way that he will be able to see through it to the ground.

The wind spreads ripples across the water's surface, blowing with the current. Baz enters the water when I am about halfway through. Like me, he has a walking stick; however, unlike me, he does not have his pack. He must have decided it would be safest for it to be in Moz's car. Baz is using his walking stick more as a probe, feeling the ground and probing for unevenness and potholes, whereas I am using my walking stick as a crutch for support, almost like an extra leg for additional balance.

The plant growth that would have been beside the road is mostly drowned, with only a few branches sticking up here and there. The thick bush starts about 15 m from the road, with lots of trees and underbrush. The trees are in blossom. There is also all manner of greens in the trees, ranging from bright lime greens through to the sombre greyish-greens of the eucalypts in a Frederick McCubbin painting. The greens are accented by the colourful butterflies, dragonflies and damselflies. A blue Eastern Billabong damselfly darts near the surface. An orange and brown Australian Painted Lady butterfly floats off to my left, riding the breeze with several yellow butterflies. A Blue Skimmer dragonfly darts about, followed soon after by a red Wandering Percher dragonfly. Occasionally, I see clumps of 30 or 40 insects dashing about the surface of the water like molecules in a science experiment. Water striders ski across on their long legs and whirligig beetles zip this way and that, all in a frenzy. Two small red perchers drift within arm's reach. Another zips around to my left before it alights on the uppermost portion of a grass stem submerged except for centimetres above the water. A Blue Skimmer flies close by before disappearing into the shadows. With all of this life surrounding the water, it crosses my mind that a crocodile might have floated from the north with the floodwaters.

When I have been in the water for ten minutes, a four-wheel drive approaches. The men in the vehicle are workers from the Shire of Bourke. They both wear yellow fluorescent vests.

'What are you doing out here?' It is not the friendliest tone I have heard in the past 11 days.

'Walking from Hungerford back to Bourke.'

'Oh, so you're the guys doing the walk, are you?'

'That's us.'

'How has the water been?'

'It is a frustration, but it is fine enough to walk through. It just slows things down and adds another level of difficulty though, which is hardly something that is needed.'

They nod their heads, as if in understanding. 'What's the bloke in the car going to do?'

'You'll have to ask him.' Officially, the road is closed.

'He won't get through, will he?' It seems more of a command than a question.

'I don't know anything about that but you'd be surprised what you can do when you set your mind to it.' I know of what I speak.

'Is it a four-wheel drive?'

'Yeah, but you'll have to talk to him to find out more about his car.' Cars do not interest me and neither does the prospect of a couple of shire officials telling me that the walk stops here. I do not bother to ask them what they are doing. I assume they are checking the depth and difficulty of the crossings to see if the road should be opened or closed. 'I better keep going fellas, or I won't be able to get going again.'

When I emerge from the water, I wait for Baz.

'I need a shit,' he tells me.

'Congratulations.'

'But I don't have my pack.'

'What does that matter? Are you planning to drop it in there?'

'No, but my toilet paper and shovel are in there.'

I laugh. 'Sounds like you have a problem, Baz.'

'I want to get it done before Jane comes across.'

'I can't help you, mate, but there are plenty of leaves around.' I don't have my toilet supplies with me either because I have no plans for wading through swamps to find a bush dunny.

I wait with Baz for Moz to make the crossing. Given this is the last water crossing for the day; I plan to dry my feet and put on a clean pair of socks and my dry Merrell hiking shoes. As with all of the crossings, Baz has removed his shoes and socks and crossed in thongs.

We see Moz is still talking with the shire workers. 'Let's get this show on the road,' I say to Baz.

'I need a shit.'

'Yeah, I heard that somewhere before.'

We see that the workers' vehicle departs. Jane remains in Moz's car, but Moz starts to walk carefully and cautiously towards us, a walking stick in each hand as he wades through the water.

'What the hell is he doing now?'

We watch his slow approach.

'Come on Big Fella.'

'He can drive through,' says Baz.

'It doesn't seem any worse than the other day when he got through.'

'No, it doesn't.'

The water level marker beside the road shows the depth to be the same as when we crossed on the outward leg. The washout in the middle is the most dangerous spot, but we think that was already washed out when we crossed on Day Five. Perhaps the Bourke workers have planted a seed of doubt in Moz's mind.

'He's gonna take forever to get over here Baz.' Even without the time I spent talking with the Bourke Shire employees, it took me almost 15 minutes to walk across. 'Maybe those other blokes told him he can't drive through. He'll have to go back all the way around. Either way, it's going to take forever for him to get to us.'

'Come on Moz,' Baz yells across the water. 'The car can make it.'

I am staying out of that. All along, my position has been that it is Moz's car—Moz's prized possession—and that whether he wants to make water crossings or not is entirely up to him. I told him that when we reached this crossing and stopped on the day before we started the walk. Although it meant we could not continue to put out our water supplies, I told him the decision was entirely his.

Moz is making slow progress. 'I can't wait here forever. I'm going to keep going, Baz. Do you think he'll mind?' Moz might *like* my opinion on whether or not he can make it but what I think does not matter so there seems no point in waiting. If Moz has to go back, it will be a long way around. He will have to go all the way back to Hungerford to catch the road south and around and then pop up somewhere in front of us. If he is to go back, for all I know, he will probably have to go 200 or 300 km to get to where we are now.

'No, you just go, Gregga.' Baz knows of my determination to keep moving so as to avoid seizing up. He cannot continue with me because, even if he did have water and his pack, he is wearing thongs. 'I don't mind perishing out here alone,' Baz adds.

'Good.' I laugh.

'I need a shit.'

I laugh again. 'I'd love to stay around and see how that turns out for you, but I'm gonna go. See ya, Baz.'

'See ya, mate. I need a...'

I don't wait for him to finish his sentence. I think I know where he is going with it.

The road makes a long bend and I walk the remaining hundred metres or so of bitumen before the Track becomes rocky and difficult. My old wet Nikes are showing wear and tear and have no cushioning left. When we get to Bourke, the shoes will find their way into a rubbish bin. My feet have been holding up well, but this stretch knocks them around badly. Some of the rocks are as big as my fist and many of them are sharp edged. They slice at my shoes and prick at my blisters. I push across to the side of the Track in search of a smoother route.

Now that I have left the creek behind, there are no gum trees on either side of the Track. A few small, scribbly saplings stand here and there but nothing offers much in the way of shade.

I look back but see no sign of Baz emerging from the bend around which I passed after I left him. Most often on the way to Hungerford, Baz was out in front. Because I am not stopping to rest, most often on the return leg, I have been out in front of him. It has been good to walk with him through most of this morning.

≈

At about 12.45, the Bourke Shire blokes drive past me, heading back towards Bourke. They wave but do not stop. Just a few minutes later, Moz and Jane catch me. I am relieved to see Moz successfully completed the crossing, although Baz tells me later that when Moz emerged from the car he was shaking like a leaf.

'Those Bourke Shire blokes told me I shouldn't cross,' Moz says. It was just as I suspected.

'What were you supposed to do?'

'Go back and go all the way around, I guess.'

I think of the fast water and serious wash out at Back Creek. 'Returning over Back Creek would have been just as bad though wouldn't it?'

'That's what I thought. If I went back, I'd have to go over that *and* go all the way around. I decided going forward couldn't be worse than going back.'

'I am glad you got through.'

'Me too.'

'I'll bet.'

Moz gives me a couple of cold hamburger patties for lunch. After he turns back towards Baz, I gobble down the patties and then retrieve some dental floss from my pocket. Were someone to come along now, I would make quite the sight, walking along in my grubby clothes in the middle of nowhere, blithely flossing my teeth. People might think I have rats.[2] Being out here, maybe I do.

Just before 1.00, I pass through the eerie ghost town of Yantabulla... again. All I can hear is the wind and the singing of the cicadas. A sign informs me that 'Yantabulla was once a prosperous town of nine houses, a Hotel, Store, School, Police Station and Cordial Factory.' The sign also says that Cobb and Co. once had a 'changing station' here. Those days seem long distant in the emptiness of the town now.

Away to the east, a windmill leans like the tower at Pisa. The debris of human occupation is a lasting scar on the land—iron, tin, steel, wire. It is all here and will be for a thousand years, along with the discarded tyres and plastic bottles, to say nothing of the shells of the forgotten automobiles.

The singing of the cicadas penetrates my mind like a torturer drilling into my skull. When the FBI is engaged in a siege, they pump noise into the site and try to drive their enemies crazy. David Koresh and his Davidian mates at Waco might have thrown their hands in the air if the FBI had tossed in a few hundred cicadas. After living underground for several years, the cicadas have come to the surface and realise they have only a narrow window of opportunity to breed. Little wonder that the males sing their hearts out.

With the piercing sound of the cicadas burrowing into my brain, I push through town. Nothing moves in Yantabulla except for long-forgotten washing fluttering on a clothes line. It is like a scene from a post-apocalyptic movie. Only the cicadas, deep underground at the time of atomic detonation, have survived.

I see the sign to the cemetery. John Stephenson told us at Warroo that his grandmother is buried there. I wonder if Baz is still planning to explore the cemetery when he gets here.

When I emerge from Yantabulla, I round a bend and pass a sign indicating that Willara Crossing is 61 km to my right. Had Moz been forced to go back and around, he would probably have emerged at this point. I come to yellow signs in the middle of the road notifying travellers of the road conditions. Road closed. Detour to Willara Crossing. Nobody is supposed to be going where we have been.

I am now on a sandy stretch, where the sand is soft yet compacted—perfect for walking. The Track is clear of rocks and feels soft and cushy underfoot, the red carpet extending for kilometres before me. It is warm, as it always is, but the temperature is bearable. More than anything, I feel sticky. The cicadas are serenading me and, at least for a time, I can forget about the difficulties of this undertaking. I have a long, straight and open expanse in front of me. It is only because there is a crest in the far distance that I cannot quite see forever.

≈

Henry knew that the good things in his life had been drowned in alcohol. Bertha did not want her children to see their father in his drunken state and insisted that he not visit the children.[3] Henry realised that the cost of his drinking was much more than the loss of whatever money he managed to earn from writing. He had lost those things that were most dear to him.

I have known too well, God help me! to what depths a man can sink,
Sacrificing wife and children, fame and honour, all for drink.

Four steel fence droppers, some twisted lengths of wire and a couple of yellow hazard signs are not about to stop us early on the return journey towards Bourke.

I walk through the gate to where we have been invited to stay at Warroo Station.

Having endured a tough afternoon on the Track, Baz collapses on a bed after arriving at Warroo.

Keeping my weary feet elevated while resting in the shearers' quarters at Kia-Ora on Day Ten.

Green Tree Frogs are my favourite frogs. During the trek, however; occasionally they pushed the friendship hard.

Baz and I pick our way across a flooded creek where the road has been washed away.

The rocks make for difficult walking on blistered, battered feet.

The Track seems to stretch forever as Baz and I trudge on until, as Henry Lawson said, 'speech has almost died.'

As I cross Cuttaburra Creek on Day Eleven, I feel like Moses ordering the waters to part. Alas, they do not.

Baz and I push on through the floodwaters—just one of the 17 water crossings we had to complete on our journey.

One of the buildings in the deserted Yantabulla ghost town. The piercing sound of the cicadas just adds to the eeriness of the place.

Amazingly, the phone still works and so Baz calls his family from the deserted ghost town of Yantabulla.

As I lean against a distance marker on Day Twelve, I know that I have less than 100 km to go to finish my undertaking.

Writing in my notebook while I wait for Baz to catch up to me outside of Fords Bridge on Day Thirteen.

William Johnson 1915 portrait of Henry Lawson. Image used with permission of the State Library of New South Wales. (Image no. A4219057)

Baz does what the sign tells him to do as we enter Fords Bridge on Day Thirteen.

Baz walks beside me in silence, his presence providing for me the support and encouragement that, at this time, his words would not.

Despite the cloud cover, no rain falls to settle the dust on the dry Track back towards Bourke on Day Fourteen.

When I get back to Sutherlands Lake on the return journey, I discover that the water has turned green.

My favourite time of the whole journey. Baz reads Henry Lawson stories to me as we sit around the campfire on the last night before our return to Bourke.

This Eastern Bearded Dragon is just one of the unique creatures of the harsh Outback country where everything must be tough to survive.

Bourke seems not to be drawing any nearer as Baz and I find the final day of our tramp an endless slog.

The flies are a constant companion as I haul my stinking body along the torturous Track.

Even at the Back O' Bourke Exhibition Centre, I refuse to sit down for fear that I will not be able to get up again to finish my quest.

Smiles all round as Baz and I celebrate our achievement alongside a portrait of Henry Lawson at the Port O' Bourke Hotel.

≈

A breeze lifts dust and pushes it left to right in front of me, sprinkling the trees to the west of the Track. As dust blows in my eyes, it makes me think of the possibility of a vehicle driving by. Every time that happens, it raises a cloud of dust. On what side of the road will a car be travelling? I think hard—as hard as my weariness will allow. I am buggered if I can remember. I know that cars travel on a different side of the road here to the ones in Canada. I try to recall which side the steering wheel is on in my car in Canada. I know the information is in my mind somewhere, but the storage files of my brain are covered in red dust. With my muddled, confused brain, this is hard work. I think. I think. I think some more. If a car comes over the crest of the next hill I might be run down if I am on the wrong side of the Track. I know this is an easy one to figure out. I think hard. I can do this. I think some more. Let me see—when the car is in the shed at my house in Canada, where is the steering wheel? All of this thinking is making me weary. I give up. If a car runs me down, I guess I deserve it.

The nasal, yet flutelike, repetitive call of Grey Butcherbirds breaks the monotony of the endlessly droning cicadas. The butcherbirds again follow my path from the safety of the Mulga Trees until they encounter a scalding Yellow-throated Miner. The miner aggressively chases the bigger birds away. A kangaroo watches me cautiously from the scrub.

After a long stretch of Mulga Trees, I reach a tall and straight Desert Bloodwood. As with most of its species, the branches do not extend away from the trunk very far and so it does not offer much in the way of shade. It is a beautiful tree nonetheless and it catches my eye in such a way that I break my rhythm to take a photograph. The reddish scales on the trunk add colour to the otherwise subdued surrounds of the Mulga.

I go further and my eye is drawn to a Black Box gum. The tree is a good 20 m in height and so it reaches well beyond any of the Lignum shrubs and Mulga Trees around it. I decide to snap a photo of that too. Almost immediately after I depress the shutter, six Major Mitchell Cockatoos explode with a short series of screeches from their hiding place amongst the linear leaves. The sudden noise and movement startles me. The orange, yellow and pink crests wave at me as the birds bob their heads in raucous excitement.

At 2.35 pm, I reach the H90 marker—Hungerford 90 km. I could sure use some rain to cool me down. The billowy clouds threaten, but nothing comes to cool my head except the sweat trickling into my eyes. I lean against the sign and take a self-portrait photograph. With my whiskered sunscreen-blotched face and grubby shirt, I look ragged and haggard. Beneath my bandana and the brim of my Tilley hat, my eyes stare at the camera unblinkingly—hollow and haunted by my self-inflicted suffering.

≈

Henry has been criticised as a pessimist who 'sat down...and let himself be miserable,' looking only for 'the worst side of everything.' Certainly, he was embittered by the failure of his marriage. He and a mate were walking home one night when they passed a house where a marriage celebration was occurring. Henry's mate happened to know the happy couple quite well and so he said that he and Henry could join the party.

'"No," [Henry] answered firmly, and chuckled. "It's too late to save him, anyway."'[4]

≈

As I continue along the Track despite all the pain that wracks my body, I am getting back my full measure of self-esteem. Not everyone could do what I am doing. Indeed, very few could. To keep going in the face of all this pain demands mental toughness. In the face of a similarly daunting endurance event, I actually did lose all hope for a time. It pains me to think back on that time in my life. As I struggled with the effort of earning my Doctorate in literacy education, I could see no way out. I floundered in the absence of support, leadership, or assistance from the professors who served on my dissertation committee. I was continually stripped of any vestiges of self-esteem that I might once have possessed. As I confided to my wife, tears running down my cheeks, my relationship with my committee was all about power but

I had no power at all. I was at a complete loss. With so much invested into achieving my Doctorate, I even lost my will to live. Eventually, I got to sitting in my office with papers piled on the desk beside me, spending as much time thinking about how I might take my own life as thinking about how I might successfully complete my dissertation.

With my thoughts having turned to suicide, I rightly decided that my dissertation was not worth the trouble that it was causing me. After much soul searching, I finally informed my wife that I could not continue with my dissertation.

'I can't do it,' I said.

'But you have to.'

It was a dangerous response. I had turned to Jennifer for a way out from the black hole in which I was enveloped. Unaware of the seriousness of the situation, Jennifer did not grant me the free passage I sought. I had struggled for months to find a way to tell her that I could not finish my dissertation. Here, when I finally managed to squeak out the words, her response was that I *had* to finish. She offered to provide whatever help she could, but I knew that no assistance she could provide would be enough.

We both cried.

'Don't worry about it,' I said. 'I'll take care of it.'

I can only guess how Jennifer interpreted those final words before I left the house to return to my office. What I meant was that I was going to kill myself.

I sat in my office and cried until I thought that I could cry no more.

I did not want to do this to my wife and daughters—to leave them without a husband and father—but I could see no way out.

My leaders at work had been watching me, trying always to provide the right balance of support and encouragement. As I sat in my office crying, I remembered the advice of the Acting Dean, Jon Young. Jon encouraged me to speak to Doctor Wayne Serebrin. Wayne is the University of Manitoba professor most closely aligned to my own areas of study. Jon encouraged me to ask for Wayne's help. However, he was *always* buried beneath work and so I did not want to add to his burdens with my own.

'What have you got to lose?' Jon asked. I had already told him I could not finish my dissertation. 'Even if Wayne cannot help, you've lost nothing in asking him.'

I sat and cried and thought some more and cried some more.

I thought of my beautiful wife and my wonderful, happy daughters.

Slowly, uncertainly, I rose from my chair, opened my office door and walked three offices down the hallway. As usual, Wayne was busy.

'I wonder if I can talk with you when you get a minute,' I said.

I returned to my office but Wayne soon arrived.

'What can I do for you?' he asked.

I closed the door behind him and ushered him to a seat.

I looked across the crowded confines of my office and into his caring eyes.

I knew that he had problems of his own—like everyone else in the building; he was trying to do too much extra work on top of all of the work that was already too much for him to do. I did not want to add to his burden, but he was my last chance. If he was too busy, I was dead and my daughters were fatherless. I looked at Wayne but, knowing the enormity of what I was going to ask, and knowing the enormity of the potential consequences of his response, I could not form words. Tears roll down my cheeks. My face crumpled. I buried my face in my hands and sobbed as though I were a little child. I have never sobbed like that before and I do not think I will ever sob like that again. There was no stopping the tears. I rose unsteadily to my feet, opened my office door and, unable to speak, I pointed for Wayne to leave.

≈

Henry believed he did not get the support from his country that he needed, and that he felt he deserved. 'My advice to any young Australian writer whose talents have been recognised,' wrote Henry, 'would be to go steerage, stow away, swim, and seek London, Yankeeland, or Timbuctoo—rather than stay in Australia till his genius turned to gall, or beer.' Henry knew of what he spoke. His own genius had turned to beer. Henry's thoughts were not just of beer though. Henry continued, 'Or, failing this...to study elementary anatomy, especially as it applies to the cranium, and then shoot himself carefully with the aid of a looking-glass.' After penning a farewell note to Henry, a friend had recently committed suicide. Now, Henry was thinking of killing himself.[5]

≈

Following his time in my office, I wondered what Wayne was thinking. Obviously, I placed him in an extremely uncomfortable position and for that I am sorry. I shuffled down the hall to his office and apologised for my breakdown. I asked if we might talk. He came to my office and this time I managed to ask him for the help that I needed. Wayne graciously agreed to look over my dissertation. Having done so, he then provided me with detailed, supportive feedback. He helped me to consider things in different, more sophisticated, ways.

After Wayne's feedback, the next draft of my dissertation was considerably improved. Seven years after beginning my Doctoral programme, but just a little over a year after receiving Wayne's feedback, I passed the oral examination and my dissertation was completed.

I remember the funeral of my colleague, Jim Welsh. One of his former graduate students said that, upon the completion of her Master's studies, she had approached Jim about the possibility of doing her Doctorate.

'I think that you can do it,' Jim said, 'but you need to know that a Doctorate comes at the cost of a part of your soul.' I understand perfectly what Jim was saying. My Doctorate almost cost me everything—body and soul. It was during that dark period that I attended church for the last time. Although I had been a Sunday regular for years, I had too much to worry about in this life without the added burden of also worrying about an afterlife. It has been over four years since I have been inside a church and I have no plans to return.

I have always prided myself on being mentally tough. Getting to the stage of surrender with my dissertation stripped from me my self-esteem. Eventually finishing my Doctorate restored some of what was lost. Toughing this out—this Bourke to Hungerford hell—is restoring what is still missing. As my watch ticks through the afternoon of Day Eleven, I know that I am a hard son of a bitch and make no mistake, Custer may have been a pussy, but I am not.

≈

Separated from his wife and his children, frequently incarcerated for failure to pay for the maintenance of his family, and struggling with the debilitating disease of alcoholism, Henry's life slipped into darkness. In the gloomiest period of his life, Henry was suicidal. On 6 December 1902, Henry threw himself from the cliffs at Manly. *The Sydney Morning Herald* reported the incident in the following manner:

> *ACCIDENT TO MR. HENRY LAWSON*
> *Shortly after 10 o'clock on Saturday morning, a fisherman named Sly, while walking along the cliffs at Manly, noticed a man lying near the water's edge. Sly climbed down a path which is used by fishermen and found that the man was Mr Henry Lawson, poet and story-writer. He was quickly carried to the top of the cliffs, and Dr Hall, who was summoned, found that Mr Lawson was suffering from a broken ankle, a lacerated wound over the right eye, besides other injuries. It was ascertained that Mr Lawson had fallen over the cliffs, which at that place are about 80 or 90ft high. He was conveyed to Sydney.*

While Henry was in hospital recovering, George Robertson sent some reading materials to help pass the time. 'I wasn't a success as a flying machine, was I?' Henry asked ruefully.

Although the newspapers reported Henry's injuries as the result of an accident, in a letter 14 years later, Henry left no doubt that it had been a deliberate attempt to end his life. He wrote that his struggles 'ended in that mad (?) attempt at suicide near North Head, Manly. Since then...it has been one bitter black struggle all through.' Although it was not published during his lifetime, Henry also wrote a poem about the incident. Despite being entitled, 'Lawson's Fall,' the poem indicates the 'fall' was a jump.[6]

After surviving the cliff-top drop, Henry found himself in hospital with 'forty-nine wrecks beside myself.' Of his broken ankle, Henry wrote, 'I can honestly say that I scarcely felt it at all after the first night.' He was in far greater pain mentally than physically. Some newspapers reported that Henry was dead. Those newspapers printed flattering obituaries. When Banjo Paterson asked Henry what he thought of the tributes, Henry replied that, 'after reading them, he was puzzled to think how he had managed to be so hard up all his life!'[7]

≈

I walk past a lake but, unlike on the outward leg, I do not cross to the water's edge to fill my hat and wet my bandana. It is only 20 or 30 metres off the Track, but I do not feel like walking even that far out of my way.

With the road closed and Baz and the support vehicle somewhere behind me, it is lonely and quiet and I have not seen another soul for three hours until Moz and Jane eventually catch up to me. 'How are yer poppin' up?' Moz asks for what seems the hundredth time since we left Bourke.

They tell me I am 10 or 11 km in front of Baz. I would rather be where I am than where he is. Moz says that when Baz got to Yantabulla, he decided to explore the town and the cemetery.

'Did he find John Stephenson's grandmother in the cemetery?'

'Yeah, Baz found that. Baz called home to his family from Yantabulla too.'

'I wondered about that.'

'Baz also used a toilet in Yantabulla.'

I laugh. 'That sounds about right.'

'We had a good rest in Yantabulla.'

'I'll rest when we get to our campsite.'

'Yep.'

'How hot is it Moz?'

'Um.' He looks at the temperature gauge inside his car. 'It is 35 degrees.'

Bloody hot. 'I figured it'd be somewhere about that.'

'We have lots of water, Grog,' he says. 'Do you want me to tip some over you?'

'That'd be good, mate.' It beats walking off the Track to get to a lake.

Moz parks the vehicle and walks alongside me holding a flagon of water above my head.

After he gives me a shower, Moz says, 'We should make camp soon.'

It would be nice to stop but, despite my weariness, I am travelling well. It is only mid-afternoon and I had not planned to stop yet. All the same, if Baz really is ten kilometres behind me, he will not get to where I am until 5.30 pm.

I tell Moz that Baz had earlier said that he wants to get to the Bourke 120 km sign and then go three or four kilometres beyond that.

'Jane and I will drive ahead to find that sign and see how far away it is then.'

'No worries.'

Moz returns to the car and he and Jane drive away.

They return a while later and tell me that they could not find the B120 sign.

'We even drove ahead to the next sign and then doubled back and measured to where the sign should be but it is not there. We'll go back and see how far back Baz is and find out what he thinks about where we should stop. Then we'll come back and let you know what our plan is.'

'No worries, Moz. When you come back though, can you make sure camp is at least a kilometre or two out in front of me? That way, I can wind down as I get towards the end.' It is not possible for me to explain my thinking to him but, psychologically, I want to feel the triumph of seeing the car ahead of me and knowing that I am getting to the end of another long day.

I reflect upon another successful day without stopping to rest. That makes three days in a row now. The first three days of this return leg have been taxing but not nearly as traumatic as the first three days of the outward leg from Bourke. I feel good and I feel strong, albeit weary. Nonetheless, it will be nice to get to camp tonight and to lie down and put my feet up.

I have only a faint shadow in front of me, slightly to my left. Although behind a solid cloud at the moment, the sun is over my right shoulder. However, up above there is lots of clear blue sky. Back towards Hungerford there is a big cloudbank. The same is true ahead of me towards Bourke. At this stage, however, there is nothing that looks likely to produce rain. Tony was right this morning. There will be no rain today.

The cicadas continue to sing.

≈

Henry's alcoholism and depression was fuelled in part by what he felt was a lack of support from Bertha. He claimed to have been brought to his low ebb 'because of a woman's work.' Banjo felt Henry's claim was justifiable. 'I remember Lawson's wife telling me that she was quite happy because Henry was "working" again,' Banjo recalled.

'What's he working at,' Banjo asked. 'Prose or verse?'

'Oh, no,' Bertha supposedly replied. 'I don't mean writing; I mean working. He's gone back to his trade as a house painter.'[8]

≈

I pass beneath the powerline that runs through the Bush back into Lake Eliza to my right. Despite my earlier convictions, I have lately started to think that maybe I will pass this way again one day. If so, I will have to venture into Lake Eliza to camp. When Henry passed this way, he was told that 'You can't do better than to camp / To-night at Lake Eliza.' I expect the lake is full of water at present but, then again, that is what Henry thought too. When he heard about Lake Eliza, he 'thought of green and shady banks, / [he] thought of pleasant waters.' As it turned out, when Henry camped at the lake, it was bone dry. 'I hope that I shall never be / As dry as Lake Eliza,' he wrote.

Moz and Jane return.

'Baz is about six kilometres behind you.'

Six kilometres? What happened to ten or eleven? Is the little bugger running down the Track?

'Okay,' I say.

'Are you okay, Grog?'

'Yep, I'm good.'

'Baz said we should look for a spot to stop fairly soon. We'll go a couple of kilometres up the road and find a good spot.'

'No worries.'

≈

I do it long and hard over the last half an hour or so of the day's journey. My left shoulder is aching and my neck and back are killing me: 'I'm stung between my shoulder-blades—my blessed back seems broke; / I'm too knocked out to eat a bite—I'm too knocked up to smoke.' My legs feel as heavy as lead. Although my tactic of keeping moving is getting me down the Track, it is a hard slog to go 30 km without a rest, particularly after doing 40 the day before. Fortunately, the pain of the blisters has largely subsided—either that or I have just gotten used to it. Henry figured when you are on the Track, you just get used to it until your spirit is broken and you lose all hope of anything better than a life tramping.

Now there's nothing for it but to tramp, tramp, tramp for your tucker, and keep tramping till you get old and careless and dirty, and older, and more careless and dirtier, and you get used to the dust and sand, and heat, and flies, and mosquitoes, just as a bullock does, and lose ambition and hope, and get contented with this animal life, like a dog, and till your swag seems part of yourself, and you'd be lost and uneasy and light shouldered without it, and you don't care a damn if you'll ever get work again, or live like a Christian; and you go on like this till the spirit of a bullock takes the place of the heart of a man.

I am hot and tired by the time I reach camp. This is the third day in a row that I have got in before Baz, but on our previous two evenings at Warroo and Kia-Ora there was nothing for me to do to prepare things for him. Now I have the opportunity to do a few things to help Moz get Baz's tent and sleeping gear set up for his arrival.

I potter around and try to help but I feel so buggered that I know I do not contribute a great deal. My back aches when I lift Baz's pack from the backseat of the car and pain shoots up my back when I bend down to hammer a few tent pegs.

'What's that noise that keeps clanging?' I say to distract myself from the pain.

'There's a windmill over there,' Moz says, pointing to a spot a couple of hundred metres away and across the other side of the road. 'It is an old public watering hole. I thought it might add to the authenticity for you.' Moz laughs. He knows I am a slave to authenticity. I can see the windmill above the small surrounding trees. The clanging and clanking of the tail vane and blades suggests it has stood there for a long time.

I fill a couple of bottles of water and retire to my tent. I feel bloody hot. I strip down in the hope of cooling off. My body is red with heat rash and my back feels worse than at any time over the past 12 days. I lay flat and hope for the pain to ease away from my spine.

I'm lyin' on the barren ground that's baked and cracked with drought,
And dunno if my legs or back or heart is most wore out.

I hear a vehicle roll to a halt outside and hear Tony Marsh's voice. He is back from Bourke. Moz walks over and starts talking with Tony. I know that I should go out for a chat. I know that I should ask how he got on with the doctor. I also know that it will take me forever to get some clothes back on inside this tiny coffin tent. I am completely naked and by the time I manage to get dressed and to crawl out of the tent, Tony will probably be gone.

After a while I hear Tony's truck pull away.

'That was Tony Marsh,' Moz says from outside my tent.

'Yeah, I heard.'

Baz gets in an hour-and-twenty minutes after me and half an hour after Tony leaves. He is tired and footsore.

'How are you, Gregga?'

'Rooted.'

'He didn't even get up to say g'day to Tony.' Moz says it good humouredly, but I recognise he is having a shot all the same. I do not mind though. I should have got up, but Moz does not know that I am lying here in all my heat-rashed naked glory.

Baz is of a mind that we have 49 km to go to get to Fords Bridge. 'It'd be nice to do more than 30 kilometres tomorrow and then have less than 20 to do to get into Fords Bridge the day after tomorrow.'

'Yep,' I agree with a succinctness borne of weariness.

'Maybe we should aim to do 32 or 33 tomorrow.'

'Yep.' I am not arguing. I am resting.

The windmill keeps clunking and clanging. It is unceasing—a noisier, less bearable version of the daytime cicadas.

'What do you want for tea, Greg?' Moz gives me the tinned options.

'I'll take the Stockman's Lamb Stew, thanks.' It seems like a good option, although I would prefer to sleep than eat.

When the stew is heated, Moz brings it to my tent.

'Bloody hell, you look after me, mate.'

'Yeah, I know.'

I eat the stew and, although it is good, it only makes me hotter. Looking out through the tent mesh, the sun still seems high in the sky. It is taking forever to set this evening. I am broiling as I lie in the tent, rubbing my aching left shoulder. Sensing my discomfit and without saying a word, Moz thoughtfully moves the car right beside my tent to block the sun.

'Thanks a lot, mate. Gee, that's bloody thoughtful.' I am sincere. He is very considerate and this makes him an invaluable person to have as support.

After they eat, Moz and Jane go for a walk to investigate the windmill. Baz is laying on his swag outside his tent, his feet elevated on an esky. He is in shorts and a Jackie Howe singlet, reading from Henry Lawson. If Henry is looking down at us, he would love to see that.

'Listen to this, Greg. I am reading "Stragglers". "To live you must walk. To cease walking is to die."'

'Ain't that the truth.'

I drift in and out of sleep.

After a time, Moz and Jane return with the news that they saw a goanna at the windmill. I look up at the sky through my tent and see a huge clown face in the clouds. He is laughing at my nakedness.

At 8.49 pm, I retrieve my camera and take another self-portrait. My face is red and hosts a grizzled beard. My green eyes still look haunted. Somewhere inside those eyes seems to be the broken shell of a broken man.

I fall asleep to the clanging and bashing of the windmill.

DAY TWELVE

Henry Lawson's Drinking

I might come along there, dusty and tired, and ragged and hard up and old, some day, and be very glad of a night's rest at the Lost Souls' Hotel.

- The Lost Souls' Hotel, *Children of the Bush*, 1902

We saw no more, but we knew that there were several apologies for men hanging about the rickety bar inside.

- In a Wet Season, *While The Billy Boils*, 1896

Tony was right about there being no rain through the day and about it falling tonight. Although I hear the others moving about to put up their tent flies, I cannot be buggered doing so. After all, Tony said there would not be much in it. Light rain trickles through the mesh of the tent from about 10.00 until 11.30 pm. Even though I am weary, it is difficult to fall back to sleep with rain splashing on my face. Although it is only light, it falls for a while and so water forms in the bottom of my tent. I pull a t-shirt from my pack and wrestle in the confines of the tent to get the damp t-shirt over my head and down over my shoulders. Eventually, I give up the fight and just drape it over my upper body like a small blanket. I find my hat in the darkness and rest it on the side of my head to stop water splashing on my face and fall back to sleep until I am awoken again by more rain. Once again, the rain is only light and through the top of my tent I can see the stars twinkling between the clouds. I hear the crashing of the nearby windmill and check my watch. It is 1.15 am.

I wake again at 5.15 and think with regret that I have to start getting ready soon. It is still dark outside though and I figure I might steal a few more minutes of sleep before I begin to move. In fact, I fall into a deep sleep. I wake with a jolt only when Jane starts the car right beside me. When I gather my senses and regather some composure, I see that it is 6.30 am. I feel exhausted before the day has started. As with Henry, 'my body craved...for another hour's sleep.'

I look through the open mesh of my flyless tent and ask, 'Where's Baz?'

'Oh, he left a few moments ago,' Jane says.

Moz adds, 'We thought you must have needed the sleep so we figured we wouldn't wake you.'

I retrieve damp socks from my pack and discover that my shirt and pants are also wet.

After staying inside my tent since shortly after getting into camp yesterday afternoon, more than 14 hours earlier, I finally emerge from it. I hobble across to the esky in my thongs, take a seat and pull on my shoes.

This is the first time someone has left while the other person slept. Usually we hear one another getting ready in the morning. Baz will be far down the road before I get going.

I rush my preparations and take to the Track at 7.15 am. I would love a pee but with Jane at the camp, I figure I should hit the Track and do the job once I get around the bend. Moz and Jane finish packing up my tent as soon as I leave, having already packed up everything else before I awoke. I am not feeling very good. My long and hard slogs over the previous three days are taking their toll on my body and I feel weary even as I set out. I hope that my feet and legs will loosen up quickly because I am feeling worse than at any time on this return journey. It is going to be a tough day.

As I think of the cold drinks that await me at Fords Bridge tomorrow, it reminds me that I need to 'take a leak'. I turn and see that Jane is on the Track behind me. Bugger it.

Given Baz's 45 minute head start, I cannot see him. There are crests in the road ahead and I expect he is in the distance somewhere beyond one of those crests.

I pass the windmill, still clanging and bashing as it has done all through the night. A sign indicates it is the Boongunyarrah Bore at Mother Nosey Spring. I wonder who Mother Nosey was. Henry occasionally considered his mother-in-law a nosey, troublesome woman. Unable to endure Henry's drunkenness, Bertha took the children and rushed to her mother's side. Henry turned his anger to his mother-in-law, writing of Bertha, 'She is now at her mother's....Her mother is also insane. I want to get her away from there.'

I pass the H100 sign. It is a 100 km back to Hungerford; more if you go via Tony Marsh's place. One hundred kilometres—when I was in Hungerford, there were people who no doubt expected I would not get more than a couple of kilometres down the Track. Maybe I too was one of those people.

The sun is only about ten degrees above the horizon, but I am already warm. There is a solid band of cloud in front of me stretching from high in the sky towards the horizon. There are also patches of light cloud behind me. Unfortunately, there is no cloud to my left, where the sun is rising in the east.

With the morning sun shining brightly, the colours of the world around me are lovely. The stalks of the waist-high Native Millet either side of the Track are a soft, light green. Above, the golden tips of the seeds seem to sparkle as they blow in the breeze. 'Things always look brighter in the morning,' Henry said. 'More so in the Australian Bush, I should think, than in most other places.' The morning light is sublime and so I stop to take a couple of photographs of the grass. It extends a long way back from the road on each side before giving way to the lime-green of the Silver Cassia shrubs, the grey-green Tangled Lignum and Mulga and, eventually, the darker green of a sprinkling of Black Box eucalypts. I am surrounded by beauty at every turn.

Moz drives up. 'Is there anything that you need, Grog?'

'No. I am doing fine,' I say. Despite how poorly I feel, surrounded by the allure of the morning, there is nothing that I need.

'Jane's going to walk to the breakfast stop.'

I see her behind me. Okay, there is one thing that I *do* need. I would love a piss.

≈

In the last decade or two of his life, Henry was reduced to living what one biographer termed 'a pitiful travesty of life'. Shuffling around the streets and alleys of Sydney a prematurely old man, Henry was reduced to begging.

After trying to take his own life just before Christmas in 1902, the day after Boxing Day two years later, Henry had a rare chance to be a hero. He was at Circular Quay when he saw a young woman put down her five-month-old baby and suddenly jump into the water. Henry and another man quickly leapt into the waves behind her, pulling Ettie Thrush to safety.

Under the heading, 'ATTEMPTED SUICIDES: BRAVE RESCUE BY HENRY LAWSON,' a newspaper reported the incident.

EXCITING SCENE AT SYDNEY NORTH HEAD.
Sydney, December 27.
Shortly after noon today Ettie Thrush, a married woman, attempted to commit suicide by jumping into the harbor at Circular Quay. She walked to the edge of the North Shore wharf, and, after placing her baby on the staging, jumped into the water. The Australian poet, Henry Lawson, without waiting to divest himself of his clothing, sprang into the water, and supported the woman until she was drawn on to the jetty. The would-be suicide was removed to the Sydney Hospital. When asked why she jumped into the water she replied, 'I have got my troubles.'[1]

Afterwards, the woman wanted to meet the man who had saved her life. Thinking of his own problems and his own suicidal tendencies, Henry was reluctant. 'I can't help her,' he wrote to a friend. 'You might.'

≈

At 7.50, I pass a decrepit ruin of a windmill on my right. The wheel is broken into sections and hangs like the head of a disgraced schoolboy. The rusted tower leans to the south, as if trying to edge its way toward a cooler clime. The windmill looks a lot like I feel. One day soon, the wheel will separate from the tower and fall completely.

I notice long trails of ants. Is this a sign that rain is on the way? The nymph of an Australian Wood Cockroach slowly edges its way across the sandy Track too. The spiky, spiny legs are fascinating. It is as if the cockroach carries medieval spiked war clubs.

I move forward and pass a sign indicating that any stock on the road belongs to Youngerina Station. I wonder who 'young Erina' is or was. She is probably old Erina by now—probably ancient, long-deceased Erina. I still need a pee but the Track is so straight and open that I cannot get the job done without Jane seeing what I am up to, not that it would worry me. I am just thinking of her sensibilities so I continue to hold tight.

I notice the rounded shapes of a pair of Brown Goshawks flying above the Track towards me. They pass close overhead, clearly interested in my presence. Drifting back over the top of me a second and third time, they eventually come to rest in a nearby eucalypt. I stop and watch them, dark amongst the shadows of the leaves. One of the birds whistles from its perch, 'Kee-kee-kee'.

Just two or three minutes later, a Superb Fairy Wren seduces me with its lyrical song. With the contrasting royal blue and navy blue throat and head, it has a physical beauty to match its delightful song. If it were not for the fact that I am approaching a bend that offers the chance to empty my bloated bladder, I could stop and watch the bird all day. Instead, I hurry around the bend and, hidden from Jane's view, I relieve myself at the side of the road, an hour-and-fifteen minutes after leaving camp with a full bladder.

Boy, did I need that.

≈

On one occasion while working for a newspaper in New Zealand, Henry managed to wrangle his way into reporting the opening of a local brewery. The editor's instructions were clear: The report was to be ready for Wednesday's edition. Wednesday came and went without an appearance either of Henry or his report. When Henry did appear, the editor said, 'I thought I told you I wanted the report for Wednesday's paper.' Henry's response was, 'You did not state which particular [Wednesday].' When the report eventually did appear, it explained Henry's tardiness: 'The Mangatainoka Brewery was opened one day this year. It was a gigantic success and ended in oblivion.' Despite his sharp sense of humour, the story hints at things more sinister. Henry's on-going alcoholism was playing out in the public eye.

≈

I pass the dry bed of Youngerina Creek. Some of the creeks are bone dry, while in other places there is flooding. I wonder why that is, and how it can be. This is a strange place, this Outback. The rocks in the creek bed near the Track suggest that, at some times, the water must roll down here swiftly.

At 9.45, I reach Moz who is parked beside the sign indicating it is 110 km back to Hungerford. With the additional kilometres walking in and out of Kia-Ora behind us, we are more than halfway back to Bourke.

Moz has a tiny fire going. 'When Baz stopped here,' he says, 'I boiled the billy so he could have a cup of tea with his breakfast. While I boil the billy for Jane I will cook you some toast for breakfast if you like.'

I would prefer not to stop, but the lure of Vegemite on toast cooked over a campfire is too great for me. I fill my hydration bladder and water bottles while Moz starts on the toast.

'Take a seat while I get this toast for you, mate,' Moz says, pointing towards the folding chair.

'I'm okay thanks,' I say, remaining on my feet while I watch him work.

I can do a lot of things badly, but there is one thing I can do well—cook toast on an open fire. As I watch Moz hunched over the tiny flame, I can see that I should do the job myself or it will take forever and, even then, I will end up with scorched bread rather than cooked toast.

'I can take over if you want, thanks mate. You can get the cups ready for your tea with Jane,' I say as diplomatically as I can, pointing back to where Jane is approaching.

Moz hands over the skewered piece of bread and I break the single-pronged twig that the bread is on and use it to build up the fire. I kneel down with a two-pronged stick to toast the bread my way, but I do not sit.

'Baz set off along this track here,' Moz tells me, pointing to a sand bed trail behind the trees. 'He figured it would be nice and shady.'

'I'll give it a go,' I say.

Although this constitutes more of a stop than a pause, I stay on my knees. I am still fearful of taking a seat and not being able to get up.

'You have done about 12 kilometres,' Moz tells me. 'Baz said he'd like to do about 33 today.'

Jane arrives just as I finish spreading Vegemite over my three pieces of toast. I bid farewell and depart with my toast in hand.

'We'll make sure your water is full when we catch up to you. Then we are going to head in to Bourke.'

'No worries.'

I do not stay on the trail behind the trees for long. It is shady, but the sand is too soft and it is a slog to walk through. With each step I take I can feel sand kicking into my shoes. After only 40 or 50 m, I rejoin the main Track and I can see Baz on a crest almost two kilometres ahead of me. The going here is not bad. It is firm but not hard and so it beats slogging through the sand on the trail in amongst the trees.

As I approach a steep crest, a flock of 60 or more Galahs pass overheard, wheeling about in the sky and screeching as they go. Galahs are so common that they are taken for granted but they really are a stunning bird. With their vibrant pink and soft grey colouring, they are eye-catching and beautiful. I take delight from the visual spectacular as they fly above me.

The sun is high now, out in front of my left shoulder. There is not much cloud cover and so I feel the heat. With the rain from the previous night, it is sticky and humid. Fortunately, a slight breeze offers some relief and I enjoy watching it sway the uppermost branches of the trees.

Moz and Jane catch up to me in the car. 'We're off to Bourke to pick up my new tyres now.'

I top up my water bladder and my bottles.

'Have a good trip, you two,' I say. 'Drive carefully.' Moz drives these unsealed roads like a bat out of hell.

'We'll leave some crackers and tuna for your lunch near the Hungerford 120 km sign. If I am right, that will be at the old bridge,' Moz says, consulting his notes. There is an old wooden bridge at Coonbilly Creek. The Track diverts around the creek now. Presumably, the bridge cannot hold the weight of cars. On the 31 December, we buried water beneath it. 'Do you want anything else left there with the water?'

'Maybe throw out a lolly or two, please.'

'We'll do that.'

After Moz leaves, I find myself thinking about Darcy Niland's fabulous book, *The Shiralee*, and the wonderful television mini-series adaptation starring Bryan Brown as Macauley and Rebecca Smart as Buster. I think particularly of the scene where Macauley first goes to the country and is getting his supplies before he takes to the Track.

If I remember correctly, the storekeeper gives Macauley a hat and Macauley says, 'I didn't ask for a hat.'

The storekeeper responds, 'You're gonna need a hat.'

That is certainly the way it is out here. You need a hat. As always, I am wearing my Tilley hat from Canada. Although it is serving me well, I continue to have the Tilley augmented by the blue bandana Bronwyn gave me at her cowgirl birthday party in October. Together, the hat and bandana keep the sun from my head and neck and enable me to keep going.

'You're gonna need a hat,' I repeat. Under the fire of the Outback sun, you need your 'face half-hid 'neath a broad-brimmed hat / That shades from the heat's white waves.'

I must have about another eight kilometres to get to the lunch spot. From there, I will have another 11 or 12 km until the end of the day. I am weary and my feet are sore. It is going to be a long, tough afternoon. Even though breakfast is still in my belly, I decide that I am going to stop and rest at lunchtime. It will be my first proper stop for this whole return leg. If my 'recovery' is like it was after breaks on the outward leg to Hungerford, I will not stop again before we get back to Bourke. Those torturous afternoon sessions after lunch stops still drive fear into my heart.

A vehicle passes me on its way towards Hungerford. The driver does not stop. All I get is a wave and a cloud of dust. I pass the lifeless shell of what would once have been an adorable little billy goat. Other than the four little trotters and the skeleton, all that remains is the white hair and, presumably, the skin beneath that hair. There is no head though, and no body or innards. The poor little bugger.

I approach a sheet of corrugated iron upon which someone has used red paint to make a sign that says, 'Youngerina'. The sign brings back painful memories of my time at this spot when I originally tramped to

Hungerford in 2009. At one of my lowest, pain-filled points, I had lain down beside this sign and suffered. Back then, the sign stood in a patch of rocky, hard-baked ground devoid of vegetation. The photograph of me lying there in my misery in the bowels of hell later brought to mind Henry saying, 'I thought it was a pity that a chap couldn't lie down on a grassy bank in a graceful position in the moonlight and die just by thinking of it.' Now, the sign is surrounded by tufts of seed-tipped grass. Things are so different out here in 2011 than they were in 2009—the drought is now just a black memory.

≈

Although not related to Henry, Will Lawson was inspired to become a writer through the influence of his namesake. One time at the local pub, Will Lawson, the writer, was mistaken for Henry Lawson, the great writer. People insisted on buying Will, who was 'Henry' to them, several drinks. He later told Henry about the misunderstanding and asked Henry if he was upset about the case of mistaken identity. Henry responded that he was only upset thinking about all the free beer that he missed.[2]

≈

The previous hour has been a slog. Despite the clouds, there has been no rain and I am hot and bothered and extremely thirsty. It is very humid. In addition to my weary legs, my left shoulder is throbbing like a toothache and my neck is sore.

Where the bloody hell is the next road marker? When I cooked my toast while Moz boiled his billy, it was before 10.00 am. At that point, there was a sign indicating I was 110 km from Hungerford. That meant I was 105 km from Bourke. As such, the next kilometre marker I come to should be the B100 sign—100 km to Bourke. I have been walking for over an hour-and-a-half and I have not reached that sign yet. At this rate, it will take me forever to get to the lunch stop, which is another five kilometres past the B100 sign. After so many long, unbroken stretches over the previous few days, I am desperate to rest.

I struggle on. The Track here is strewn with rocks. It is the worst kind of walking surface. Rocky and hard. Rocky and sharp.

The Track is endless but, in contrast, my reserves of energy are at an end. I am spiritless, flat, and dejected.

≈

In response to criticism of his drunkenness, Henry grew defensive. Despite evidence to the contrary, Henry argued that he and his friends did not drink to the excesses often attributed to them. 'I had been a total abstainer for two years before [going to] England,' claimed Henry. 'We of the old literary *Bulletin* school in Australia never drank to anything near the extent they gave us credit for. The trouble was we never hid it,' Henry wrote. But then, in the very same paragraph, Henry conceded 'we might have been "drunkards".' Despite that admission, Henry felt he could have been much worse. Henry was proud that he and his friends were 'never blackguards' and that they 'never dealt in lies, false pretences, or dishonesty.' Henry also defended his relationships with family, saying, 'I was a good son—one of the best—and, in spite of all the lies they told, a good, kind, considerate, and indulgent husband and father.'

≈

Oh, what joy—in the distance I see the old bridge over the dry Coonbilly Creek. That is where Moz said he would leave our lunch and lots of water. The B100 sign I have been looking for must be missing. Not seeing the sign gave me the impression I had not yet gone five kilometres from where I had breakfast. In reality, I have actually gone almost ten! What a relief! I am not travelling nearly as badly as I thought. With a renewed spring in my step, I battle on to the bridge, arriving there a few minutes before midday.

'G'day Gregga,' Baz says.

'Baz? What are you doing here?' He is sheltering under the old bridge. I thought he would be many kilometres further down the road. I am more than a little pleased to catch up to him. It has been a lonely morning and I can use his company.

'That was a tough mongrel of a stretch,' I say.

'What can I get you? Do you want some water? Are you ready to eat?'

'Oh, I think I just want to sit and relax for a minute.' I lower my backside onto some big rocks in the shade beneath the bridge. It feels nice to remove my hat and bandana.

'You look like Grandpa Simpson again,' Baz laughs.

He has his hat and shirt hanging from a nail on the underside of the bridge, which makes him look like Tarzan.

He retrieves the water flagon from the hole in which it was buried in the sand.

'Here you go mate,' he says, handing me the water.

'Thanks.'

'Take this too.' He hands me a tin of tuna and what remains of a packet of rice crackers.

I open the tin and start to dig out some tuna with a cracker. It tastes bland and uninviting.

'I'm just about sick of tuna, that's for sure.'

Baz laughs. 'Too right.'

'Do we have any lollies?'

There are mosquitoes in the shade and so I retrieve Aerogard from my pack and spray it on my hands and in my hair.

'I might have a spray too, please,' Baz says. He hands me a juice box and I hand him the insect repellent.

Baz tells me that he passed the morning compiling a fantasy dinner guest list. 'Steve Waugh is at the top of the list.'

'That sounds about right.'

He opens his notebook and reads me his list.

'Allan Border, Tommy Hafey, Gary Ablett senior, Lance Armstrong, Muhammad Ali, Bob Geldoff, Nelson Mandela and, despite the fact that it is now too late, Mother Theresa and Burke and Wills.'

It is a great list. Surely Henry should be on there somewhere though.

'Did Tony stop on the road and talk to you yesterday evening, Baz?'

'Yeah.'

'How was he?'

'The doctor said he has bruised ribs.'

I think of the massive purple bruise all the way down Tony's side. 'Brilliant. I wonder what that cost.'

'You didn't talk to Tony, Gregga?'

'No. I had just got into my tent and stripped off when he arrived. I was moving so slow that I knew it would take me forever to get dressed. I figured by the time I got out, he'd be gone.'

Baz understands. He knows how slowly I can move.

'I wonder if Big Moz and Janie's Got A Gun will bring us back anything from Bourke,' Baz says.

'I hope so. That'd be good.'

'What do you want?'

'Anything except effing tuna!'

We share another laugh, but I am serious.

'I saw a huge flock of goats earlier,' Baz says. 'They had great big, long horns. There were 40 or 50 that crossed the road right in front of me. Did you see them?

'I have seen one or two goats today, but that's all.'

'I also saw a lot of feral pigs, including lots of piglets.'

We sit and eat lollies together, tossing the container back and forth between us.

'You stink,' says Baz.

'Thanks.'

'Don't you have a change of shirts?'

Wouldn't I have worn it by now if I did? 'This one is still clean.'

Baz disagrees.

'You needn't think that you don't stink too,' I say.

'No, I don't—not old Two Shirts Bazza.'

I figure I am in a better place to judge than he is. I take a gum leaf I have been carrying in my pocket for several days and grind it in my hands. Now, *that* smells good.

With Moz and Jane in Bourke, it is important that we make sure we carry enough water to get us through each ten-kilometre stretch to where our next buried water site is located. We load up fully and return to the Track at 12.30 pm.

My legs and feet feel tight and painful when we get going again.

'You better go on ahead, Baz. I think I might be moving pretty slowly this afternoon.'

≈

At 1.10 on Day Twelve, I am in bright sunshine and suffering in the heat with my shadow almost completely beneath my feet.

Baz is probably only 800 m in front. He is not moving very well this afternoon either.

I wish that it would rain. Up ahead there are dark clouds. Off to my right, there are dark clouds. Away to my left, there are dark clouds. When I look back, there are not only dark clouds, but it actually looks like it is raining across a broad swath not too far behind me. There is also a thick black cloud above me but somehow the sun manages to elude its cloak of darkness. There is cooling, refreshing rain or rain clouds everywhere around me but for the very spot where I am. To cool off would be fantastic. Light rain would be ideal. The wind blows dust at me from the left. Can that same wind not blow one of those clouds over the top to dump water on me?

The temperature would have to be mid-thirties again. The water that I have with me is not only heavy but it is warm and not particularly refreshing. If I find a lamp along the Track and manage to get a genie to grant me a wish, a little rain will be at the top of my list.

'Come on Huey. Send 'er down!' I cry out…into the emptiness.

≈

Baz stops in the middle of the Track in front of me. In the distance, I can see him scratching into the ground. He waves at me and points towards the side of the Track, then turns and continues towards Bourke.

When I eventually get to the arrow he scratched into the ground, I wander in the direction in which the arrow points. I can see that Baz has stopped up ahead to watch and see if I can find whatever it is that I am supposed to be looking at.

Hanging from a Mulga Tree is a hive of thousands of bees. I back away carefully.

'I didn't think you saw it,' Baz says when I catch up. 'I thought you were going to walk straight into the hive.'

≈

We get to the B90 distance marker at 1.45 pm. I lean against the metal pole and Baz takes a photograph. I look almost as buggered as I feel. Near here, there are ruins of Old Kerribree Station. Henry stopped here for a handout from William Walter Davis, who owned the Kerribree run. The squatter kindly gave Henry as much food as he could carry and also sent him on his way with money. In a childhood accident, one side of Davis' head was burned and so, although he always wore a red wig, he was known by the nickname, Baldy Davis. Baldy's generosity convinced Henry that not all squatters were bad people. Henry's experiences in the shearing sheds at Toorale had already taught him that not all workers were good people. In return for Baldy's kindness, Henry used him as the model for a short story entitled 'Baldy Thompson'. 'The track by Baldy Thompson's was reckoned as a good tucker track,' Henry wrote.[3]

≈

We cross the bridge over Kerribree Creek. The cows and the bull are at the creek again, but this time they are down on the flatland, away from the bridge. In need of a rest, we venture down to the water, thinking it will be nice to dunk our hats to cool off. And finally, after threatening to do so for hours, it begins to rain!

'How's that, Baz? I have been hoping for rain all day and as soon as we get to water, it decides to rain then—seems like a bit of a waste to me.'

We watch the bull on the other side of the creek. 'Have a look at the sack on that thing,' Baz says. I cannot miss it.

The bull seems to be enjoying the fresh green grass. 'Old Ferdinand has a tough life, eh?' I say, facetiously.

Cows move about, grazing contentedly.

'...With all of his bitches,' Baz adds.

We watch a heifer and a young calf splash their way across the creek under the bridge.

'If the calf doesn't make it through, you'll have to go in and save it, Baz.'

Luckily, the calf makes it. It finds its mother after crossing the creek, leaving the heifer to wander off in search of another playmate. Baz and I move on before the heifer decides to play with us.

≈

At 2.45, Baz and I encounter an Eastern Grey Kangaroo with a joey in its pouch. The kangaroo seems not to be bothered by our presence and we are able to approach closely before it decides on a languid retreat. We walk another 20 m or so and catch up to the mother and baby again. Baz takes some photographs as the angular mother kangaroo cleans its forearms without any indication of fear of humans. Eventually, the kangaroo hops away but it seems more so out of boredom with us rather than perceived threat from us.

We move on.

'The next water dump is not far away,' Baz tells me.

As my water runs low, I get increasingly hotter and drier through the afternoon. Baz tells me again that the next water stop is not far away but I think he is being optimistic about the pace at which we are travelling. I dream of getting to our water spot, having a big drink, replenishing my hydration pack and then pouring any excess water over my head.

'I need a rest, Baz. Even just a short one will do. When we get to the South Kerribree mailbox, if there is shade there I am going to lie down.' We have just gone past the North Kerribree turn off and I know the South Kerribree turn off is only about a kilometre further down the Track. When we get there, however, the long grass around the mailbox looks neither safe nor inviting. I do not fancy lying down in there with the snakes.

'Maybe there'll be a better place further ahead.'

We see a large tree ahead and hope that it will offer good shade. When we get to the eucalypt, it is tall but, as with all of the trees out here, it offers only mottled and patchy shade.

I lie down anyway and drink what remains of my water.

I encourage Baz to continue without me, but he decides to wait. Lying down flat on my back, I stare up at the bright sun through the gum leaves. I watch as the moving clouds cover the sun and I hope for more rain.

In my muddled, exhausted mind, the movement of the clouds makes it seem as if the tree is about to crash down upon me. 'I had a dreamy recollection,' wrote Henry, thinking back to his childhood home beneath a large tree. 'That tree haunted my early childhood. I had a childish dread that it would fall on the tent.'

'I reckon the next water dump is only about ten minutes up the road,' Baz says.

'I reckon you're dreaming. I think it is at least half an hour away.'

We rest for 15 minutes and then carry on.

We have only gone another five minutes when Baz asks, 'Is that the sign up ahead?' Our next water stop should be beside the H130 sign. Baz takes out his binoculars and looks down the Track. 'That's it.'

Baz hands me the binoculars. My spirits soar. Water! At last!

We rush ahead as fast as our exhausted legs will carry us.

Baz gets there first. 'There is no water here.'

'Son of a bitch.'

Baz and I dug up the water here and drank it all on the way out and, much to our surprise, Moz has not replenished the dump before he and Jane drove into Bourke. It is after 4.00 pm and I have not seen Moz or Jane since they left for Bourke at 10.20 this morning.

Having looked forward to getting to this water dump for so long, I have a sense of what Burke and Wills felt like when they got back to Cooper Creek and no one was waiting for them. I feel totally deflated and sink to my backside in the dirt thinking of one of Henry's poems: 'And, oh! it's a terrible thing to die of thirst in the scrub Out Back.'

'There are some apricots here,' Baz tells me.

'What?'

'There is a tin of apricots in the hole.' Baz retrieves the tin from the hole where the water had once been buried. 'We didn't eat these when we passed through the other day.'

I think of the Dig Tree.

I do not feel like eating. I am desperate for a nice cold drink. I might cook if I do not drink soon. Maybe the juice in the tin will help.

'There's only one problem,' Baz says—famous last words.

'What's that?'

'We don't have a...'

I know where he is going. '...Bloody can opener.' I finish the sentence for him.

Baz decides he will get the tin open anyway. He takes his metal hand trowel and starts bashing away at the top of the tin while I sit in the dirt and watch, depressed.

To my surprise, he not only hacks the tin open, but he does it without chopping off a finger.

I retrieve my fork from my pack and eat a couple of the apricot halves. They have been in the sun since Day Three. Still, they are wet, and so is the syrup that I gulp down.

When I have had my fill, I return the tin to Baz. 'You have the rest, mate.'

'I wonder how far away Moz and Jane are.'

I do not know. 'I am long past ready to stop for the day.'

'We might as well keep going until they arrive,' says Baz. I want to take his trowel and kill him with it, but I am too exhausted.

I say nothing.

'I'm going to keep going,' Baz says.

'I'll wait here and cool down a bit,' I say.

'See ya later,' Baz says.

Baz has only gone 20 or 30 m when I haul myself to my feet and start to follow him. 'The dumb bastard will die out there alone if I don't look after him,' I say to myself, hurrying to catch up.

We struggle on for another half an hour, Baz out in front of me.

It occurs to me that the temperature where I live in Winnipeg is likely minus 30 or 40 degrees.

After what has been a hell of a long day, we see a car approaching from the east. As it draws closer, we see that it is Jane and Moz.

Moz has the new tyres on his car and seems pleased. Jane, on the other hand, has a headache after she accidentally walked into the uplifted rear hatch and knocked her head. Later, she tells us of her conversation with Moz on the return trip from Bourke.

'Gee, I've got a headache. My head is throbbing.'

'Gee, I like my new tyres,' was Moz's reply.

Needless to say, Moz is in the doghouse. I do not complain about the empty water situation. The poor bugger has enough on his hands. In any case, Moz redeems himself when he gives me and Baz ice-cold Powerades and a delicious, cold Cherry Ripe.

'Moz, I could kiss you.'

Moz turns the car and drives a little ahead of us where he and Jane set up the camp.

When I get in, I fill three bottles with water and crawl into my tent. I gulp down the water and then strip. My butt rot feels bad. My crotch rot looks horrible. We have completed almost 35 km today and I am buggered.

'Grog, I picked up a couple of newspapers in Bourke,' Moz tells me. 'I thought you might want to read them.'

'Thanks mate.' I cover my crotch rot as Moz comes over and passes me the newspapers.

In *The Sydney Morning Herald*, there is an article by Peter FitzSimons, an author and historian. I have a number of his books in my home library back in Canada. Here, FitzSimons writes about how to get published.

'Listen to this,' I call out to the others, 'Peter FitzSimons says, "If you want to write, you have to have something to say that people will give a stuff about. And they will be more likely to give a stuff if you've done things they haven't."[4] That's FitzSimons' advice on how to get published. Well, no one has ever done *this* before.' With only three days to go, Baz and I will be the first people ever to retrace Henry's footsteps all the way from Bourke to Hungerford and back. 'It sounds like I might have a book.'

As I continue to read the newspapers, I am shocked to learn of the enormity of the floods to the north of us in Queensland. Most of all, I am shocked by the rapidly rising death toll in and around Toowoomba. It brings into perspective the difficulties we are experiencing with our water crossings. Things are so very much worse for the people up river.

Later, I emerge from the tent when tea is ready. There are tonnes of ants at this site and we spray our feet with insect repellent to try to keep the little buggers off.

When I am resting after tea, it starts to rain. Wouldn't you know it? I baked all day long and the only time it rained was when we reached Kerribree Creek and did not need rain. Now that we have finally stopped and are able to rest, it rains again.

When I put the fly on the tent to keep the rain out, no breeze can circulate and I practically roast in the still air inside. Given the long, thin dimensions of the tent, I feel like I am inside one of those sliding drawers that they have at a crematorium.

At 7.45, I watch a fly crawl up the side of my tent. It has only four legs and three of them are on one side of its body. Still, it seems to be coping. Seeing the fly directs my thoughts to the great Canadian, Terry Fox. After losing a leg to cancer, he set out to run across Canada to raise money for Cancer research. I guess that I sit somewhere between a four-legged Australian fly and an inspirational Canadian hero. Regardless of where I sit on that continuum, I draw inspiration from the two extremes. If they can continue, then surely so can I.

DAY THIRTEEN

Henry Lawson and an Old Mate in Leeton

You take a drink, and look over your glass at Tom. Then the old smile spreads over his face, and it makes you glad—you could swear to Tom's grin in a hundred years. Then something tickles him—your expression, perhaps, or a recollection of the past—and he sets down his glass on the bar and laughs. Then you laugh. Oh, there's no smile like the smile that old mates favour each other with over the tops of their glasses when they meet again after years. It is eloquent, because of the memories that give it birth.

- Meeting Old Mates, *On the Track*, 1900

There was a something of sympathy between us—I can't explain what it was. It seemed as though it were an understood thing between us that we understood each other. He sometimes said things to me which would have needed a deal of explanation—so I thought—had he said them to any other of the party. He'd often, after brooding a long while, start a sentence, and break off with 'You know, Jack.' And somehow I understood, without being able to explain why.

- The Babies in the Bush, *Joe Wilson and His Mates*, 1901

With rain falling occasionally through the night, I keep the fly on the tent, which ultimately creates an oven. I open the fly at the entrance but then the rain falls in my face. Between the heat and the rain, it is difficult to fall asleep and my problem is compounded by the large rock digging into my shoulder. I am forced to one side or the other, but this pushes me off the mat and onto the little twigs and grit I have inadvertently carried into the tent. Naked because of the heat, the dust and grit sticks to my back and makes me itchy and miserable.

It is little wonder that I am awake early and so decide to get ready and get going as soon as I can. I will rest better on a soft bed in Fords Bridge than lying here beside the Track.

≈

Late in 1915, knowing how badly off Henry was for money, Archibald called together Henry's friends and they arranged a job for him in the town of Leeton in the New South Wales' Murrumbidgee Irrigation Area. The once arid landscape was being transformed into productive agricultural farms. Officially opened in 1912, the government wanted to attract farmers to the area. Consequently, it was Henry's job to write promotional materials that would hopefully entice settlers. Henry felt equal to the responsibility. 'Anyone who has read my works and known something of my earlier life would know that I'm qualified for the work suggested,' he wrote to his prospective employer while the deal was being negotiated. Before he had even seen Leeton, Henry had in mind the types of things he would write. 'The articles need not be as dry as the Drought itself. I would probably turn out something novel and readable,' he said.

When he did see Leeton, Henry immediately noticed the absence of a hotel. He sat down and penned his 'First Impressions of Leeton'.

> *Yes. The barber's shop is here—and a sociable barber's shop it is—and the grocer's shop, draper's, butcher's, baker's, news-agent's, ironmonger's, etc., etc.; and very cheerful and kindly people they seem. And the dentist, chemist, doctor, and schoolmaster—reserved men for the most part, I should think.... The ministers are here—quiet men who know a great deal and say little, I expect; and the policeman is here. It must be an awful dull job for a policeman in a town like this. The churches, creeds, my old friend the Salvation Army, and my older friends the Ungodly are here. Yes, all things are here that are in most country towns—and more;* but, lo and behold! the pub is not here.

Henry was accompanied to Leeton by his long-time housekeeper and companion, Mrs Isabel Byers.[1] She quickly improved the appearance of Henry's new home and started a garden. Henry was proud to report her progress: 'She has all her vegetables up and the place well stocked with poultry of the best breeds. Also the flower garden in first-class order.'

Henry was less impressed with her pets than he was with her garden, complaining of a dog she bought with her that Henry considered 'a fowl-killer and a man-killer—a horse and sulky chaser, and, all together, a dangerous dog.' Henry, however, had the last laugh. 'I lost him with the aid of two mates and a two-gallon jar of beer,' he reported with satisfaction. Henry hardly put forth his best efforts to find the lost dog either, posting a note saying:

PUBLIC NOTICE
TO ALL WHOM IT MAY CONCERN
Lost, a Dorg named 'Charley'. The Man wot finds that Dorg and Brings him back to me, will get the Biggest Hidin' a Man ever got from the undersigned.
Yours truly,
Henry Lawson

Living in Leeton was a man who had published in the *Bulletin* under the name, Jim Grahame. Grahame had been amongst the first settlers to the Murrumbidgee Irrigation Area. It so happened, that same man was also known as Jim Gordon—the same Jim Gordon who was Henry's old mate from the Bourke-Hungerford Track. The two had not seen one another for 23 years.

Upon learning of Henry's arrival, Jim penned a letter to his old swagman mate, albeit 'not without some hesitation' for Jim wondered 'how the applause of all of Australia and other parts of the world might have affected [Henry].'

With a letter dated, 'Damn the Date,' Henry responded enthusiastically. 'I was delighted to get your note....Drop in any day you're in and we'll get acquaint again, and arrange a time for a long chat, either here or at your place.'

Jim wrote of their reunion:

That evening I walked up the street in the hope of meeting him, and saw him coming down the hill....Of course I knew him at once. I do not think anyone would fail to recognize Lawson after once seeing him. But what surprised me was that he knew me when we were yet ten yards apart. What a hearty, silent handshake we had! And great peering into faces, and looking up and down. Then he insisted I must go with him to his bungalow. I went, and soon we were all good friends.

Henry was anxious to get to know his old friend once again, but nonetheless unsure and uncertain too. They once had been the best of mates. What would things be like between them after so many years apart?

My mate, James Grahame.... We first met in Bourke some twenty-five years ago, and thus we share two pasts, so as to speak; but we were very young men then, those pasts are boys' pasts; and being but recently re-mated we haven't got to speak of those pasts yet. There's a certain shyness about the matter, if you understand, which may or may not deepen as those twenty-five year pasts are cleared up. They'll have to be cleared up first, and it is mainly for that reason we have come out into the gnarledest, wildest, weirdest, oldest bush in Australia, to 'camp on the river' for a night or two under the pretence of fishing.

Our faces and voices have already grown re-familiar to each other, and we seem to have changed not at all. Not to each other.

Leeton was a dry area and, although Henry slipped outside to imbibe, Jim felt that having him back in the country away from the hustle of city life was good for Henry. 'As the week's passed, the colour came back to Lawson's cheeks,' Jim wrote, 'and his step was lighter and his chuckle heard more often.' Jim was also pleased with the quality of Henry's Leeton writing. After the publication of 'By the Banks of the Murrumbidgee,' Jim wrote a congratulatory letter to Henry, saying, 'It is splendid, excellent; I know, for I was there! It was written by the same hand that made the name of Henry Lawson famous in the [eighteen] nineties.'

≈

'Is Moz moving yet?' Baz asks from inside his tent when I am almost ready to leave.

'No sign of life yet,' I say.

'Seeing as it is going to be a short day, Moz was saying last night that he is planning for us to have breakfast here before we set off.'

'Oh.' I know nothing of this plan. I want to get into Fords Bridge as soon as possible so that I can feel like I am having another rest day lounging in the pub. I expect we have only 16 km of toil today and I want to hit the Track and knock it over as soon as I can.

'If he is not awake yet, it might be a while before breakfast is ready.'

'Yeah, I'm not waiting Baz. As soon as I am ready, I'm hitting the Track.'

I slop sunscreen across my face and hands and then fill my water bottles.

'See ya later, Baz. I expect you will catch me pretty quickly. If you don't and I happen to get near Fords Bridge before you, I'll wait for you a kilometre or two out of town so that we can walk in together.'

I am on the Track at 6.45 am. Although the old Nike shoes have served me well, I am wearing my Merrells for the first time on the return journey. With just a short day ahead, I want to wear the Merrells to see how they feel. If they do not create more blisters, I will also wear them on the final two days, where their extra cushioning will be useful when I get back to the hard bitumen surfaces.

Even in the first 200 m, I notice that the Merrells do not feel as comfortable as the Nikes. My feet are, however, still loosening up and so I push on in the hope that once I am warmed up, things will feel better. I pop a couple of painkillers into my mouth and take a swig of water to wash them down. I have less than 100 km to go until Bourke, but I realise that if things go badly with my footwear today I could be stopped in my tracks.

Although the sun is only just emerging above the clouds sitting low on the horizon, it already packs a punch. It is another reason that I want to get to Fords Bridge early. I can always eat once I am in front of a fan, rather than sitting around eating breakfast out here and then having to sweat through the heat to arrive at Fords Bridge in the midday sun.

Despite my early start, I have not seen any lizards or animals this morning and I have seen only a small handful of birds. I hear the caw of a crow as I climb towards a crest on the Track but, other than that and the long rows of ants crossing the Track, there are surprisingly few signs of life.

I take the portraits of Henry from the pocket of my tracksuit pants.

'What do you make of all this, Henry?' I say to myself.

Although I have received no otherworldly visitations, from what I know of Henry from photographs and the physical descriptions I have read, I can picture Henry on the Track, tall and angular but, like me, at times a broken man. I can see him walking beside me with his full waterbag. Jim Gordon said that Henry was fussy about the appearance of his swag. Jim's slovenly swag used to annoy Henry, who felt there was a *right* way to fold and carry a swag and that was the way it should be done.

> *The swag is usually composed of a tent 'fly' or strip of calico (a cover for the swag and a shelter in bad weather...), a couple of blankets, blue by custom and preference, as that colour shows the dirt less than any other (hence the name 'bluey' for swag), and the core is composed of spare clothing and small personal effects. To make or 'roll up' your swag: lay the fly or strip of calico on the ground, blueys on top of it; across one end, with eighteen inches or so to spare, lay your spare trousers and shirt, folded, light boots tied together by the laces toe to heel, books, bundle of old letters, portraits, or whatever little knick-knacks you have or care to carry, bag of needles, thread, pen and ink, spare patches for your pants, and bootlaces. Lay or arrange the pile so that it will roll evenly with the swag (some pack the lot in an old pillowslip or canvas bag), take a fold over of blanket and calico the whole length on each side, so as to reduce the width of the swag to, say, three feet, throw the spare end, with an inward fold, over the little pile of belongings, and then roll the whole to the other end, using your knees and judgment to make the swag tight, compact and artistic; when within eighteen inches of the loose end take an inward fold in that, and bring it up against the body of the swag....Fasten the swag with three or four straps, according to judgment and the supply of straps....To the top strap, and lowest, or lowest but one, fasten the ends of the shoulder strap (usually a towel is preferred as being softer to the shoulder), your coat being carried outside the swag at the back, under the straps.*

I can envisage Henry out here, baking underneath the sun, wishing for the next place to stop and cool down. I do not know of the world of spirits or any supposed afterlife, ghosts and haunting, but if there be such things, I wonder if Henry ever does venture out here on the Track. Certainly, his experiences here had a huge influence on him as a writer and, I believe, as a person. I think of his time in and around Bourke during 1892-1893 as being amongst the happiest times of his troubled life. I think that it was here that he really enjoyed mateship. His ethos around mateship was heavily formed and refined during his time with Jim Gordon but also with other Bourke entities like his Union mates, Billy Wood (the secretary of Bourke's Amalgamated Shearers' Union), Tom Hicks Hall (secretary of Bourke's General Labourers' Union), Donald Macdonell (who became secretary of Bourke's Australian Workers' Union when the ASU and GLU amalgamated), and the kindly Giraffe, Bob Brothers. I think that, should such things be real, in the spirit life Henry would occasionally venture onto the Track to spend time with an old mate, looking forward to the opportunity at the end of a wearisome day to gather around the campfire to yarn.

Henry might enjoy a stretch like the one on which I now find myself, with colours in the sky and a wide-open Track before me. Taking to the Track again was something that Henry and Jim talked about when they were together in Leeton. Despite the torture of the Track, Henry thought fondly when he looked back on those youthful, energetic days of old. One day Jim and Henry explored the rubbish tip outside Leeton. As they did so, they spied a large flock of goats dining on the refuse. Excitedly, Henry grasped Jim's hand. 'The last flock of goats we saw,' said Henry, 'was out at Hungerford, on the Queensland border in the days when the billy boiled.'[2]

They made plans but a return to the Track never eventuated. If spirits exist, I think Henry might like to return out here, at least for a short jaunt—perhaps on a cool day.[3]

≈

Although they did not return to the Track, Henry and Jim did spend time together in the Bush around Leeton. One day they went out to gather a load of firewood. 'What a day that was for him!' Jim recalled, '[Henry] had not forgotten the way to swing the axe. We boiled the billy for dinner, resting it on two small pieces of wood and building the fire all round it. That pleased him immensely. It was the right way, he said—the bushman's way.'

Because of his experience as a Bushman, Jim was employed to collect tree seeds for the Irrigation Commission. Henry enthusiastically volunteered to accompany him on any ventures that would involve camping overnight. On one such trip, Jim was especially weary and slept soundly. He awoke to find it had been raining heavily but he was 'snug and dry and warm' because Henry had kept the fire going all night. What's more, above Jim, 'stretched tightly on frail sticks,' was one of Henry's blankets. Henry's other blanket covered Jim's body. Jim realised that all night Henry 'hadn't even lain down,' but had spent the entire night covering Jim from the rain. Henry was 'drenched to the skin' because 'he had waited and watched through the night, stoking the fire, and tirelessly attending to the crazy awning.' When Jim expressed his gratitude, Henry replied, 'Ah! But then nothing matters between mates.'

For old time's sake, at Christmas in 1916, Henry and Jim arranged to spend two nights camped beside the Murrumbidgee River. 'What a time we had in making our preparations!' Jim recalled. 'Whatever he was in other ways, as a camper, Lawson was the king of optimists.' Jim wrote that when the two set up camp, Henry felt 'he had found a haven of calm and peace.' It 'pleased Henry immensely,' for it made him feel that he was 'once more in the bush, and doing what we used to do.' With that theme of *auld lang syne* in mind and with Yanco Station less than a mile from where they were camped; Henry and Jim decided to walk to the station to see if they might beg a little tucker, as if they were back on the Track to Hungerford. When the homestead door opened in response to their knock, Jim retreated to the safety of saying simply, 'This is Henry Lawson.' The famous writer and his mate were immediately welcomed inside to sit down and wait while a feast was prepared in their honour. After their meal, the two men returned to their camp beside the river. 'Henry's spirits were as blithe as those of a boy, while he appeared almost as active as one too, as he scrambled up and slithered down the banks, or climbed along dangerous logs that overhung the water.' But Henry was not a boy. As they sat around their campfire, Jim could see how much the day's activities had taken out of his mate. Jim wrote that as Henry gazed into the fire, 'I knew then that the day had been too much for him. The trip had too vividly recalled the old days...and, as he squatted there, looking such an old, tired man, I felt a great pity for him.' Jim asked what Henry was thinking and Henry replied, 'Old songs I used to sing. I would like to write now.' Then, knowing that his days were numbered, Henry added, 'You know Jim, I haven't many more writing years left.'[4]

≈

When I get to the crest of the rise, I cross a cattle grid and pass a sign indicating that I am now travelling through the Green Creek Station. It was somewhere on this property that Moz pulled out of the tramp on Day Three. A Nankeen Kestrel hovers above me on rapidly beating tan-coloured wings, watching me from a height of less than 20 m. Drifting this way and that on the breeze, the kestrel seems perplexed by my presence. As I move down the Track, the kestrel shadows me. Maybe the bird is curious to know if, as I walk along, I might kick up a few grasshoppers or perhaps even a mouse. After several minutes pass, the bird loses interest and drifts away over the sparsely vegetated paddocks to my right. Over those bare fields, I now notice there are a number of Nankeen Kestrels. Two hover together in one spot. Another one floats on the breeze and, behind it, a fourth and fifth bird explore the emptiness for food. Further back, in the distance, there appears to be three more, drifting together easily and effortlessly. Up ahead, I see two Black Kites fly low over the Track, their dark plumage and forked-tails obvious even from a distance. Suddenly, there are birds aplenty.

A little after 7.30, I come across a Red Kangaroo on the road. It struggles to scramble away from me but it has obviously been hit by a car and is unable to rise. It seems so very much alive, yet its fate is obviously sealed. I fear the roo has a long day of suffering before it eventually dies. I look around for someone who might help, but I am alone.

It is easy to see that the animal in the dust is terrified. It scratches and claws at the ground in agonised fear. 'It's alright, mate. It's alright,' I say gently, but I know that it is not. At best, all it can hope for is a speedy death. The more likely outcome, however, is a long, painful one. I look around for a stick that I might use as a club and am relieved when I cannot find anything suitable. The roo seems so alive that I worry I will make a mess of it if I try to finish it off.

Will the birds at least wait until the kangaroo is dead before they pick at its eyes? Will a fox or a Dingo wait for it to die? I doubt that. Out here, everything is about survival. To wait is to increase the likelihood of losing the meal to another scavenger.

Sadness wells inside me and I feel a cloak of gloom descend. I can tell that my presence is adding to the kangaroo's grief so I decide that the best thing to do is to leave. The kangaroo is such a beautiful creature. The fur on its broad back is the distinctive rust red, but is grey on its neck and much of its head. On the powerful hind legs, the fur is so light as to be almost white. The animal's thick tail is rust red at the base but then the colour transitions to grey and then tapers through shades of brown to a light tan at the tip. It has black spots on its cheeks. The fur at the base of each large, pointed ear is white and distinctive white streaks run from below its mouth to almost join with the white patches immediately below its black eyes. I look into those eyes and see hopelessness. The eyes break my heart. With nothing I can do to help, I move off slowly, filled with sadness.

I have only walked another eight minutes when I see a kangaroo bound across the Track. I can tell that it is an Eastern Grey Kangaroo because of the black at the end of its tail. Seeing a healthy, hopping kangaroo—of any species—lifts my flagged spirits.

I have not gone much further when I see four foxes together on the right-hand side of the Track. I stop to watch them from my distance of 20 or 30 m. Two of the foxes are very playful and so I suspect that it is a family with pups—maybe a mother and father and two pups or, probably more likely, a mother and three pups. Suddenly, one of the foxes detects my presence and bolts through the long grass and into the Bush to the right. To my surprise, the two playful foxes continue to wrestle, seemingly oblivious to the departure of one of their companions. I move forward and the fox not involved in the play suddenly becomes aware of my approach. It makes a coarse, yet high-pitched, bark-like noise and explodes into movement. Rather than turning and taking the departure route of the first fox, it bolts across the Track and disappears away to the left. Shaken into action by the bark, the two playful pups take off too, but they go in different directions—one turns and disappears through the grass to the right, while the other one seems confused by the sudden drama and it runs along the Track straight towards me, closing the gap between us to only ten metres. It eventually sees me and makes a hasty departure through the grass and into the Bush.

A short time later, Moz pulls up beside me in the car.

'I guess you saw that kangaroo.'

'Yeah. It's bloody sad, eh? The poor bugger.'

'He must have a broken pelvis.'

'Yeah, he's buggered but, to me, he seems a long way from dying.'

'I'm going to go ahead to Fords Bridge to see if I can find someone with a gun to put it out of its misery.'

'Yeah, that'd be good, Moz.' I hope he can find someone but I do not fancy that anyone at Fords Bridge will go to the trouble of coming out here and, in their eyes, wasting a bullet on a kangaroo that will die anyway.

I reach the Green Creek mailbox where Moz had initially sheltered from the sun on Day Three. Even though it is now ten days later, I can see the spot where the grass is broken and pressed down from where Moz was lying when I approached and gave him my water bottle. It looks like one hell of a 'snaky' spot.

I continue on, scanning the swamp to the left for birds or animals. Purple Swamphens wade through the reeds, pushing forward with the aid of the bright red shields atop their foreheads. I see a newborn chick following close behind its mother. I can hear thousands of frogs croaking and, in the din, I realise that today I have not heard any cicadas. A Straw-necked Ibis flies over the water on the left. Five Australian White Ibises stand in the middle of the Track ahead of me, but they take flight as I draw nearer. A frog hops across the Track in front of me, perhaps relieved by the departure of the long, curved bills of the ibises. Dozens of dragonflies patrol the swamp like squadrons of tiny military helicopters searching for a hidden enemy.

As the morning has progressed, the clouds have built up so that the cover is consistent across the sky now, albeit still light and not suggestive of rain. The wind is blowing regularly as well. It actually might be the coolest day of our entire walk—ironic that it should fall on the shortest day of our journey too.

I hear a vehicle coming towards me and then see Moz's car approaching from the direction of Fords Bridge. He has not been gone long. 'I couldn't find anyone at the pub,' he tells me. 'No one is around.'

'No?'

'I found some bloke outside and he told me just to give the kangaroo a tap on the head.'

'Yeah, well, I wondered about that when I was there but it is very much alive, Moz. It'll take a hard tap to kill it. I don't think one tap will do it either. I thought I'd make a bloody mess of it, even if I could find something to do it with.'

Moz and I both obviously wish that we could do more but we cannot. Somewhat dejectedly, Moz drives back towards the injured kangaroo.

Three more foxes cross the Track in a line in front of me, passing from the left to the right. When I get to where they crossed, I see all three foxes again. They are sitting on the bare claypan that extends to the south and the east. The foxes see me and run.

At 8.30, I begin what I hope will be the last of the water crossings that I have to complete before we get back to Bourke. This is our 17th water crossing for the journey. I saw a snake near here on the way out and so I keep my eyes peeled lest I stumble across another one.

The crossing is only short—perhaps 30 m—but I recall that it was a source of great frustration on the outward journey after leaving Fords Bridge on Day Three. My mindset is different now and the crossing is merely something that has to be done.

White-faced Herons fly past while I am in the water. Their plumage is as grey as the stratus clouds that have descended around me.

Moz stops his car on the other side of the water and gets out.

'I'm too late,' he calls out. 'I brought back your thongs in case you wanted them.'

'Oh, I've just worn my shoes through. I'm okay thanks. On the way out, this crossing actually was the only one where I took my shoes off. Thanks anyway.'

Moz seems to be taking that all in. After a moment, he says, 'The kangaroo is dead, Grog.'

'Really?' I am surprised to hear that news.

'It died while Baz was there with it.'

'Good. I thought it was in for a long day.' I am relieved to hear that the roo is dead.

'Baz checked it for a pouch just in case it was a female with a joey.'

'Yep. Good.'

'Baz pulled the kangaroo off the side of the road too.'

Good for him. Good old Bazza. He has made me proud yet again.

≈

While he was in Leeton with Jim, Henry wrote a letter to a theatre director working on dramatising some of Henry's work. In the letter, Henry identified the real life people upon whom his characters were based. Although Henry conceded that one of his most commonly reoccurring characters, Mitchell, was a composite of several people, his letter suggests that one of the people upon whom Mitchell was based was Jim Gordon. 'There were many other Mitchells...In fact, there is one living here now,' wrote Henry. Henry further indicated that Mitchell was based on Jim Gordon in 'By the Banks of the Murrumbidgee' when Henry wrote, 'There used to be two young fellas knockin' about Bourke and west-o'-Bourke named Joe Swallow and Jack Mitchell in those old days.' Joe Swallow, of course, was one of the pseudonyms Henry often employed for himself.

≈

I see a truck travelling towards me.

'Are you okay?' the driver asks, a look of concern on his face.

'Yeah, thanks. I'm one of the blokes walking back from Hungerford.'

'How's it going? Alright?'

'Better now that we are getting closer to Bourke, that's for sure.'

'I guess the bitumen up ahead will make things easier for you when you get there.'

'Well, it is pretty hard on the feet. I prefer the sand for walking.'

'Do you need anything?'

'I'm fine, thanks. Thank you for stopping.'

He drives off and I continue along the Track.

I think of my wife and wonder what she is up to at this time. I start to sing John Williamson's *Raining on the Rock*. 'I'm wishing and I'm dreaming that you were here with me.'

My singing is interrupted when I see another truck travelling towards me from the direction of Fords Bridge. The driver swerves as if to run me over. He is joking with his mate who sits beside him. The truck rolls to a stop.

'Are you okay?' asks the driver.

'Yeah, I'm good, thanks.'

'What are you doing out here?' He is a young Aborigine and now that he knows I am okay, he has a broad smile.

'My brother and I have walked to Hungerford and now we are walking back to Bourke.'

'Wow. You're crazy, man.' The driver looks around me. 'Where's your brother?'

From the look on the driver's face, I cannot be sure if he thinks I have killed my brother or merely imagined him. I laugh.

Henry was asked a similar question when he was tramping.

About the first thing the cook asks you when you come along to a shearers' hut is, 'Where's your mate?' I travelled alone for a while one time, and it seemed to me sometimes, by the tone of the inquiry concerning the whereabouts of my mate, that the bush had an idea that I might have done away with him and that the thing ought to be looked into.

'Oh, my brother is back behind me a little ways,' I say. 'I got an early start this morning but you'll pass him back behind me somewhere.'

We chat for a little longer before the driver says to his companion that they had better get going. 'We've got work to do.'

So have I—continue edging towards Fords Bridge.

When I get to within a couple of kilometres of town I start to look for a place to sit and wait on the roadside for Baz to catch up. Were it not for him stopping and waiting with the injured kangaroo, he would have caught me by now. I do not want to make things unnecessarily difficult for myself though and, depending how far behind me Baz is, I might have a long wait. If it proves tough to get going again after waiting, I only want a short hobble to get to the pub. I keep pushing on and do not stop until I have only about a kilometre left.

I find an open spot on the right-hand side of the Track and sit down in the dirt to wait. I detach my cup from my pack, pour myself a drink and then I take out my notebook and scribble some notes to pass the time.

Moz appears in his car and stops for a chat.

'Baz is probably about two kilometres back,' he tells me.

'No worries. I'll just wait here for him and we can walk in together.'

'Jane is going to walk all the way into Fords Bridge. I might take the car in and then walk out and meet her.'

It will be Moz's first time back on the Track since he withdrew on Day Three. 'Good for you.'

While we are chatting, another car pulls up.

The woman who emerges from the vehicle is energetic and animated.

'I am so glad that I found you,' she says. 'I was hoping I wouldn't miss you.'

'G'day,' I say.

'I'm Annette Parker. I'm from the pub in Fords Bridge.'

'I'm Greg and this is Sean.'

'Hi Greg. Hi Sean.'

'You might not want to come too close,' I say self consciously. 'I am looking forward to a shower at the pub.'

'I bet you are.'

'So, you're Scott's wife then, eh?' I ask, trying to put the pieces together amidst the whirlwind of her enthusiasm.

'Yeah, that's right. I missed you the other day when you went through but Scott and the girls told me all about it.'

'Yeah?'

'How are things going for you? Have they got any better?'

'Yes, much better. It gets better the closer I get to the end.'

'I heard you were in a very bad way. Matilda and Julia said that you were in trouble. Julia said "his feet are not working." You know how the Germans talk—how they would say it.'

'Well,' says Moz, 'that's a pretty accurate description actually.'

I nod my agreement. 'Is Julia one of those backpackers working at the pub, is she?' I am still trying to put the pieces together.

'Yes, Julia and Matilda.'

'And you said they are from Germany?' I was too tired to sort all this out when I went through on the way out.

'Well, Julia is from Germany and Matilda is from France.'

'Okay.'

'But things have been better on the way back than on the way out, have they?'

'Yeah, much better. I think it must be downhill all the way from Hungerford. Only two more days to go now.'

'Well, I better get on to the pub, I guess. Now I can say that I saw you out on the road.'

'Yeah, well, we'll be in at the pub soon. We'll see you again there.'

Annette drives away, her pretty face beaming at having seen us out here. Moz sets off in his car behind her.

At 10.00, Baz wanders along the Track to me.

'Hi Gregga,' he says.

'Baz.' I nod in greeting. 'How are you?'

'Good. You are travelling well, Gregga.'

'Things are good.'

'Let's go.'

Stiffly, I get to my weary feet. I realise how comfortable I had been, sitting there in the dirt. Baz is moving quickly and, after waiting for him for almost half an hour, now I have to hurry to keep up.

We talk about this morning's injured kangaroo.

'Poor bugger.'

We walk on towards Fords Bridge.

'Cop a load of this,' Baz says.

Up ahead, emerging from a bend in the road, I can see Moz walking out towards us. He is dressed in white trousers, a white t-shirt, a white legionnaire's hat and he seems to have materialised from the Bush.

'He looks like a ghost.'

'No, he looks like an escapee from a lunatic asylum,' says Baz.

Moz walks talk and erect. I wonder what he is thinking and how he is feeling being back out here.

'G'day Grog.'

'G'day, mate,' I say as we pass one another. His eyes are hidden behind dark sunglasses.

I pass on by, not really knowing what to say.

Baz and I continue ahead while Moz walks back towards Jane.

We get to a blue sign that says, 'Please stop and drop dust.' It could mean a variety of things. Baz decides to take it literally, stopping to bend and pick up a fistful of red sand. I photograph him as he lets the sand run through his open fingers. The sleeves of his dusty navy blue shirt are folded to three-quarter length. His white shorts are stained red from sand and his hat is fastened to his head with its strap. His face sports a grizzly beard and moustache. Although his eyes are hidden behind dark glasses, he looks happy.

At 10.20, we step onto the veranda of the Warrego Hotel. An old 'soak' from the pub is on the veranda to greet us.

'You made it back,' he says.

'We made it back.' I am pleased to have got to the pub so early. We have been rewarded with an unscheduled rest day.

The present Warrego Hotel was built in 1913 and is the oldest surviving mudbrick pub in Australia. The previous pub at Fords Bridge was a Cobb and Co. changing station. That pub was named the Salmon

Ford Hotel and would have been the one that stood here when Henry passed through.

We go out the back to where Annette is attending to laundry and check that it is okay for us to go to our room. 'I'm not sure if the rooms have been made up yet,' says Annette. 'I don't know if they have had time to do it.'

When we get to our room, we see that it has not been touched since we slept in here 11 days earlier.

I gather the things that I need and then head out the back to use the toilet and to take a shower. There are frogs and mosquitoes everywhere.

By the time I emerge, refreshed and clean, Jane and Moz have reached the pub. We gather our dirty clothes and Jane puts them in the laundry.

Baz and I venture into the bar, where Baz attends to his journal. I prop myself on a barstool and down a series of ice-cold drinks.

'Just let me know when you blokes would like some lunch,' Annette says.

We decide to wait until Moz gets out of the shower before we order.

'I feel like I'm done, Gregga,' Baz says as he sips from a cold beer. 'It feels like we're at the end.'

'Don't get too carried away yet, Bazza. There are still a lot of miles between us and Bourke.'

'Yeah, I know, but I feel like it's over. With this long rest we have here, I find it difficult to think that there are still two days of walking to go.'

I will not allow myself those thoughts. I know a lot can still go wrong. I also know that, even if things do not go wrong, the 68 km between Fords Bridge and Bourke still need to be walked and 68 km is never going to be an easy distance for anyone, let alone two people who are already worn down and exhausted.

Caps and stubby holders hang from a clothesline behind the bar. A framed photograph of a rugby team adorns one of the pillars that hold up the ceiling. Beneath that photograph, a steel-jawed rabbit trap yawns wide open, as if tired of waiting for some fool to put his fingers in there. Under the trap someone has pinned a photograph of a man in a yellow shirt dangling a snake by its tail. The snake is as long as the man. What appears to be a grass skirt hangs from the ceiling in front of one of the glass-door refrigerators. The skirt does not look like it has been worn by anyone in the last several decades.

Baz is scratching at his legs. His heat rash has been driving him crazy for much of our time out here and, no doubt, he has mosquito bites on his legs too. 'I am sick of wiping blood from my legs,' Baz says. 'Gregga, do you ever make yourself bleed?'

I laugh, although not as loud as my wife would laugh if she had heard that question. 'Jennifer says I am like a psychopathic killer the way that I make myself bleed. Dad does it too, you know. It's a Bryan thing.'

'Yeah.'

'I wonder what Jennifer and the people in Canada are thinking now. I wonder how they think we are going.'

'Do you really think they have given you a second thought?'

Baz is stirring. He loves it. I ignore him. 'They would think we are still out on the Track. They wouldn't imagine we are in a cool room sipping ice-cold drinks.'

When Moz emerges from his shower, I decide that for lunch I will go with chips and gravy again. They were perfect on the way through on Day Two and so I am happy to stick with a winner. My food is delivered to where I sit at the bar and so I eat it there although the other three eat at a table near what is labelled the Marriage Counselling Corner. Beside the sign a hard hat, a rolling pin and a pair of boxing gloves hang.

'You did pretty well on the quiz last night, Moz,' Baz says.

Moz and Jane both seem to know what Baz is talking about, but I have no idea.

'What quiz is that?' I ask

'Oh, I was doing some questions from the newspaper last night.'

'Really? I didn't hear a thing.'

'You were too busy snoring.'

'It was after the car pulled up,' Baz says.

I actually had heard a vehicle pull up on the road beside our tents. They must have seen Moz's car beside the road and wondered if we were in need of assistance. When no one seemed to be moving, I had called out into the darkness, 'Are you going to attend to them, Moz?'

'Are you all okay?' I heard someone ask.

'Yeah, we're fine thanks,' Moz responded.

'Why did you ask *me* to attend to them, Grog?' Moz asks me now.

'Well have a look at you. You're six foot eight and a man mountain, mate. You're far more imposing than me when strangers stop by in the dark.'

'Great thanks. It could have been a killer and you're sending me out to investigate.'

'Well, *I'm* hardly going to scare anyone away.'

'There was another car later,' Baz says.

'I only heard one,' I say.

'No, there was another one,' Moz says. 'It was weird. They didn't say anything. It was kind of unsettling.'

'Yeah, I agree,' says Baz. 'I thought that it might be Ivan Milat or Bradley John Murdoch stopping.'

On the one hand, I am pleased to have slept through a visit from the Backpacker Murderer or Peter Falconio's Outback killer. On the other hand, I would like to wake up and have a fighting chance if it ever comes to that.

'You blokes have been watching too many horror movies,' I say.

After a delicious lunch, Moz and Jane decide to go into Bourke to visit the Back O' Bourke Exhibition Centre. Although Baz agrees to go to Bourke with them, it does not seem right to me to get into a car and go for a drive before the tramp is completed.

'I'll stay here and rest,' I say.

When the others leave, I retire to my room. I lie on my bed and read from Henry before I settle into a snooze.

≈

All good things must come to an end and in Henry's life that most often happened sooner rather than later. As his employers began to wonder if they were getting value for their money, Henry grew tired of life in Leeton. The ban on alcohol in the Murrumbidgee Irrigation Area was wearing thin. 'Christ didn't believe in prohibition,' Henry reasoned, appealing to the Bible for support as one means of alleviating his growing thirst in what he described—despite all the water—as 'the driest and thirstiest' place imaginable.

Leeton's prohibition rule was always going to be problematic for Henry. He found the absence of a pub 'uncanny [and] supernatural,' saying 'that a Place is not natural without the Pub.' Henry penned a short note to Jim, complaining 'I'm the Commander of the Army of the Fed-ups in Leeton.' After a little over a year-and-a-half in Leeton, Henry was heading back to Sydney.

≈

I venture into the bar and I see that the women backpacking from overseas are working.

'How are you feeling?' they both ask.

'Yep, I'm good.'

They are both attractive and smile freely. I had been too exhausted to notice when I was here on Day Two. Julia is taller and darker, while Matilda's English seems stronger. Baz told me that one did not have a very strong command of English but they both seem good to me.

'How do you both like it out here?' I ask. It must be very different for them, so far from home and, I suspect, in a place so very different from home.

'It is okay. I like it mostly.'

'I like it too.'

Julia is in and out of the bar, attending to her laundry so I speak primarily with Matilda.

'Where do you live?' Matilda asks.

I tell her that although I am an Australian, I live in Canada.

'I love Canada,' she says. 'I want to go back there.' She was in Canada a year earlier and tells me how much she enjoyed the experience, wishing only that she had time to stay longer. She plans to go back one day.

'What do you think of the heat out here?' I ask.

'It is too hot for me.' She laughs.

'What about the mosquitoes.'

'Oh, I hate the mosquitoes.' She shows me some lumps on her arms and gives them a scratch. She tells me about swimming in the Warrego River on New Year's Eve. I can imagine there would be a million mosquitoes for every drop of blood.

I find the two women's stories interesting. Henry's wife said of him that he 'was interested in every human being whom he met.' Once upon a time, I was like that too. Talking to Matilda and Julia—with their occasionally faltering English—reminds me of my days backpacking in Europe when I met Jennifer. I used to like meeting people but have become reclusive with age. Had I then been as introverted as I generally am now, I would never have met Jennifer, let alone won her affections. I was different back then. People change over time. I have withdrawn and become insular. 'There were times when I really wished in my heart I was on my own,' Henry wrote, and I know what he means. Henry referred to himself as 'painfully shy' and that description fits me well. However, I do consider one of the highlights of this trip has been the people that I have met and talked with along the way.

Jane has noticed a difference in me and says to Moz, 'Greg is a different person this second week. He has really come out of his shell.'

'How are your feet now?' Matilda asks me, pulling me back to the present. 'When you were here I felt so bad for you and I wondered how you were going to get to Hungerford?'

'There were times where I wondered the same thing.'

'I read about what you are doing when a report appeared in the newspaper just before Christmas,' Matilda says. 'It must be so hard.'

'You've got that right.'

Matilda is a happy, friendly person. I like the freckles on her pretty face. She tells me she is 23 and that she is visiting Australia for one year. 'Julia is out here for six months,' Matilda says when Julia re-enters the bar. 'She has to get back to see her boyfriend.' She laughs.

Baz comes in to the bar, having just returned from the trip into Bourke.

'How was it?' I ask.

'Yeah, it was really good.' He tells me about the Outback Centre and the things on display there.

'How was the drive in and out?' Baz and I have been talking about Moz's driving.

'Hairy.'

'Fast?'

'I was hanging on so tight that I had to pluck my fingernails out of the upholstery when we got to Bourke.' He holds up his hands like claws and then pulls them backwards, as if removing his grip from the seat in front of him.

When the backpackers leave, Baz tells me Julia sleeps with a knife under her pillow.

'How the bloody hell do you know that?' I laugh.

'Scott was telling me. He went with another bloke to pick them up and when they were driving back here, she told them about staying in a hostel in Sydney. Apparently, some bloke snuck to her bed one night and she woke up and just said to him, "I sleep with a knife under my pillow." That got rid of him.'

'I'll bet it did.'

A bloke comes in to the bar and asks for a packet of cigarettes.

Matilda retrieves a pack from behind the bar and looks up the price on a sheet of paper.

'Eighteen dollars,' she says.

I just about fall off my chair. She must be mistaken. I have never been a smoker and so I have no idea how much they cost but I cannot believe that a person would pay that much money for a packet of cigarettes.

The man hands over the money.

'Can you believe that, Baz? Eighteen dollars for a packet of cigarettes!'

Baz thinks about it for a moment. 'Eff 'em.' Like me, he thinks smoking is a ridiculous habit.

Scott approaches us at the bar. 'Do you blokes want to be in a pool comp?' he asks.

Baz dobs us in to play before I even have time to think about the meaning of the question.

'Five bucks in and the winner gets $40 and the runner up gets $20.'

'Put us down. We're in,' Baz says.

'Yeah, no worries, Scott,' I say, reluctantly. After Scott has left, I turn to Baz, 'Bugger ya, Baz. I haven't had a pool cue in my hand for ten years.'

'Me either, but when in Rome...'

I picture the two of us in togas. 'We're gonna make fools of ourselves.'

≈

In the pool competition, most of the participants are from the roadwork party staying in caravans in the car park outside the hotel. Baz is drawn to play first. His opponent has a beard like Ned Kelly. If Baz wins, he might get shot.

Baz breaks. He looks like he knows what he is doing and he starts well and gains an early lead, sinking a number of balls before Ned Kelly has put even one in a pocket. Throughout the early days on the Track, Baz was always way out in front of me but, on this return leg, he has lost some of his energy and has often been behind me. Things pan out the same way in his game of 8-ball. He leads early but he is gradually overtaken and eventually goes down.

I am drawn to play next. As with my time on the Track, I start slowly. Things look so bad that I am desperate to sink merely one ball before the other bloke cleans up. The longer the game goes though, the better I play. By the end, I have a shot on the black and should win the game. Unfortunately, I fluff it and, in what I hope is not foreshadowing of what is to come out on the Track, like Baz, eventually I am beaten too.

'There's ten bucks we won't see again, Baz.'

The music has been getting louder and louder and it looks and sounds like they are settling in for a big night out here. We watch a couple of games and then Baz says that he is going to bed. I get my toothbrush and head outside to brush my teeth and get ready for bed too. It would have been nice to win the pool competition but it is even nicer to go to bed. When we hear the next morning that they still did not have a winner at 1.00 am, Baz and I are glad to have been first round losers.

'They're playing some good music,' I say to Baz as I lie in the darkness, trying to fall asleep. Australian Crawl's *Reckless* is followed by Mental As Anything and *The Nips are Getting Bigger*. I fall asleep but, sometime later; I am awoken by the sound of Neil Diamond singing with a loud and drunken accompaniment every time the song reaches the chorus of *Sweet Caroline*. It is going to be a long noisy night.

DAY FOURTEEN

Henry Lawson's Final Years

He was an awful object by this time, wild-eyed and gaunt, and he hadn't washed or shaved for days.

- Telling Mrs Baker, *Joe Wilson and His Mates*, 1901

I saw at once that he was a very sick man. His face was drawn, and he bent forward as if he was hurt. He got down stiffly and awkwardly, like a hurt man, and as soon as his feet touched the ground he grabbed my arm, or he would have gone down like a man who steps off a train in motion.

- No Place for a Woman, *On the Track*, 1900

My day starts with getting massacred by mosquitoes while I am sitting on the toilet seat. I shower then venture into the kitchen. While I am preparing breakfast, Matilda wanders in to join me.

'Have you tried Vegemite, Matilda?' I ask while I am putting my beloved spread on my toast.

'Yes. It is quite good. I like it.'

Good on you. 'Who won the pool tournament?'

Matilda tells me that she went to bed before it was over. 'It was late!'

Baz comes into the kitchen to say goodbye. He was up and about before me and is ready to leave. Several minutes later, however, I see him still poking around in the bar, taking photographs. I also see him talking on the telephone.

I fill my hydration bladder and two water bottles. Once I have everything in readiness, I say goodbye. I move outside, where the sun is already hot. It is 8.45 am—a very late start to what will be a long day.

Shortly after I take my first steps along the Track, I arrive at the Warrego River where it is extremely verdant. Little wonder Matilda got covered in mosquito bites when she took a New Year's Eve swim here. There are hundreds and hundreds of Tree Martins darting about the bridge, presumably dining on the surronding insect smorgosbord. The martins' white rumps flash as the birds twist this way and that. I cross the bridge, enjoying the site of so many martins filling the sky about me. There are also several dozen on the ground on the other side of the bridge. To see so many birds is, for me, one of life's great joys.

Ten minutes after I cross the bridge, I venture off the Track to visit a grave that marks the final resting spot of someone who is, at least to me, unknown. A wire fence has been erected around the lonely gravesite. Someone has threaded onto the fence what is now a tattered and faded bouquet of plastic flowers. As I leave the grave, I notice fresh footprints in the sand. I assume Baz made a similar detour to pay his respects and satisfy his curiosity when he passed earlier this morning.

When I return to the Track, I notice that my shoes have picked up several prickly burrs so I stop to remove them before proceeding.

'Are you okay?' Asks a driver in a one-tonner that has pulled up alongside me.

He is an older gentleman, travelling alone.

'Yeah, I'm fine thanks. My brother and I are walking back to Bourke and I've got a shoe full of burrs bothering me.'

'Oh, yeah, I heard about you blokes being on the road.'

I wander closer to the vehicle. 'Do you live out here, do you?' I ask.

'I'm from back at the turn off at Youngerina. You would have passed that.'

'Yep. I remember the sign for Youngerina alright.'

'I just thought I should stop and make sure you are alright. You don't see a lot of people on foot out this way. I better keep going though. I have an appointment in Bourke at ten o'clock. I'm not sure if I'll make it.'

I look at my watch. It is 9.15 am. 'You better get cracking.'

It feels good to know that, although I am on foot, I am within a distance that can supposedly be driven in less than an hour.

Up ahead, I can see that the road workers from the pub are in full swing. A water tanker showers the road as it rolls before me and a couple of huge Caterpillar graders slowly creep their way over the wet surface. The graders' blades push big piles of red sand before the machines, filling depressions and knocking the tops from high spots. In their wake they leave behind a smooth soft surface that is tailor-made for walking.

≈

Late in 1917, Henry left Leeton and returned to Sydney. During these days before television and movie stars, Henry—being a famous writer—was considered a celebrity on the streets of Sydney. Yet, at least in the mind of one friend, the hustle and bustle of crowded city streets were lonelier places for Henry 'than the solemn immensities of the sunburnt plain' where Henry knew and understood himself and his companions. In Sydney, people made strange requests of him. A salesgirl in a shop Henry frequented asked for his opinion on one of the poems she had written. A friend said, 'It was a poor thing...but Lawson thanked her for the privilege of being allowed to read it.'

Strangers would often stop Henry and ask him to make a note in their autograph albums. Bemused, but generous with his time, Henry would happily oblige, often making the entry:

'Mary had a little lamb—
And I had steak and onions.'

≈

Remembering the song to which I fell asleep last night, I wander down the Track singing Mental As Anything's *The Nips are Getting Bigger*. There are few more catchy songs:

Started out, just drinkin' beer.
I didn't know how or why
Or what I was doin' there.
Just a couple more
Made me feel a little better,
Believe me when I tell you
It was nothin' to do with the letter.

The song seems perfectly in keeping with Henry and his trials with drink. In a letter to the *Bulletin* newspaper in 1903, he wrote:

Dear Bulletin,

I'm awfully surprised to find myself sober. And, being sober, I take up my pen to write a few lines, hoping they will find you as I am at present. I want to know a few things. In the first place: Why does a man get drunk? There seems to be no excuse for it. I get drunk because I'm in trouble, and I get drunk because I've got out of it. I get drunk because I am sick, or have corns, or the toothache: and I get drunk because I'm feeling well and grand. I got drunk because I was rejected; and I got awfully drunk the night I was accepted. And, mind you, I don't like to get drunk at all, because I don't enjoy it much, and suffer hell afterwards. I'm always far better and happier when I'm sober, and tea tastes better than beer. But I get drunk. I get drunk when I feel that I want a drink, and I get drunk when I don't. I get drunk because I had a row last night and made a fool of myself and it worries me, and when things are fixed up I get drunk to celebrate it. And, mind you, I've got no craving for drink. I get drunk because I'm frightened about things, and because I don't care a damn. Because I'm hard up and because I'm flush. And, somehow, I seem to have better luck when I'm drunk. I don't think the mystery of drunkenness will ever be explained—until all things are explained, and that will be never. A friend says that we don't drink to feel happier, but to feel less miserable. But I don't feel miserable when I'm straight. Perhaps I'm not perfectly sober right now, after all. I'll go and get a drink, and write again later.

Henry Lawson

Save for the year-and-a-half that he spent in Leeton, for much of the last decade of his life Henry was often seen stumbling through the streets of Sydney drunk. He also earned a reputation for constantly begging money that would remain with him no further than the nearest bar. Not everyone begrudged Henry's begging though. One newspaper editor felt that giving money to Henry as a token of their friendship, 'bestowed some indefinable honour upon me.' Proud was the editor to know Henry Lawson.

≈

When he appears, Moz gives me his customary greeting. 'How are you popping up?'

'Good. I take it Jane is walking again, eh?'

'Yep.' He smiles. 'Grog, you know that chick you were chatting with last night?'

'Matilda? Yeah.' I can tell by the look on his face that he is going somewhere with this.

'Well...for authenticity, I suppose like Lawson, you were "humping Matilda" last night.'[1]

'Clever Moz, but the thought never crossed my mind.'

'Yeah, right.'

It is early morning and the sky is filled with clouds, but already I am feeling the heat.

'What's the temperature, Moz?'

'It is 29 degrees.'

Warm enough already.

≈

As much of a celebrity as was Henry, not everyone wanted to see him. Smelling of stale beer and tobacco, he became a nuisance around the Angus & Robertson offices. In order to get some work done, George Robertson would often hide when Henry showed up.[2] As their star author, Henry's drunken presence around the building did not reflect favourably on the business. Robertson got so fed up that he began to pay Henry's housekeeper, Mrs Byers, to stop Henry from going to the office.

Henry's drunkenness was turning more and more of his friends away from him.

≈

As I talk into my voice recorder at 9.45, I say that I am 'buoyant, optimistic, hopeful and even confident' given that it is Day Fourteen. Tomorrow this hellish experience will be over.

The current stretch is one to be enjoyed before I get to the rocks and, eventually, back to the bitumen. There are graders and a water truck working on the road and this makes it even easier to enjoy the red carpet because they compact the sand so that it is not so soft that it is slipping beneath my feet and working its way into my shoes and socks. The graders also make the surface smooth and obstacle free. The Track here is about as good as it gets—flat, clear, and soft but firm. I enjoy the easy walking while I can.

A car zips past me and I notice that the driver and his passenger are both wearing fluorescent safety work vests. To my surprise, once the car gets ahead of me, its starts to slalom, gouging ruts into the surface. I guess the driver thinks he will give himself and his work mates more to do. A jeep passes. The man driving was in the pool tournament last night.

I see Moz ahead, returning from having checked upon Baz. He smiles and waves as he passes and drives back to see how Jane is getting along. He has not been gone long when he drives back to me again. He must be enjoying driving back and forth on this graded road.

'How are you popping up now, Grog? Have you had fun humping Matilda this morning?'

'I'm good, Moz. I was just working on a voice recording. I think I have made about 70 so far—usually 5 or 10 minutes each, but some are 15 or 20 minutes.'

'Yeah, good. I know you are writing notes too and so are Baz and Jane. When I was walking I never bothered to document anything because I was just hoping it would hurry up and be over.' He smiles, but I know the pain behind that comment better than anyone.

'It would have been interesting to have other people's thoughts too, eh? It would have been interesting to have people record what they thought as they saw us or as they went by at different stages along the tramp.'

Moz thinks a moment. 'I think I can sum up what the other people thought, Grog.' He pauses for impact. '"You're effen mad! " Tom Bloggs in Fords Bridge: "You're effen mad." Joe Bloggs in Hungerford: "You're effen mad." Tony Marsh at Kia-Ora: "You're effen mad."'

We laugh.

'Yeah, you're probably right. I think most people think we're insane.'

In addition to, or as a result of, his alcoholism, Henry's later life also involved battles for his sanity. Amidst the pressures of trying to write for a living, alcoholism coupled with the eccentricities of his genius saw Henry in and out of the Mental Hospital and the Reception House for the Insane attached to the Darlinghurst Gaol. One time, a poet friend perhaps doubting his own sanity commented to Henry that a general opinion is that poets must all be crazy. Henry 'turned the thing over in his mind' and then whispered comfortingly into his friend's ear, 'Don't let that worry you, Ted. It's them that's mad, not us.' Another time, a friend asked Henry why he had been inside the asylum. Fed up with what he saw as a life of persecution and criticism, Henry replied, 'I went there to escape from the damned lunatics outside.'

Moz drives ahead again and another truck passes. The driver toots and waves but also blows a cloud of dust into my face and eyes. When the dust settles, I can see Baz's footprints on the freshly graded road and so I step onto them and I am surprised to discover that the distance between each of my footprints is further than the distance between Baz's footprints. I am taller and have longer legs, so can obviously take a bigger stride, but for much of this tramp mine has just been a short shuffle. I try to figure out how far it takes for me to make a full step less than Baz. I walk along, counting. 'One, two, three, four, six, eight, ten, twelve, fourteen, sixteen, eighteen.' In the space where I have taken 18 steps, Baz took 19 steps.

Another vehicle approaches. I see it is one of the Country Energy vehicles. The driver shows his consideration by drifting across to the other side of the road as he passes and, by so doing, reducing the amount of dust that flies into my face. I wave in thanks.

A Toyota vehicle approaches from the rear. Although it is mainly the workers going to and fro, there is a lot of traffic on the road this morning. It seems strange after days of seeing almost no one. The Toyota driver gives me a friendly wave but, as with most of the vehicles this morning, the driver does not stop. By now, they all know who we are and what we are doing.

I have noticed that my water tastes weird. I wonder if the tube of my hydration pack is dirty or if it is because of the tank water from Fords Bridge. I am thirsty, but my water is unappealing. It seems almost to have a woody taste to it. What I would give for a glass of clean water with cubes of ice bobbing in it.

I am still on the red carpet but when I get beyond the road works, the sand is soft and moves beneath my feet due to the absence of the water truck. This is a long stretch of red carpet, but it might be the last. Soon enough, the Track will be punishingly hard and unyielding.

≈

After leaving Leeton, Henry missed his old mate. While on a visit to the country, he wrote Jim a letter. Jim had been writing poems[3] and Henry advised him to play two newspapers against one another in order to secure the best payment. Henry also said Jim should submit his best materials to the *Bulletin*. Anything that the *Bulletin* rejected, Jim could then submit elsewhere. Henry encouraged Jim to think about a compilation of his best materials into a book. 'I paved the way with Mr Robertson of A & R,' Henry wrote, explaining that he had shown Angus & Robertson some of Jim's best writing. 'Your stuff is all good, some splendid,' Henry enthused, 'It throws all mine on similar lines on to the Men's Hut rubbish heaps altogether. But then you had all the experience and I next to none—except for six months in Bourke and beyond.' In the letter, Henry said that where he was staying there was 'a pub next door, but, you know,' wrote Henry, 'all the interest seems to have gone out of life since we parted.'[4]

In Sydney in June 1921, Henry suffered a mild stroke. It led to just one in a long line of hospital stays as Henry was in and out of hospital through the final years of his life. Henry was shocked by the state of one hospital and the patients within, describing his surroundings as 'the dreariest place in Australia [and] the most depressing.' The wards were 'shamefully, appallingly overcrowded'. He wished, perhaps, that he and his fellow patients could have 'died amongst men who understand.'

Henry realised that most people did not understand him or his troubles. 'They drank and threw money away,' he could hear people scoffing snidely about him and the other patients. Henry's response was that life could get so bad that 'the periodical spree was all [some people] had to look forward to.'

Because of his celebrity status, Henry often received visitors bearing gifts. One time when he was hospitalised, his friend Roderic Quinn planned to visit. Quinn asked those who had already visited Henry what he needed. He was advised to take nothing, for Henry had received numerous gifts and had all that he might possibly want. When Quinn visited, he was surprised when Henry asked him for some tobacco. 'But [everyone] told me you had heaps of tobacco and forests of cigars,' Quinn said.

'They told you that, did they?' asked Henry. 'Well they shouldn't have.'

'Why Henry?' Quinn asked.

Henry discretely indicated his fellow patients and then whispered, 'Because there are others.'

Every gift that he received, Henry shared.

As Quinn made a hasty departure to purchase some tobacco, Henry called out for some bananas as well. 'You know how I love bananas.' In fact, Quinn had never known Henry to eat a banana in his life. Henry also told another friend, Walter Jago, that he was 'very fond of bananas.' Jago later learned that every banana Henry received he gave to children in the hospital.

≈

I pass a big goat dead on the side of the Track. There is no obvious damage to the animal and, as it is not yet smothered either with flies or maggots, it cannot have been dead for long. Except for a thin line of black that extends along the ridge of its back, it is tan over most of its upper body and flanks. Dark black hair covers its underbelly and it is also black from its knees down to each hoof. A long black beard extends from its chin. Above, each horn is more than a foot in length. It must have been a beautiful animal when it was alive.

I crest a rise and round a bend and find Baz and Moz sheltering in the shade beneath some trees.

'Would you like a cordial, Grog?'

'That would be nice.' With my water tasting strange, I am dry and thirsty.

We watch an Emu walking towards us from the other side of the Track.

'The walking has been good this morning, Baz.'

'Yeah, it's been nice on the feet,' Baz agrees.

'So you enjoyed "humping Matilda" this morning, did you?' Moz teases.

Moz and Baz both laugh.

'I'm going,' I say, skulling down the rest of my cordial. 'You buggers are going to have me divorced before long.'

I have not gone more than ten minutes when, at 11.25, I take my last few steps on the red carpet and transition onto the bitumen. I see that the black top stretches a kilometre or so to the next crest and, presumably, beyond. As I step off the sand, I figure that from this point forth, I have either bitumen or hard stones. Now, my feet are in for another battering.

I move to the side of the Track, where it is softer than on the bitumen. Unfortunately, however, there are lots of rocks on the side and so it is difficult to walk on. I feel caught betwixt and between—do I want to walk on the rocks at the side where it is softer, or the hard surface where it is flat and even? I edge back onto the bitumen.

I notice that the cicadas are singing again today. One gets so used to them that their noise blends into, and becomes a part of, the background. The temperature must be well into the thirties and unfortunately the water coming from my hydration bladder still does not taste good. Although there is cloud cover, the cirrus clouds are soft and wispy. There does not appear to be any relief in terms of showers or shade for the remainder of the day.

To be on the bitumen suggests we are getting closer to Bourke. I figure I have done about 12 km so far today. That must leave me about 56 km to go. Having already done almost 400 km, what remains does not seem very much. There is still a lot of work to do though. With the hard surface, the hot sun, big rocks that threaten a twisted ankle, a heavy load on my weary back, and snakes and such, a lot can still go wrong. I cannot celebrate yet. It is important to remain mentally tough.

When Moz and Jane catch me, I ask them to drive ahead one kilometre. I am interested to know how many steps I might take on this journey. From an agreed starting point, I begin to count my steps and Moz begins to measure a one-kilometre distance. I am aware that I am walking better today than at almost any time on the outward leg to Hungerford but at least my calculations will serve as a rough guide. With my notebook in hand, I record a tally mark for each one hundred steps. In one kilometre, I take 1308 steps. Working with an overall distance of 450 km, that number suggests I am going to take approximately 588,600 steps between leaving the Port O' Bourke Hotel and returning to the Port O' Bourke Hotel.

Forty minutes later, I come to a line painted across the road by workers for some purpose or another. On the way out, Baz mentioned he had found a 400 m distance marked on the road and I see that is the point where I am now. Using the 400 m distance, Baz calculated his overall number of steps. I know I just counted my steps, but just to be sure I will check against this marker and count my steps again. I take 545 steps over the 400 m distance. This suggests I will take 613,125 steps over the course of this journey. Given the two totals—one a little under and one a little over—it seems reasonable to conclude that I will take about 600,000 steps!

Back on the bitumen, it doesn't take long for me to notice that the hard road is troubling my left Achilles tendon. I decide to have a little rest. As I make my way towards the Mulga Trees for a place to sit down, I see a stunningly beautiful Splendid Fairy Wren. The range of blues across the body is, indeed, splendid. With its tiny size and ornate colouring, the bird seems so delicate, yet it must be tough and hardy to survive out here. As with many times during the past two weeks, when I sit down, I see a Red Jewel Bug. The bold red shield of the bug is undeniably stunning. To the rear of the shield, triangular metallic green spots are bordered by dark violet lines. In amongst the dust and the heat, not all of the beauty out here is subtle.

Baz eventually catches up to where I sit and so I get up and join him on the Track. Over the next half an hour, I grow increasingly hot and the unpleasant taste of my water discourages me from drinking. As Henry said, 'The heat is bad, the water's bad, the flies a crimson curse.'

'How does your water taste today, Baz?'

'Hot.'

'Well, yeah, of course, but mine tastes like shit too.'

At 1.15 pm, we leave the Track and climb the fence to explore the ruins of an old hotel at Lauradale. It is one of a number of spots along the Track where a hotel stood at the time of Henry's tramp. On the way to Hungerford, I did not have the energy for such explorations. Amongst the broken red bricks and jagged tin, the metal pipes and the splintered wood, a decrepit windmill still stands, although only the tail vane remains at the top of the stand. The wheel has tumbled from its roost. The rusty joints and seams in the round water tank suggest that, like me, the tank is well past its prime. When I see a series of thick wooden poles standing upright beside the tank, I realise that, like the wheel of the windmill, the tank has also tumbled from its roost.

Where they might once have led to an inviting bar, three concrete steps now lead to nothing but thin air and broken debris. Today, one can walk up those steps and stumble off the top because they do not lead anywhere. I wonder how many inebriated people stumbled from those steps when they did lead somewhere.

After taking some photographs, Baz peeks down a well that is beneath the windmill and discovers an Eastern Blue-tongue Lizard down there. The lizard clings to the floating end of a big log that is largely submerged beneath the water in the well. It is the biggest blue-tongue I have ever seen. It must stretch more than 60 cm from the tip of its tail to the point of its nose. It is not moving but, with its head held high, it is clearly alive. About two-thirds of the length of the lizard is dry but its rear legs and tail are underwater.

'Do you think he can get out?' Baz asks.

'It's hard to know if it wants to be there or not.'

'It might be catching insects and staying cool in the water.'

'Who knows?'

I expect there is plenty of food there. We see several frogs and a couple of toads that are also down the well, presumably hunting insects and other water invertebrates. The walls of the well are entirely vertical. Whether the lizard wants to be there or not, it is hard to imagine it being able to climb out when it eventually decides it would like to.

'We should see if we can help it,' says Baz.

The difficulty of a rescue operation is that the lizard is at least two metres below the ground surface and added to that, we have no idea how deep the water might be. The ground surface is uneven and crumbly too. From where I stand, it looks as if Baz will be doing well if he doesn't fall into the well. He retrieves a thin but long metal pipe from the ruins of the pub and pokes it down into the hole. He tries to fish the lizard out with the pipe. Both Baz and the lizard are in awkward spots though and I worry that Baz will end up clonking the lizard on the head. The lizard seems to share my fears and it wobbles off the log and into the water. It seems to swim okay and there are plenty of things for it to grip onto so I do not see it being in any immediate danger of drowning. I go to the ruins and pick up a long wooden plank. I drop that into the well and lean it against the side.

'Maybe it'll be able to crawl up that,' I say.

After several minutes of unsuccessful rescue attempts, we decide that we have probably done all that we are able to do to help. Perhaps the lizard will be able to crawl up the things we have lowered into the hole. Perhaps the poor bugger is perfectly happy where it is anyway.

From the Lauradale ruins, it is only a kilometre to Sutherlands Lake. There are many birds at the lake and we use Baz's binoculars to watch them going about their business. I am surprised to see how green the water is. I submerged my head in that water when I was boiling on Day Two. I do not remember it looking like this.

'Maybe I made it turn green,' I suggest.

Whether it is because the temperature is cooler or whether it is because the water is green, I do not dunk my head today.

Back on the Track, two trucks pass by in quick succession, burying us beneath storms of dust. Baz and I continue on through the dust, walking side by side. We continue until 2.40, when we reach the place where Moz and Jane have set up for lunch.

Moz has a small fire going. 'It looks like we have extra food, so we can have a hot lunch if you blokes like.'

Given that we are in the process of completing one of my life's goals in retracing Henry's footsteps, somehow we get to talking about Baz's childhood fantasies too.

'Baz used to love Celeste and Judy Green. They were models on *Sale of the Century*.'

'Judy Green used to be my kindergarten teacher,' says Jane.

'What? Really?'

'Yeah, she was a teacher. She also won Miss Australia.'

After lunch, I say to Baz, 'Imagine that. Jane's kindergarten teacher was Judy Green.'

'Yeah, I'm not sure about that,' Baz says dubiously.

'What do you mean?'

'Well, I think there was another Judy Green who was Miss Australia but that's not the one we were talking about.'

Sure enough, Judith Green won Miss Australia in 1987, but the Judy Green who every evening would brighten our teenage years by draping herself across a car was a different Judy Green. Baz has a good handle on these things—he knows his sexy models.

Moz and Jane have gone ahead to find a suitable campsite. Knowing that we still have a long distance to make it to Bourke tomorrow, they have decided to go as far as they can before they hit the plains, where there is nothing but long tussock grassland and nowhere to set up our tents.

As we walk together after lunch, Baz and I see various wildlife, including a small Painted Dragon. At one point, Baz stops to photograph something and, as I push ahead, a snake slithers across the Track and into the grass before me.

Within half an hour of taking to the Track after lunch, Baz decides that he needs urgently to answer the call of nature.

'Well, is it any bloody wonder?' I ask. 'While I chose to have spaghetti, you went for baked beans!'

'Yeah, and as I ate them, I remembered I don't really like them anyway.'

Baz leaves the Track and makes off towards the Bush in a hurry. Even if I want to, I cannot move that fast. I hear a few curses and oaths and turn back to see Baz tangled and stumbling over a fence.

'Come and give me a helping hand,' Baz calls.

All I can do is laugh.

'If you don't come back,' I say, walking on ahead, 'I'll let your wife know that you were last seen heading into the Bush with a trowel and toilet paper in your hands.' That should make him recognisable on any Missing Persons posters. I hear the distant roar of an approaching vehicle. 'Here comes a busload of Japanese tourists,' I call out.

≈

During one visit to hospital, Walter Jago found Henry seething at a newspaper article in which Henry was portrayed in an unfavourable light. The article was supposedly the product of an interview that Henry claimed did not even occur. Henry duly prepared a scathing written response in which he exposed the journalist's lack of honesty and showed it to Jago. The letter would destroy the journalist's career. It so happened that Jago knew the journalist and knew also that he was having a difficult time making ends meet. When Jago informed Henry of this, Henry proclaimed, 'The poor devil' and immediately tore the letter to shreds.

≈

Eventually, Baz and I make it into camp a little after 5.00 pm. We are on the edge of Seventeen Mile Plain, amongst the last of the trees before the plain opens before us. We are at the same spot where we camped at the end of Day One. Moz has taken the spade from the back of the car and he and Jane have hacked paths through the brush and burrs to make it easier for us to walk about our campsite. By the time we get there, Moz is sweating almost as much as he was during the heat of the first few days on the Track. Jane is also hot and even sunburnt. The two of them have obviously been working very hard and Baz and I appreciate their efforts.

'It is like the "Yellow Brick Road",' Baz says effusively, following the cleared path to our tents.

Seeing the sweat pouring from his brow, I ask Moz about today's temperatures. Despite the cloud cover that existed for much of the day, it got to 38 degrees.

Referring to the weather, Jane says to Moz, 'If the second week had been the first week, you would have made it.'

Moz says nothing to suggest disagreement but I am not so sure about that. It is possible he would have made it but it is easy to say that from the comfort of a car. Yes, it has been cooler in the second week and today has been one of the coolest days but it still reached 38 degrees. Given the cloud cover, I suspect it has seemed much cooler from inside the car than has really been the case out on the Track. One also need not suppose that battling the heat is the only war that needs to be waged. If this was just about withstanding heat, we could all have sat in a sauna.

The ground here is clayey and small clumps of it stick to the soles of my shoes. I cannot help but drag some of it into the tent with me when I enter. Although my feet are tired from walking on the hard surfaces, the Merrells have served me well today. Because it is our last night and I am planning to sit around the campfire with the others, I leave on my shoes. Despite the cleared paths, it is a prickly spot and I do not fancy walking about in the dark later in thongs.

Although it is still hot, we have a larger fire tonight than any other night. Moz has cleaned out the hole that we dug to bury water for use at the end of Day One and transformed it into a deep fire pit. Jane retires to her tent. From the look of her sunburnt face and neck, I imagine that she is not feeling well.

We watch the sun set on our last night out here together, knowing that the approaching curtain of darkness portends the close of this act of our lives.

As the sun set on Henry's life, he thought often of his old Bourke mates. He was excited to have been reunited with Jim, for those friends from Bourke remained always in Henry's heart. In 1911, Henry had learned of the serious illness of Donald Macdonell, with whom he had shared Christmas Day in 1892. Macdonell was one of Henry's companions trying to escape the Bourke heat sitting in his underpants. On learning of Macdonell's illness, Henry reached for his pen and wrote:

Donald! They say you're dying!
God grant that it be not true;
But the sweep of the mulga's sighing,
Donald, in memory of you.

As we three sit around the fire waiting for our food to heat, we watch several Emus wandering through the long grass in the paddock on the other side of the Track.

'Bloody hell it'd be funny to watch Baz riding an Emu,' I say. In the good humour of knowing that we are almost finished, the idea tickles my fancy. 'I can just see him hanging on for dear life.'

'Why would it have to be Baz though?'

'Well, he's the size of a jockey. You and I are too big, Moz. Besides I remember the look on his face when we were kids and I watched him ride his pushbike down the side of the Sugarloaf.' The Sugarloaf is a big hill in Mount Gambier where Baz and I passed many wonderful childhood days in exploration and adventure. 'The bike was going so fast that Baz couldn't possibly jump off, so he just hung on for dear life. There was terror written all over his face. Bloody hell it was funny.' Baz stayed with the bike until it crashed into a fence at the bottom of the hill and he went flying over the handlebars—and the fence—and crash-landed with a thump. 'Baz would just hang on to that Emu with all that he's got.'

Baz decides to humour me and participate in the conversation.

'Well, inevitably I'd get bucked off, right?'

'Not if you held on tight enough.'

Moz interjects, 'Now, this is the sort of philosophical discussion that is missing from the world.'

We all enjoy a good laugh, knowing that this is our final night out here together.

'Nope,' I say. 'There is no way the Emu would buck Baz off—unless it ran into a fence, that is.'

We eat our tea, but Moz is up and down attending to Jane because she cannot get comfortable. From what we can hear, it seems Jane has been out here one night too many. Baz and I are amazed by how patiently Moz responds to the situation and to Jane's discontent.

Baz says to me, 'Gregga, you'll be pleased to know that my appreciation of Lawson as both a man and as a writer has grown considerably on this trip.'

I *am* pleased to hear that. As a result of that growing appreciation, the bond that I share with Baz is even stronger.

As the sun slowly goes down on our last night on the wearisome Track, I enjoy sitting around the camp fire, listening to Baz and Moz read aloud to me from collections of Henry's short stories. It is a scene that I think Henry would love. Henry's friend, Jack Moses, wrote of a time when the two men were visiting Gundagai. In the evening, Henry suggested a walk along the Murrumbidgee River. As they walked, they were drawn towards the flickering light of a distant campfire and presently arrived at a drovers' camp. The two strangers were invited to join the circle around the fire, where they partook in an evening of swapping yarns and singing old Bush songs. Not knowing who was in their company, a drover eventually arose and proceeded to recite one of Henry's poems. Henry would not allow Moses to reveal his identity. Henry 'said nothing,' Moses recalled, 'but I knew from the satisfaction that I read in his eyes in the firelight, that this was his best reward—to be loved and appreciated by his bushmen.'

While Baz and Moz drink billy tea, I hang my head wearily but listen to every word they read. Gee, Henry is a great writer. Despite the late night in Fords Bridge, I stay up late again tonight and 'we sits an' thinks beside the fire, / With all the stars a-shine.' For me, these few hours tonight are the best hours of the entire journey: sharing laughs with two great mates; listening to Henry; sitting around a campfire in the Aussie Bush. If I die now, I go happy.

DAY FIFTEEN

The Death of Henry Lawson

They thought of the far-away grave on the plain,
They thought of the comrade who came not again,
They lifted their glasses, and sadly they said:
'We drink to the name of the mate who is dead.'
And the sunlight streamed in, and a light like a star
Seemed to glow in the depth of the glass on the bar.

- The Glass on the Bar, *In the Days When the World Was Wide and Other Verses*, 1896

They followed me with flattery
In the days when I was brave—
But for those who have been true to me
I'll strike back from the grave!

- The Afterglow, *When I was King and Other Verses*, 1905

I want an early start in the morning so the only thing that I remove when I lie down to sleep is my shoes. Along with everything else, I decide to sleep with my socks on my feet. This will hold the bandaids in place and, presumably, mean that I will not need to attend to them in the morning. In any case, I reason that if my feet get knocked around tomorrow, it does not really matter. After tomorrow, I am putting my feet up to rest—maybe forever.

I dream through the night. In one dream, my old mate, Peppi Bueti, is my companion when I get to Bourke. I do not know what happened to Baz. Maybe he is still out in the Bush, attending to the baked-beans-fuelled call of nature. When I finish the trek with Peppi, my mother is there to greet me. She leads me to a bath but I am surprised to discover the dirty old tub is full of leaves and grime.

'If that's not good enough,' Mum says, 'you can go without.'

Initially, I decide to go without. Feeling sweaty and grimy; however, I eventually decide that, although it is obvious no one has bothered to clean it for decades, it is my job to clean the bathtub.

≈

On 1 September 1922, Henry visited the offices of the *Bulletin*. According to friends, he looked 'white and shaken' and said that he was 'very tired'. That evening, Henry suffered a seizure and was assisted to bed.

The following morning, Henry felt uncomfortable but propped himself up and proceeded to write regardless. Mrs Byers called a doctor to Henry's Abbotsford residence and he examined Henry at 9.00 am. After the doctor left, Henry recommenced writing in bed. Henry's discomfit intensified and he rose from bed but collapsed in a corner of the room. At 10.00 am, Mrs Byers found him there. She rushed to get the doctor back but by the time he returned Australia's greatest writer was dead—the cause of death: cerebral haemorrhage. He was just 55 years of age.

≈

When I awaken, I realise that it is finally Day Fifteen—the final day. Although I feel like celebrating, I also know that with a 35 km day ahead of me, there is still much to be done. Having slept in my clothes, I need only to venture outside to slide on my shoes.

As I fill up my water bottles and hydration bladder, I notice that the flies are thick—more so than on any previous morning. I ask Baz to take a photograph and, despite the fact that his close-up photograph does not encompass all of my back, I count 114 flies on the back of my shirt. It is not even 7.00 yet.

I move onto the Track, enjoying the cool of the overcast morning and feeling pleased to have gotten an early start on what promises to be a long day. Baz returns to attending to his feet when I leave camp. Moz and Jane are both in their tent, apparently still sleeping.

The Track beneath me is strewn with rocks. I have Seventeen Mile Plain and Walkdens Plain to negotiate today and so as I leave the last of the trees in which we camped, I know that from this point forth, there is little shade. The absence of shade, the long distance, and the hard surfaces are going to make for a tiring day. I swallow a couple of painkillers for I want to arrive at Bourke feeling well enough to enjoy my accomplishment.

To my sides, I see half a dozen or so scattered trees, but as far as I can see in front of me, there are no trees. A rabbit runs through the field to my right and a kangaroo bounds along the fence line. The fence is a good 30 m or more from the Track but it is close enough that I am afforded a good view of the powerful roo and its easy movements.

When I hit the bitumen, one of the first things I find is a hawk squashed on the road, its wings, tail feathers and one leg are all spread in different directions as if, in death, the bird is stretching to reveal its size and grandeur. It is too late for it to intimidate anything now though.

The lush green thigh-length Mitchell grass extends to the edge of the bitumen so that, even forgetting about shade, there are not likely to be many spots to sit and rest unless, of course, I am happy to wade into the long grass and plonk myself down amongst the snakes.

Stratus clouds cover the entire sky. The clouds are grey and, although not very heavy, there is the prospect of rain. That said, like Henry, I have learned not to expect rain no matter how the sky might look. Henry said that 'out here, it can look more like rain without raining, and continue to do so for a longer time, than in most other places.' As with all things to do with life on the Track, Henry was right.

After half an hour of walking, I am surprised to come to the H180 sign. I was under the impression that we had 35 km to do today of which I must already have done at least 2 km. Yet, with Hungerford 180 km behind me, I still have 35 km to get to Bourke from this point. Bloody hell. Once we get to Bourke, it might still be a couple of kilometres to get to the Port O' Bourke pub too. The day is going to be longer even than I thought. I am now looking at 40 km today.

The flies from the campsite this morning have stuck with me and they seem to be calling to their friends to join them. This might be the most flies I have had with me for the entire journey. I spread my insect netting over my head and carry my hat, bandana and a bottle of water in my hands. I come to another dead animal squashed into the bitumen surface. From the dark colour of the flattened corpse, my first thought is that it was once a boar but then I notice feathers and decide that the mess before me was once an Emu.

A couple of Emus bounce through the paddock to my left, running at high speed and dissolving into the distance, wishing for all the world that they could fly I bet. According to one Aboriginal legend, the reason that Emus are flightless is because of the trickery of Brolgas. One day, Emu saw Brolga's graceful dancing and asked Brolga for dancing lessons. Brolga quickly folded her own wings against her back and responded that she would be happy to teach Emu how to dance. However, because Emu would never be graceful with such long wings, Brolga offered to clip them. Poor, gullible Emu allowed her wings to be cut and, to this day, they remain short and useless. Realising she had been deceived, Emu grew angry but Brolga simply unfolded her own long wings and flew away to safety.

I watch a flock of eight more Emus as they bounce recklessly through the paddock to my left. One of the Emus separates from the others but, after finding itself all alone, the Emu changes direction and, in the distance, I watch it rejoin the running flock.

Seeing all of the Emus this morning makes me think that I am seeing the things that the ancient Aborigines would have seen out here in this changeless place. I realise, too, that I am seeing what Henry would have seen. It brings to mind what one of his friends said was Henry's favourite saying: 'Well, if things don't alter, they'll be the same.' Perhaps it was the progenitors of these same Emus that Henry had in mind when he described Emus as having 'snaky heads' and watchful eyes. Henry wrote:

Their snaky heads well up, and eyes
Well out for man's manoeuvres,
And feathers bobbing round behind
Like fringes round improvers.

I stop and see a few kangaroos hidden in the long tussock grass watching me with curiosity.

I think of the three of us walking here on Day One. It seems so long ago—another lifetime. We were all so full of hope and excitement, our spirits buoyant, albeit by the time we got to this point towards the end of the first day, we had already been slapped around the head with realisation of the enormity of the task we had undertaken. Day One set up the hell that was to follow. It was unspeakably hot and on the hard road with blisters forming, I took the type of mental battering from which I might never have recovered. Were it not for the assistance of Baz, Moz and Jane, I could not have kept going.

It is Day Fifteen, I keep telling myself. You bloody beauty! Day Fifteen. An Australian Magpie warbles beside the Track, adding to my celebration. There were certainly days where I worried whether I could still be going on Day Fifteen. On the way to Hungerford, I set my mind merely to getting *to* Hungerford. I dared not even think about having to turn around to come back across the hell through which I had just passed.

When I remove my hat to cool my head, great clouds of flies rise and buzz their protest at the disturbance. After swarming around my face, they settle on my back. I do not mind carrying them provided they stay out of my face.

I look back and see Baz about 500 m behind me, closing in on me fast. I have been moving along at a leisurely pace this morning, seeing as I have a long day ahead. I am trying to preserve my energy.

A vehicle approaches from the direction of Bourke. I move to the side of the road and the driver veers to the other side and waves as he passes in a rush of noise and a gust of wind. Back here on the bitumen, things suddenly seem to be moving so fast. I feel like I am on a fairground ride and I want to get off. In Henry's day it was the railway that represented progress and tied the Bush to the cities:

The flaunting flag of progress
Is in the West unfurled,
The mighty bush with iron rails
Is tethered to the world.

Nowadays, in a world where cars have replaced trains as a primary means of locomotion over long distances, it is the bitumen that gradually extends its fingers across the country, carrying with it all and sundry.

Just before 8.00, I come across a rabbit dead on the road, presumably hit by a car. It has not been dead long. With the traffic on the road, I realise that it will not be long before it is as flat as a pancake. In an effort to preserve some of the creature's dignity, I use my foot to roll the carcass off the road and into the grass. Had I perished out here, I hope someone would have at least done that much for me.

≈

Quiet and shy, Henry was not one for public adoration and would have been uncomfortable with the outpouring of grief and affection that followed news of his death. A State funeral followed, with hundreds gathering at Sydney's St. Andrew's Cathedral to pay their respects and perhaps thousands more lining the route to the burial site at Waverley Cemetery. Amongst the mourners were politicians and government officials and many of the city's leading professionals and businessmen. Henry's brother-in-law found this bitterly ironic. 'A week before they would have dodged...to avoid him. Now they wanted to bask in his reflected glory.'[1] Henry had foreseen this response to his death, writing:

To see 'The Death of Henry Lawson' printed
In letters tall and black and fairly stout;
To read he died a hero, hear it hinted
That all his debts were paid—the Bill Wiped Out!

Upon Henry's casket 'lay a simple bunch of native roses, and about it lay a spray of gum leaves, a cluster of glowing wattle and some bush ferns.' The Prime Minister, Billy Hughes, scrambled to be heard amongst the clamour of eulogists, saying:

He knew intimately the real Australia, and was its greatest minstrel. He sang of its wide spaces, its dense bush, its droughts, its floods, as a lover sings of his mistress. He loved Australia....None was his master. He was the poet of Australia, the minstrel of the people. He was a genius and his name will live. Australia's greatest writer has passed away.

When I think of Henry, my preference is to think of the happy times of his life. Those were not times of adoration and attention. As one female admirer reminisced at Henry's passing, 'The characteristic I most distinctly recall is Henry Lawson's modesty.' Humble in the extreme, Henry rarely read his own published work and apparently 'never quoted his own lines.' This fact was borne out at a Christmas gathering when the assembled group of friends decided to take turns singing a song, reciting poetry or telling a story or, upon failure to do so, the person would forfeit some money. When it came to Henry's turn, he quietly reached into his pocket and paid his fine.

≈

With an hour behind me, I am starting to wonder at the wisdom of leaving my socks on through the night and then not attending to my feet this morning. I hope the bandages and bandaids are still in place from yesterday. My feet are sore.

Baz catches up to me soon after 8.00 and I see that, as with me, the flies are using him to hitch a ride into Bourke.

'The flies are friendly today,' I say.

'You're covered in them, Gregga.'

'You too, mate. They are all over your back and all over your hat.'

Baz and I photograph one another's backs and, in the photo of me, there are now 85 flies just on my hat alone. Countless more sit on my back and on my pack.

The bitumen appears and disappears in turns. It is strange that the road is sealed for a stretch and then not sealed and then sealed for another stretch. In the absence of towns or of nearby rivers or creeks, there seems to be no rhyme or reason as to where the bitumen falls.

Twenty minutes later, we pass the carcass of a black boar. The empty, unseeing eye sockets stare at us. With the pointy bottom tusk extending above the top lip, the boar seems to sneer at us, angry at its fate. Even amidst the flourishing grass of the plain, there is death in abundance. I stop beside a dead tree and Baz snaps a photograph. It is the only tree in sight. The leafless branches fanning out in all directions bring to mind the squashed hawk from earlier in the morning. The tree and the hawk are as dead as one another.

'How are yer poppin' up?' Moz wants to know when he and Jane join us. Jane has nothing to say. Her mood seems no better this morning than was the case last night.

After refilling my water and grabbing a bite to eat, I move onto the raised edge of a canal to the right of the Track. Although one cannot tell from the Track, the canal is filled with water. Last time, when Moz and I were out here, it was bone dry. I return to Baz and we set off again on the Track, passing the B20 sign at 10.43 am. After almost 430 km, Bourke is now only 20 km away.

A short while after entering the final 20 km of our tramp, we are visited by a couple of men heading from Bourke in a four-wheel drive.

'We are heading to Warroo,' the driver says after learning what we are doing out here. 'You would have passed it coming out from Hungerford.'

'Passed it? Actually, we stayed there. John Stephenson, the manager, let us stay there one night.'

It turns out that the driver is the great-grandson of Roy Dunk, visiting Warroo to check up on things at the family inheritance.

'You're lucky to have John managing the place for you. He is a wonderful man,' I tell him.

At 11.00, the Track again converts to bitumen. From this point forth, we will have nothing but bitumen. We get to a cattle grid across the Track.

'Can you get a photo, Gregga?' Baz asks. 'This might be the last cattle grate before we finish.'

Baz hands me his camera and I walk across the grate and call him over. He moves too quickly for me and I miss the photo.

'You better go back and do it again,' I say. 'I'll get it right this time.'

He crosses back and comes over again.

I am working with an unfamiliar camera and I miss the photo again.

'One more time, Bazza.'

'Bloody hell, Gregga.'

He is like a trained monkey. I should tell him I missed the photo of him walking through the Dingo fence and send him back to Hungerford.

I told Baz early in the tramp that these grids become a hazard when the messages from your brain have to pass through weary legs before they get to your sore feet. Sending Baz back and forth over the grid is just plain cruel and so, on the third try, I snap what I think is a good picture.

Baz looks at my photograph on the camera's LCD screen. 'That'll do. Let's get going.'

≈

A few years after Henry settled back in Sydney, Jim visited and caught up with him at the *Bulletin* office. 'After giving his old salute,' Jim wrote, 'he sprang to meet me with his bone-crushing handshake that I remembered so well; but I got first grip, and I noticed he winced a little.' Old before his time, Henry was enfeebled and weak but that was not about to stop him from showing Jim around the city. It was the last time the two great mates were together. When Jim heard the news of Henry's death, he was heartbroken. 'When the word came that he was dead,' Jim wrote, '...sorrow came as it only came to me once before, reviving old memories and marshalling hosts of old regrets. Some that cut like a knife.'

Inspired by memories of his friend, Jim wrote a poem entitled, 'The Bush Mourns'. The final stanza of the poem reads:

The sky is dark and dull the sun,
The world itself seems cold and grey;
And rough and rugged is the track,
And long and dreary is the way.
Where sliprail takes the place of gate
And fences lie beneath the sand
Proud men are they out there today
To claim they once held Lawson's hand.

For Jim, life was never again the same. In another poetic tribute, Jim wrote:

The stars have never seemed so bright
Since Lawson walked with me.

≈

Moz and Jane rejoin us at about 11.45 am. Moz parks out in front. He alights from the car and walks back to us with bags in his hands.

'We got you each a pie and some sausage rolls.'

'Moz, that's fantastic, mate.'

And it is. They taste great and go down well. We are getting close to town when the big fella can get hot food to us.

Baz and I are both covered in flies and they are keen to share our lunches but I manage to eat mine without swallowing too many. The flies are not the only things in huge numbers. Thousands of grasshoppers spring from the long stalks of grass as we pass.

We are inside the final 15 km but we have already covered more than 20 km today. With most of today's journey on sealed bitumen, the hard surface tires my weary legs and my feet ache. I have two new blisters and each of them is giving me hell. The featureless plains seem to stretch forever and Bourke seems not to be drawing closer. This last day is a long, hard slog. Baz and I both just want it to end—we want to be in Bourke.

The unbearable sameness of it all—the plain and the walking—grinds me down. I expected to feel mixed emotions about coming to the end of this journey. To my surprise, I do not. I will just be glad when it is all over. Although I expect I will soon feel relieved and also exhilarated to have made it, for now, it is just a grind. I take one weary step before another as I plough through the distance between Bourke and me.

Shortly after 1.00, we reach the Wanaaring turnoff. Beneath that sign is an arrow pointing to Bourke, ten kilometres away. I ask Baz to photograph me as I walk past the sign. He stuffs up the shot and makes me walk back and do it again. He stuffs up the shot...again. The bugger is making me pay for sending him back and forth over the cattle grid earlier.

My new blisters make walking very, very painful. 'Gee, my feet are sore today.'

'Me too.'

'Are we really getting any closer to Bourke?'

'It seems to be taking forever.'

'I'm just ready for it to be over.'

We move further down the Track and arrive at the 'Road Conditions' sign. The road to Hungerford is open today.

'I'm going to take a rest.' I sit on the step used for changing the sign to signify if the road is closed.

'I've got to take another dump.'

'Bloody hell, Baz. That's all you do out here.'

'Yeah, but I've got to go.'

'Well, you can't do it here. I'm resting.'

'How about over here then?'

He takes a step or two and then attends to his business.

'Baz! You can't do it right there. Bloody hell, Baz.'

'Well, I had to go. It was either beside you or in my undies. It was a simple choice. I don't see what all the kerfuffle is about.'

'You're not bad, are you? I find a place to sit and rest and you go and do that to it. You are a bloody crapping machine.'

'Well, I buried it, didn't I?'

I am not stopping to confirm or deny.

We go a little further down the road and, at 1.45, Baz receives a text message on his mobile phone. For me, it is the concluding signal that we are back in so-called civilisation. We are back within mobile phone range.

Moz appears. This time he is alone.

'I thought you blokes could do with an icy pole.'

This is more evidence that we are finally out of the Outback.

With the long grass beside the bitumen, Baz and I walk on the road. Moz tries to drive alongside us for a chat as we walk, but I think it creates a dangerous situation. The road is narrow and, other than Moz, the cars on it are moving fast. I drop in behind Moz's car but quickly find that I am sucking on his exhaust.

'Maybe you could go ahead and get a photo of us as we cross the bridge,' I suggest.

When we get to the Darling River, the water level seems even higher than when we left Bourke 14 days ago. There are hundreds and hundreds of trees standing in the wide channel of the river. We cross the old bridge over the flooded Darling. I take my hat off and wave to where Moz has stopped his car on the road—on the new bridge over the Darling—to photograph us. We are almost done. It is soon time to celebrate. It feels good to take the weight off my feet as I take a seat on the bridge rail.

We leave the bridge and, at 2.35, we pass a sign indicating we are about to move inside the final five kilometres of the journey. Baz is about 15 m in front of me. Many cars pass by at high speed. The closer we get to Bourke, the busier it gets and the faster things seem to move. More and more cars pass us and each one seems to be moving faster and making more noise than the one before it.

A car stops beside Baz and someone offers him a lift—with less than 5 km remaining, and after almost 445 km!

When the car leaves, I catch up and say to Baz, 'It is nice we don't have to get up and walk again tomorrow.'

'Or ever again,' he replies.

I am glad that we do not have to face another day of walking tomorrow because my feet have taken a horrid battering today. I will be so happy when we are done.

I look down and am startled to see a snake at my feet. To my relief, when my brain catches up with my eyes, I realise that it is only someone's old discarded bootlace.

I stop to empty my bladder and I empty my hydration bladder too. There is no need for me to be carrying unnecessary weight now.

Light rain falls—just spitting. Although the road surface has not been easy today, at least the weather, albeit humid and sticky, has been mostly cooperative on this last day. Officially, Bourke's temperature for the day is 28 degrees. It is the coolest day so far.

A car and caravan pass us, heading out of Bourke for an Outback adventure. 'Yer wanter go out back,' I can imagine someone saying to them. Well, like Henry, 'I don't wanter; I've been there' and, having just got back, I have no plans of returning. There are no regrets that it is soon all over, more a feeling of relief that the suffering is about to end.

At 3.15, we reach the Back O' Bourke Exhibition Centre. Jane and Moz are there to take photographs of us. Baz sits on the ground but even at this late stage I remain on my feet, worried that if I get down, I will not be able to get back up. Baz and I pose for photographs beneath a Henry Lawson quote erected on the outside of the Exhibition Centre wall: 'If you know Bourke you know Australia.'

'We'll see you both back at the pub,' Jane says as she and Moz depart. 'You only have about two kilometres to go now, boys.'

We walk into town and pass a man washing his car in his front yard. 'I see you made it,' he says, 'You've done well. Congratulations.' I recognise him as one of the blokes in the Fords Bridge pool tournament.

'Thank you.'

'Where did you camp last night?'

'Oh, on the edge of the plains, in the last of the trees,' Baz tells him. 'About 35 km back.'

'Yes, you've both done well.'

We carry on through town, passing a service station where a woman is filling up her car.

'She's the best looking woman we have seen while we've been out here.'

I take a look. 'Bugger off, Baz. You're crazy.'

'No. Have another look.'

I have another look but I do not feel inspired to change my mind.

'Narelle was the best looker out here.'

'No.' Baz is emphatic. 'This one is much better. What Narelle had on her side was the benefit of still being young.'

I think about that for a moment. Tired and battered, I feel old. 'Being young is not a bad thing to have on your side though, Baz.'

As we pass the old Carriers Arms Hotel, Baz asks me to photograph him. He leans against the wall beside the hotel doors. I know that, like me, he can now imagine Henry passing in and out of those doors to the bar.

At 3.55 pm, we reach our starting and ending point, the Port O' Bourke Hotel. We have made it! Baz and I are the first two people ever to recreate Henry's Bourke-to-Hungerford-and-back trek.

'Congratulations, boys,' says Jane. 'Well done.'

'Congratulations, Grog. Congratulations, Baz.' Moz and Jane are both happy for us.

'Well done, mate.' Baz is happy for me.

'Congratulations,' I say to Baz, 'and thanks.'

Baz takes off his pack and lowers himself onto a step. I remain standing for the obligatory finish line photographs. I will not take a seat until I have a nice cold drink beside me.

'Let's have a drink,' Moz says. 'What'll it be?'

'Just water for me, thanks,' I say, 'with lots of ice.'

'I'll have a beer,' Baz says. He has earned it.

When Moz returns with drinks for all of us, I take the weight from my feet and seat myself on a chair on the hotel veranda. I take my portraits of Henry from my pocket and prop them on the table beside my drink.

I raise my glass. 'To Henry.'

I feel my aches and pains slowly begin to ease out of my body, replaced by an enormous sense of satisfaction at having achieved one of my life's dreams.

Somewhere in my mind, and not necessarily to the back of it, was the idea that I would get to Hungerford and that would be enough. But I never made that call. I never consciously admitted that to myself. I never decided that I would stop, even if I perhaps entertained that thought. I was able to stay strong and here I now am, sipping ice cold drinks on the veranda of the Port O' Bourke Hotel.

I am unable to adequately encapsulate the pain that I endured and the doubts that I had to confront and work through and work over during this tramp. Thankfully, my brother and I will not be called upon to storm the beaches at Gallipoli. We will not be called upon to slog through the mud of the Kokoda Track, or to crawl through the jungles of Vietnam. We will not be required to run the gauntlet of hidden roadside bombs near Baghdad, or to engage in gun battles in Kandahar. The greatest test of my courage was to continue when every part of my body told me not to return to the Track in Hungerford. This was my personal Gallipoli. This was me running towards the machine guns at the Nek. I have passed my test of courage.

'I thought you would want a bath,' Moz says, awakening me from my reverie. 'But there were no bath plugs. Old Mick is looking for a plug for you.'

'Oh, I don't know if I'll take a bath,' I say. I think of the previous night's dream of the bath full of dirt and leaves.

'Why not?' asks Jane. 'We thought you would love a bath.'

'Well, yeah, but it'll probably be dirty.'

'No, they're not. Everything is clean. Well, at least it is in the women's bathroom.'

'The men's looks good too,' says Moz. 'I checked when I was looking for a plug.'

Shortly afterwards, Candy Mick hands Moz a new bath plug from a nearby store.

'Well, maybe I'll take a look at the bath and then decide,' I say.

'Jane and I bought some Epsom salts for the bath too.'

'Thanks you two. You really have been great. I could not have made it without you.'

I hobble up the stairs to the room Baz and I share and, in a moment of solitude, I lie down on my bed to think.

I am having a difficult time knowing what I feel, let alone consolidating those feelings into coherent thoughts. It is all over and so I ask myself, 'What next?' What do I do now? For a lifetime, I have dreamed about getting to where I am now, but I did not dream of where I go from here. What is the next step in my life's journey?

Right now, that next step, whatever it is, will be a hobbled, weary one. As I lie on the bed, I look at my blistered, battered feet and recognise them as proof of what I have been through. I am mighty proud of what I have done. I am proud of Baz and what he has done. Good old Two Shirts Bazza. I know enough about the hardships of the Track not to believe that it was easy for him but there were times when he made it look easy. I am proud of Moz and Jane as members of this successful team. I recognise how hard they worked in difficult conditions to serve me and Baz and to help us keep moving down the Track. Moz made it clear that Jane does not like heat and dust and camping.[2] Rarely would one have known that though. She was a trooper through and through.

This has been an incredible achievement—one of the greatest things that I have ever done or that I ever will do. Certainly, without any shadow of doubt, it is the greatest physical achievement of my life. It is hard to compare the years of mental struggle and turmoil of my dissertation against the couple of weeks of mental struggle and turmoil here, but this was done also within the context of incredible physical pain. There was the mental anguish and feeling of helplessness—the feeling of having to climb a mountain that seemed too high—but it was also experienced in the context of severe physical suffering. The heat, distance, blisters and the load I carried all physically wore me down, but added also to wearing at me mentally too.

I have achieved something very special.

Six hundred thousand steps! Each step was either forming a blister or on a blister or, indeed, both at the same time. Despite how often I have looked at Henry's portrait over these past 15 days and thought that he would be thinking, 'What on Earth are you doing?' I also think that he would be proud of me. He would be thinking, 'Well done, son, well done.' He would be tickled—chuffed—by the thought of inspiring this journey. He may have warned me against it in his writing, but he also inspired it.

I decide to check out the bath. If I am lying down thinking, I might as well be doing it in the bath.

≈

I step into the tub and immediately start to feel better. I feel all of the pain and weariness easing out of me. It is wonderful just to lie back and to feel my agonies soaking away. It is a great, great feeling. The weight of the world seems lifted from my weary body, as if that weight is suspended by the water around my nakedness.

I feel all of the sweat and grime of the Track soak from me.

My aching Achilles tendons start to feel better as do my tired legs and weary groins. I feel the satisfaction of having achieved a dream. I set myself an extremely difficult goal—something that nobody else has ever done before—and I have achieved that goal alongside my brother. I could not ask for anything better.

I could lie here in the tub forever and be happy not ever to move.

It feels so good.

I elevate my feet on the end of the bathtub.

The skin between my toes is peeling off in layers. A big roll of skin on the toe that had been infected is peeled back, ready to fall away.

I think of Henry and recall a sketch he wrote about taking a bath. Poor Henry, tired and fast growing old, 'had been working hard all day,' to avoid the grind of writing. He 'was very dirty and tired' and decided to take a bath before bed. The bathroom was separate from the house in which he was staying and water had to be heated and carried across. Henry filled the copper in the washhouse and then started a fire to heat the water. In order to retain the water's heat, Henry rigged canvas over the bath, leaving a small space for him to step in. He rigged up a headrest and planned a long, luxurious soak.

Henry went upstairs to his room and arranged reading materials beside his bed (the *Bulletin* at the top of the pile) for when he returned from the bath. He retrieved 'pyjamas, towel, and soap'. Henry even 'arranged with the landlady to have a good cup of coffee made...ready to hand in round the edge of the door when [he] should be in the bath.' He then asked the landlady to pass in some toast. He returned to the fire beneath the copper and got a bucket and started to carry the hot water across, 'pouring it in through the opening' in the canvas where his head would protrude once he stepped into the tub. He retrieved the extra water he had boiled in pots and pans in the kitchen and went to top up his bath. Once in the bathroom, he undressed; hanging his clothes 'on the wall, till morning, for [he] intended to go straight from the bath to bed in [his] pyjamas and to lie there reading.' After an elaborate production getting everything ready for a relaxing soak, Henry gingerly put his 'right foot in to feel the temperature' of the water. Alas, after all that work, Henry had forgotten the plug.

'I'm deaf, you know,' wrote Henry, who had not heard the water escaping down the drain.

Henry put his clothes back on and went inside, disappointed to have gone to so much trouble for naught. His landlady was 'so full of sympathy, condolence, and self-reproach (because she failed to draw [Henry's] attention to the gurgling) that she let the coffee and toast burn.'

Disgusted with events, Henry 'went up and lay on [his] bed, and was so tired and misty and far away that [he] went to sleep without undressing, or even washing [his] face and hands.'

Poor old Henry.

I soak in the bath for almost an hour, glad beyond measure that Moz sent Candy Mick on an errand to find me a plug. All of my aches and pains ease out of my body like demons at an exorcism. It is quite possibly the best bath I have ever taken. It is long after I notice that my fingers are like prunes that I decide that I should get out.

I dry myself and pull on a pair of jeans and, when I go to button them up, I am surprised by how much room there is. Moz says to me later, 'There is much less of you now than when I picked you up at the airport.' Lots of me remains on the Track, but I take from the Track restored self-belief, unforgettable memories, and deeper understandings of Henry.

When I return from the bath, I discover that Baz has left another of his laminated quotes on my bed.

The quote says:

It's primarily about one thing; not giving up. Being a champion means not quitting, no matter how tough the situation becomes, and no matter how badly the odds seemed stacked against you. If you have the courage, stamina and persistence, you are a champion!

Being a champion means not quitting. There were lots of opportunities for me to quit. The situation became very tough and the odds were stacked against me. I guess that I did show courage and stamina and persistence and, as Baz described it to the *Warrego Watchman* reporters at Kia-Ora, some stubbornness. Maybe stubbornness is just another word for stamina and persistence. I feel good that, despite the pain and despite the dangers, I was able to keep going and to make the return journey. I have shown Moz, Jane and Baz what I am made of, but I also have shown myself that I am made of the right stuff. I can hang in there. I can overcome difficulties and setbacks and even seemingly insurmountable obstacles. The dent to my self-belief and my sense of self-pride that I took during my dissertation was restored in part by eventually managing to complete that task but now, to have come out here and to have completed this tramp, I feel fully restored and, for that, I have Henry to thank.

I discover later that Baz's final quote comes from the mouth of Dean Karnazes, the ultra marathon marvel who just keeps on running. In the same report where I find the quote from Karnazes, he says also, 'Dreams can come true… [but] sometimes you've got to go through hell to get to heaven.' I know what he means. Indeed, sometimes you have to go through hell and high water.

APPENDIX I

An Unpublished Henry Lawson Poem

I've tramped the wild bush over
And camped where men ne'er trod.

- Buckley's Rum, Previously unpublished, Composed 1893

I was bad for three weeks after—
But wore it down at last,
And was singing Pat O'Hara
As the Warrego I passed.

- Buckley's Rum, Previously unpublished, Composed 1893

Upon completing the Bourke-to-Hungerford-and-back tramp, I proceeded to my hometown of Mount Gambier, South Australia, where I based myself for the duration of my research study leave from the University of Manitoba. Shortly after my arrival in Mount Gambier, I published in the local newspaper a two-part account of my experiences walking in the footsteps of Henry Lawson. After those articles appeared in *The Border Watch*, I received an email from a Mount Gambier lady who introduced herself as Mrs Yvonne Swaffer. Mrs Swaffer indicated that she had in her possession a copy of an unpublished Henry Lawson poem. She wondered if I would be interested in seeing it and hearing about how it came to be in her hands. Naturally, I said I would be delighted.

On a wet Mount Gambier morning, I visited Mrs Swaffer in the retirement village where she lives and she proceeded to share with me some of her family history. Mrs Swaffer's great-great-grandfather was Isaac Foster, who was born in 1830 in Cambridgeshire, England. He immigrated to Australia in 1849 aboard the *Cheapside*. It was Isaac who built the original hotel at Hungerford in the mid-1870s. Together with family members, Isaac purchased 11 blocks of Crown Land when it was offered for sale to the public shortly after Hungerford was gazetted as a township. As well as hotel publican, Isaac served as the town's postmaster (a tradition that continues today with Moc Parker acting as the Royal Mail Hotel publican and also the town postmaster).

In August 1861, Isaac Foster's wife gave birth to a son named Thomas Goodman Foster. After inheriting the position from his father, Thomas was the proprietor of Hungerford's Royal Mail Hotel at the time of Henry Lawson's visit in January 1893.[1]

In November 1890, Thomas married a woman named Agnes. They had been married for only a little over two years at the time of Henry's visit. Agnes was born in May 1865 and was 27 when Henry visited. Henry was 25 years of age. Thomas and Agnes' first child was born in May 1892. The baby was only eight months old when Henry visited.

According to family legend, Henry wrote a poem while he was in Hungerford and he gave the poem to Agnes in return for her hospitality and because he was short of the necessary funds to pay for his drink.[2] Such behaviour was not unusual for Henry. Indeed, in the later years of his life, he developed a reputation for his willingness to scribble hurriedly composed verses in exchange for money for booze. The manner of composition is also consistent with other episodes from Henry's life where he sat down and wrote a poem largely 'off the top of his head,' as it were, albeit perhaps having previously ran the verses through his mind before he proceeded to put pen to paper. There is also evidence that he wrote many poems as gifts (or payment) to people with no intention of having them published.[3]

Thomas' and Agnes' nephew was Alfred E. M. Forster (not 'Foster' like Thomas and Agnes Foster). This same Alfred Forster was Mrs Swaffer's grandfather. I found it interesting to read Alfred Forster's handwritten pencil notes about his own 1891 trip to Bourke and Hungerford:

Arrival in Bourke the Head of the Railway, which is on the Darling River. About 6 months before our arrival the River had overflowed and the town was flooded from 3 to 4 feet of water, up to window sills.
22 hotels in Bourke.
We stayed three days in Bourke owing to the flooded country to Hungerford which is 130 miles from Bourke.
My brother was offered a good job some 30 miles down the River on a large station and he left me.

I took Cobb and Co. Coach to Hungerford.
This journey was full of adventure. Crossing flooded rivers—The Warrego—Green Creek and the Cuttaburra. We had to stay 3 days at Buckley's Hotel on this river.
Then the coachman decided to cross the river.
Riding horses were used and had to swim across. After a good wetting we arrived at Hungerford in the evening, the trip having taken 6 days.
The method of transport to the Outback was large tabletop wagons with 13 or 15 horses. Some splendid teams and also bullock teams.
Queensland Border. Rabbit proof fence.
I resided 18 months with my uncle at Hungerford, who had an Hotel and Butchering business. I see some exciting events in this Outback Country.

Although they are only notes, on several points Alfred Forster's descriptions of his time in Bourke and Hungerford are of particular interest regarding Henry Lawson. One such point is that Alfred mentions that Bourke was flooded 'about 6 months before our arrival' in 1891. This is the Great Flood of '90 that Henry wrote about in a sketch by that name, where he says that he talked with men who supposedly arrived in Bourke 'shortly after the deluge, and the only sign they saw...was a dirty mark around the doorsteps where the water had been.'

As shown in the Day Seven chapter of this book, as a result of his investigations, Henry mocked the newspaper reports he had read in Sydney about the 'great danger' the flood presented to the people of Bourke. Again, as previously mentioned, Henry concluded that 'the chief danger was that the liquor would give out.' Perhaps Alfred Forster had read the same newspaper reports that piqued Henry's interest because Alfred mentioned Bourke 'was flooded from 3 to 4 feet of water, up to window sills.'[4] Certainly, old black-and-white photographs in the collections of the State Library of New South Wales show major flooding and much of the town literally under water. *The Launceston Examiner* newspaper reported two deaths by drowning while *The Sydney Morning Herald* reported four.[5] It seems Alfred formulated a truer picture of the severity of the Bourke flood than did Henry.

It is apparent that, like Henry, Alfred travelled to Bourke aboard the railway. It seems also that, as was the case with Henry, the Bourke hotels—or at least the large number of them—made an impression on Alfred.

Although he was desirous to get to Hungerford, it is interesting to note that Alfred stayed in Bourke 'owing to the flooded country to Hungerford.' Although there would not have been Bourke Shire workers posting 'road closed' signs as they did when my brother and I were on the Track, the road to Hungerford was obviously deemed impassable.

Alfred separated from his brother when the brother was offered work 'some 30 miles down the River on a large station.' This is likely to have been Toorale—the same station where Henry and Jim Gordon worked in the shearing sheds prior to setting out for Hungerford.

Alfred wrote that he was with his uncle in Hungerford for 18 months and that Bourke was flooded six months before his arrival. The Bourke flood was in late-April 1890. Six months from that time (Alfred's arrival in Bourke) would be late-October 1890. Eighteen months from that time (Alfred's departure from Hungerford) would be late-April 1892. It seems Alfred left Hungerford less than nine months before Henry's arrival.

Another point of particular interest in Alfred's notes is that he says his Cobb and Co. journey to Hungerford was delayed for three days at Buckley's Hotel on the Cuttaburra River. This identification of a Buckley's Hotel between Bourke and Hungerford is significant with regards to Mrs Swaffer's unpublished poem, which is entitled 'Buckley's Rum'.

It is understood that when Henry walked to Hungerford there was a hotel every 20 miles (32 km) or so. That was about the distance a bullock team could travel comfortably in a day. Smaller shanties called 'Half-Way Houses' or 'Ten-Mile Inns' were often located between the hotels so that the hot traveller or his thirsty animals might be rested and revived every ten miles (16 km). A list of New South Wales Hotels and Publicans' Licences reveals the identity of a number of the hotels or shanties that once existed between Bourke and Hungerford. There once was a Yantabulla Hotel. An Emu Inn stood at 'Lauradale Lake,' which seems to be the northern end of present day Sutherlands Lake or very near to it. The old Emu Inn may now be the hotel ruins that Baz and I explored at Lauradale on the way back from Hungerford. In any case, another hotel is listed as Sutherlands Lake Hotel, at Sutherlands Lake. A Lake Eliza Hotel is listed, as are the Lake Hotel (situated at 'the lake on the Warrego Road,' in the district of Bourke), Native Dog Hotel (near Warrego River, district of Bourke), Native Home Hotel (Warrego Road, district of Bourke) and, among others, a hotel called the House that Jack Built (situated at Fords Bridge, Warrego River, in the district of Bourke).[6]

In Mrs Swaffer's poem, the first-person narrator talks of having roamed the Outback countryside. He says that one day he set out from Yantabulla and continued until he stopped for a drink at an establishment called Buckley's. The rum he drank at Buckley's left him with a terrible hangover in which his 'head was bursting, heart was throbbing / aches and cramps in every limb.' He continues on and eventually passes the Warrego River where, the narrator says, a pub was kept by a man named McPhee. As much as he is tempted to have a drink, the narrator decides not to stop for fear that McPhee might serve the same 'fearful new, and awful brand' of whiskey that the narrator drank at Buckley's.

Buckley's Rum

I've whaled the Murrumbidgee,
And been on the Lachlan too,
Baked dampers on the Darling,
And roamed the wild Paroo.
I've worked on every station
From Uranna round to Bourke,
And from there, right out to Tambo,
Cunnamulla and Kanturk.
I've tramped the wild bush over
And camped where men ne'er trod.
But the last for straight out snagging
Beat the lot, so help me God.
I set out from Yantabulla.
On the tenth of March I think

And came on, till I struck Buckley's
Where I called and had a drink.
It was the real McKay I tell you
Taking breath and voice away
And scattered all my senses
Until twelve o'clock next day.
I've been in Back Block shanties
Where the best was Farmers Friend
With a dash of Gin or Whisky
Called I think the Navvies Blend.

I've drank the Bushmens Idol
Made of Pain-Paint, mixed with Wine
And to tastes not o'er exacting
Such as Bill's and Tom's and mine
Might pass muster of a morning
But that rum of Buckley's Oh
What it was made of, do not ask me
T'was a Brand I did not know.

Lord, the pains I felt on waking
I can find no tongue to tell,
Head was bursting, heart was throbbing
Aches and cramps in every limb,
And so for days together
Past events were lost, and dim,
I was bad for three weeks after—
But wore it down at last,
And was singing Pat O'Hara
As the Warrego I passed.

There's a pub here too I'd mention,
Kept I think by one McPhee,
But I did not call on passing,
Tho' true as death claims me—

I was sorely, sorely tempted
To have one only one, I heard me say
Just the nerves to steady,
After all the miles I've walked today.

How perverse we are when passion
Rears its fierce and ugly head,
Deaf to every higher feeling,
All that's good within us dead.
Stiffled conscience, shrieks her warning.
Passion smothers back its cry.
Then there comes a passage royal
Who shall win, or who shall die.
This is true, there's no gain saying,
But bear it in mind that day
For I cannot once remember
Having stopped or once said nay—
When the tempter, stood before me
Sparkling in the clear, cut glass,
Memory fails to bring an instance,
When I let the tempter pass.

This is straight, so don't imagine
Qualms of conscience, held my hand
No t'was dread of Buckley's
Fearful new, and awful brand
That resolved me to continue
Else I knew, my day had come,
If it chanced, as well it might be
That McPhee sold Buckley's Rum.

If the poem is, indeed, the work of Henry Lawson, it is not one of his best pieces of poetry. Yet Henry's work was not always of the highest quality. Even the most passionate Lawson scholar, Colin Roderick, said of Henry, 'That there was much dross in his writing no one would deny.'[7] Even midway through his writing career, Perth's *The*

Sunday Times newspaper claimed that as much as half of Henry's poems were only ever making it into print because of the good name he had established early on.

In some places, 'Buckley's Rum' is technically flawed and some of the spelling and punctuation is unconventional. The copy of the poem that exists within the Foster family today is a typewritten version believed to have been typed by or for Agnes Foster late in her life in order to preserve the poem that was originally handwritten by Henry. Unfortunately, the whereabouts of the original document are now unknown as Agnes died in 1964 when she was 99 years of age. In making a type-written copy of the poem, it is possible that errors and unconventionalities arose at that time. What is more likely, however, is that the errors existed in the original work as penned by Henry. Before publication, Henry's work very often needed considerable editing. George Mure Black was a sub-editor of the *Bulletin* and he claimed that before his work on Henry's classic short story, 'His Father's Mate,' it was 'a very crude composition' and that 'the spelling was bad, it was not punctuated, and its strong points were overlaid with words or by a bad arrangement of paragraphs.' The editor and critic, A. G. Stephens, had worked with Henry on many projects. Stephens knew how sensitive Henry was to criticism and amendment. 'I believe he feels an editorial cut in his copy as keenly as if it were a cut in his flesh,' Stephens wrote. Stephens, however, was unapologetically certain that, where Henry was concerned, editorial intervention was necessary. He took umbrage at Henry's repeated objections to the role of an editor. 'Nobody knows better than Lawson how generously he has been helped,' Stephens wrote, pointing out to readers and, no doubt, to Henry himself, 'how the misspelt, ungrammatical but forcible copy of his early days was edited into something like decent form—form for which Lawson took credit.' Stephens marvelled at Henry's ingratitude when he remembered 'how many deformities and crudities have been removed' from Henry's manuscripts.

The 'Buckley's Rum' poem does not have the benefit of this editing. It is an 'off the top of his head' composition. Henry's published works were not only edited for conventions, but also for content and, while 'Buckley's Rum' is somewhat rough, with careful and considerate re-writing of the type that Henry often engaged in before his work appeared in books or newspapers, the poem could have been significantly improved. The kernel of the idea is a good one and a humorous one; however, the poem is occasionally stilted and difficult to read in places. As much as Henry disliked editors meddling with his work, he did concede at one time that he could guarantee the quality of very little but his actual story ideas. In another letter, he concluded with the words, 'You might as well know at once to save trouble that I don't spell—never could.' Several scholars and historians have pointed out his inability to adhere to spelling and grammatical conventions.[8] With the limitations acknowledged, I might also add that having read the poem many times and in reading it out loud, it now seems far less stilted than I originally thought. With repeated readings, the quality of the poem seems to improve.

The poem is not to be taken as a strict record of fact. In the poem, the narrator is travelling east—he sets out from Yantabulla and passes McPhee's at the Warrego River. If the poem was written and given to Agnes Foster in Hungerford, at that stage Henry would only have travelled between the Warrego River and Yantabulla in a westerly direction. Of his life and writing, Henry did say, 'most of my hard-up experiences are in my published books, disguised but not exaggerated.' To write of going west to east instead of east to west is somewhat of a disguise, but certainly not an exaggeration. It makes one wonder at just how awful the hangover from Buckley's rum must have been. No wonder it took Henry about five weeks to complete what Baz and I managed in 15 days!

I note that in 1908 *The Sydney Morning Herald* reported that a John Buckley died at his home in Yantabulla, aged 64 years.[9] When Henry passed Yantabulla, this John Buckley would have been 48 years old. John Buckley held the publican's licence to the Yantabulla Hotel from 1879-1900.

In the poem, McPhee's Hotel is said to be located at the Warrego River. This, of course, is where the township of Fords Bridge and the Warrego Hotel is located. The Warrego Hotel, however, was once known as the Salmon Ford Hotel. Thus, it is worthy of note that the license to the Salmon Ford Hotel was held by a Norman McPhee from 1881-1889 and again from 1896-1900. At the time of Henry's passing through, the licence was held by Richard Green but seeing as he only held the licence for a relatively short time, the hotel may well have been known by some locals as McPhee's, particularly as McPhee only let the hotel to Green before subsequently returning.

To the modern reader, these details can only be verified through careful research. Their existence in the poem adds credibility in terms of its authenticity but in the absence of the original handwritten poem, it cannot be possible to prove without doubt that the poem was written by Henry Lawson. However, the facts as they stand do suggest that the poem is an original Lawson work.

Mrs Swaffer put me in contact with the 'owner' of the poem, Mr Graeme Foster, who lives in New South Wales. Isaac Foster (who built the Hungerford hotel) was the great-great-grandfather of both Mr Graeme Foster and Mrs Yvonne Swaffer. Both Mr Foster and Mrs Swaffer welcomed this opportunity for their treasured family possession to 'see the light of day,' as it were, and to be published and shared with a wider audience. I thank them both for this addition to my book.

APPENDIX II

Henry Lawson Books and Booklets Published in His Lifetime

Short Stories in Prose and Verse (1894). Sydney: Louisa Lawson.

In the Days When the World Was Wide and Other Verses (1896). Sydney: Angus & Robertson.

While the Billy Boils (1896). Sydney: Angus & Robertson.

On the Track (1900). Sydney: Angus & Robertson.

Over the Sliprails (1900). Sydney: Angus & Robertson.

On the Track and Over the Sliprails (combined edition) (1900). Sydney: Angus & Robertson.

Verses Popular and Humorous (1900). Sydney: Angus & Robertson.

The Country I Come From (1901). London: William Blackwood & Sons.

Joe Wilson and His Mates (1901). London: William Blackwood & Sons.

Children of the Bush (1902). London: Methuen.

When I Was King and Other Verses (1905). Sydney: Angus & Robertson.

The Elder Son (1905). Sydney: Angus & Robertson.

Send Round the Hat (1907). Sydney: Angus & Robertson.

The Romance of the Swag (1907). Sydney: Angus & Robertson.

The Rising of the Court and Other Sketches in Prose and Verse (1910). Sydney: Angus & Robertson.

The Skyline Riders and Other Verses (1910). Sydney: Fergusson.

A Coronation Ode and Retrospect (1911). Sydney: D. S. Ford.

Mateship: A Discursive Yarn (1911). Melbourne: Lothian.

The Stranger's Friend (1911). Melbourne: Lothian.

Triangles of Life and Other Stories (1913). Melbourne: Standard.

For Australia and Other Poems (1913). Melbourne: Standard.

My Army, O, My Army! And Other Songs (1915). Sydney: Tyrell's.

Song of the Dardanelles (1915). Sydney: Tyrell's.

Song of the Dardanelles and Other Verses (1916). London: George G. Harrap.

A Letter from Henry Lawson (1917). Sydney: The Worker.

'Too Old to Rat' (1917). Sydney: [no imprint].

Selected Poems of Henry Lawson (1918). Sydney: Angus & Robertson.

ENDNOTES

FOREWORD

'One-hundred-and-something-scary': Henry Lawson (1902). The Ghosts of Many Christmases. *Children of the Bush.*

DAY ONE

1. Henry Lawson (1896). Out Back. *In the Days When the World Was Wide and Other Verses.* The actual line is that he 'paid for his sins Out Back.'
2. *The Bulletin* newspaper was one of the main organs for the publication of Henry's work. In 1892, Henry published 32 works in the *Bulletin.* In 1893, he published 22 works. With the exception of 1915 (when he published 36 times in the *Bulletin,* with much of his work reflective of wartime sentiments because of Australia's involvement in WWI), these two years were the most prolific years of Henry publishing in the *Bulletin.* Two 1893 *Bulletin* pieces worthy of particular mention are 'In A Dry Season' (detailing Henry's trip to Bourke) and 'In A Wet Season' (detailing Henry's trip from Bourke), published on 5 November and 2 December respectively. – see Walter W. Stone (compiler) (1964). *Henry Lawson: A Chronological Checklist of his Contributions to 'The Bulletin' 1887-1924.* Sydney: Wentworth.
3. After his death, one newspaper wrote, 'It is said that Lawson has painted an age and atmosphere which is swiftly passing, if, indeed, it has not already passed. If that be so, his transcript becomes the more valuable.' – *The Sydney Morning Herald* (1922, September 9). Henry Lawson. (p. 12).
4. Miles Franklin (1942, September 5). 'A Great Gift from a Greatly Gifted Man.' In Pamela Robson (ed.) (2009). *Great Australian Speeches: Landmark Speeches that Defined and Shaped our Nation,* pp. 116-120. Millers Point, NSW: Pier 9. (p. 120).
   In 1899, as an unpublished teenaged author, Miles Franklin sent a manuscript to Henry Lawson and asked him for assistance. Henry edited the manuscript and presented it to a publisher. The resulting book, *My Brilliant Career,* drew critical acclaim and was immediately a best-seller. It remains an Australian classic.
5. Manning Clark (1978). *In Search of Henry Lawson.* South Melbourne: MacMillan. (p. 47).
   H. M. Green went further and identified Henry as 'the apostle and prophet of the gospel of mateship.' – H. M. Green (1930). *An Outline of Australian Literature.* Sydney: Whitcombe & Tombs. (p. 112).
   In a similar vein, J. M. Neild said of Henry that he 'preached his wonderful gospel of brotherhood.' – J. M. Neild (1944). *Lawson—and His Critics.* Melbourne: The Henry Lawson Memorial and Literary Society. (p. 5).
6. J. T. Lang (1956). *I Remember.* Sydney: Invincible. (pp. 7-8).
   On the occasion of a Royal Visit to Australia, Henry predicted that at the various receptions, the Royals would be joined by other honoured guests who would invariably be highly-esteemed city dwellers. The people of the Bush and the Outback would remain uninvited. With this in mind, Henry wrote 'The Men Who Made Australia'. Of the forthcoming receptions and celebrations, he predicted, 'The men who made the land will not be there....For the men who conquer deserts have to work.' – *When I Was King and Other Verses,* 1905.
7. For an early declaration that Henry's work would become increasingly important to Australians, see Miles Franklin, *op. cit.* #4.
8. H. M. Green, op. cit. #5. (p. 115).
   Similarly, William Thomas Ely, Australia's Minister for Health from 1931-1932, said at the 1931 unveiling of a memorial at a hospital where Lawson spent time recuperating, that 'One has only to read the verses he has left us to feel the spirit that he conveyed in his writings; to see in the mind's eye the scenes he described; to smell the scent of the gums and the wattle and the rain on the bush;

to hear the call of the birds and the thousand and one sounds that are of the great wide spaces. His verses bring before us in clear detail life in the shearing shed, in the bush shanty, and on the track. His writings shine with the sparkle of his faith in his country; they reflect hope in the future, and cover with the cloak of charity the shortcomings of those about whom he wrote.' – *The Sydney Morning Herald* (1931, December 10). Henry Lawson: Coast Hospital Memorial. (p. 8).

Other Day One Sources

'send back for their blankets': Henry Lawson (1902). That Pretty Girl in the Army. *Children of the Bush.*

'watchword of Australia': J. Le Gay Brereton (1931). In the Gusty Old Weather. In Bertha Lawson & J. Le Gay Brereton (eds.) *Henry Lawson by His Mates*, pp. 1-16. Sydney: Angus & Robertson. (p. 15).

'greatest gift': John Tighe Ryan quoted in Colin Roderick (1991). *Henry Lawson: A Life.* Sydney: Angus & Robertson. (p. 162).

Royal Hotel a Union pub: Robyn Burrows & Alan Barton (1996). *Henry Lawson: A Stranger on the Darling.* Sydney: Angus & Robertson. (p. 32).

'Watty Bothways': Henry Lawson (1907). Send Round the Hat. *Send Round the Hat.*

'indulgent or fatherly expression': Henry Lawson (1902). That Pretty Girl in the Army. *Children of the Bush.*

'If a horse bolted': *Ibid.*

'go to the brink of eternity': Henry Lawson (1896). In A Dry Season. *While the Billy Boils.*

'second Mississippi': Henry Lawson (1900). The Darling River. *Over the Sliprails.*

'narrow streak of mud': Henry Lawson (1897, February 26). The Bush and the Ideal. *The Bulletin.*

'the drover hauled off': Henry Lawson (1896). The Union Buries Its Dead. *While the Billy Boils.*

'holy horror of snakes': Thomas Davies Mutch quoted in Roderick (1991). *Henry Lawson: A Life.* Sydney: Angus & Robertson. (p. 309).

Old North Bourke Bridge: Robyn Burrows & Alan Barton (1996). *Henry Lawson: A Stranger on the Darling.* Sydney: Angus & Robertson. (p. 63).

'Australian of the Australians': T. W. Heney. Appreciation. In Bertha Lawson & J. Le Gay Brereton (eds.) *Henry Lawson by His Mates*, pp. 170-174. Sydney: Angus & Robertson. (p. 172).

'first articulate voice': Arthur H. Adams. Stories. *Ibid.* pp. 252-254. (p. 252).

'just a hint of kangaroos': Henry Lawson (1900). But What's the Use. *Verses Popular and Humorous.*

'average Australian boy': Henry Lawson (1893). Our Countrymen. *The Worker* (Sydney). In Colin Roderick (ed.) (1972). *Henry Lawson: Autobiographical and Other Writings 1887-1922.* Sydney: Angus & Robertson. (p. 20).

*Hereafter referred to as 'Roderick *Autobiographical Writings*'.

'awful desolation': Henry Lawson (1897, February 26). The Bush and the Ideal. *The Bulletin.*

'lamp of hope': Henry Lawson (1901, December 14). Out Back. *The Queenslander.*

'make smoke round our camp to keep off the mosquitoes': Henry Lawson (1902). The Ghosts of Many Christmases. *Children of the Bush.*

DAY TWO

1. In a poem written by a close friend, the friend wrote:
   Soft, Lawson spoke of boyhood days,
   　　Of childish love and hate,
   Of struggles that his father had,
   　　In his long fight with Fate.
   'My father's heart was gold,' he said,
   　　For Lawson loved his sire.
   – Jim Grahame (1940). When Lawson Walked With Me. *Call of the Bush*, pp. 49-50. Melbourne: Bread and Cheese Club. (p. 49).
2. Louisa Lawson note to A. G. Stephens after Stephens reviewed Henry's first book, which had been published by Louisa. Quoted in *Barrier Miner* (Broken Hill, NSW) (1935, December 19). Struggles of Lawson: Letters tell of Early Trials. (p. 6).

As much as Louisa is here trying to emerge as an independent identity and not just the *mother of* Henry Lawson, there can be no doubt that Henry Lawson was also the son of Louisa Lawson. As Roderick said, 'From her activities Lawson drew inspiration. Her arguments became part of his mind.' – Colin Roderick (1960). *Henry Lawson's Formative Years (1883-1893).* Sydney: Wentworth. (p. 21).
More recently, Louisa seems to have been granted her wish with a number of biographers choosing her life as the subject for their works. One biographer begins with the assertion, 'Louisa Lawson was a formidable woman' and then sets about portraying Louisa as Louisa and not just as the mother of Henry. See Sharyn Pearce (1992). *The Shameless Scribbler: Louisa Lawson.* London: Sir Robert Menzies Centre for Australian Studies, University of London. (p. 1).

3. For 'The Chieftainess' see Henry's composition 'Men Who Did Their Work, or, The Books That I Like,' Composed 1917. In Roderick *Prose Writings*, pp. 775-778.
One female biographer of Louisa says, 'Louisa has been airily and summarily recorded as being to blame' for the strained relationship with her son. The female biographer considers that to be an unfair and biased representation of the situation born of the fact that most of Henry's biographers are men. See Lorna Ollif (1978). *Louisa Lawson: Henry Lawson's Crusading Mother.* Adelaide: Rigby. (p. 17).
4. 'In form and feature, character and mentality, Lawson was a copy of his mother with male emphasis.' – A. G. Stephens (1922, November 1). Henry Lawson. *Art in Australia (3rd series #2).* (8 pages unnumbered [p. 5]).

Other Day Two Sources

'not glorious and grand and free': Henry Lawson (1893, November 18). Some Popular Australian Mistakes. *The Bulletin.*
'wandered off in the quest of gold': *The Sydney Morning Herald,* (1922, September 4). Henry Lawson: Death Announced: Poet and Prose Writer. (p. 8).
'a most industrious man': *Ibid.*
'died before I began to understand and appreciate him.': Henry Lawson. A Foreign Father, manuscript published in *The La Trobe Journal, 70*, Spring 2002, pp. 35-40.
'pick and shovel over his shoulder': Henry Lawson (Composed 1903-1906). A Fragment of Autobiography. In Colin Roderick (ed.) (1984). *Henry Lawson: The Master Story-Teller: Prose Writings.* Sydney: Angus & Robertson. (pp. 714-755).
*Hereafter referred to as 'Roderick *Prose Writings*'.
'Father was always gentle': Henry Lawson (1900). A Vision of Sandy Blight. *On the Track.*
'a remarkable woman, with many graces of character': *The Sydney Morning Herald,* (1922, September 4). Henry Lawson: Death Announced: Poet and Prose Writer. (p. 8).
'selfish, indolent, mad-tempered woman': Henry Lawson letter #335 to George Robertson, Leeton, New South Wales, [January 1917]. In Colin Roderick (ed.). (1970). *Henry Lawson Letters 1890-1922.* Sydney: Angus & Robertson.
*Hereafter referred to as 'Roderick *Letters*'.
'ridges on the floors of hell': Henry Lawson (1893, November 18). Some Popular Australian Mistakes. *The Bulletin.*
'excruciating ear-ache': Henry Lawson (Composed 1903-1906). A Fragment of Autobiography. In Roderick *Prose Writings*, pp. 714-755.
'On sunset tracks they ride and tramp': Henry Lawson (1896, December 19). The Swagman and His Mate. *The Town and Country Journal.*
'The local larrikins called him "Grog"': Henry Lawson (1896). Bogg of Geebung. *While The Billy Boils.*
'sickly stream that looked like bad milk': Henry Lawson (1902). The Ghosts of Many Christmases. *Children of the Bush.*

DAY THREE

1. *The Canberra Times* (1927, September 2). Our Authors. Literature in Australia. Past and Present. (p. 8).
In one history of Australian literature, the *Bulletin* is said to have 'fostered the robust manhood of

Australian poetry.' – Elizabeth Perkins (1998). Literary Culture 1851-1914: Founding a Canon. In Bruce Bennett & Jennifer Strauss (eds.), *The Oxford Literary History of Australia*, pp. 47-65. Melbourne: Oxford University Press. (p. 47).
Upon his death, one newspaper considered it a 'happy coincidence' that Henry should write at the time that Australians were adopting this new view of themselves: 'Most of Lawson's work is animated by the divine fire, and it has achieved a popularity unique in Australia. How are we to account for the strength of his appeal? Perhaps it is because the time and the man were in happy coincidence. Lawson's most active and fruitful years covered a period which, in some ways, marked a definite epoch in Australian history and in Australian literature. He expressed that newly-developed and growing Australian sentiment which thirty years ago or so was just beginning to become articulate.... Again, Lawson was in the forefront of a literary movement which broke new ground in Australia, and exploited the material to be found within our own gates. Hitherto Australian writers had in the main ignored the possibilities of this field.' – *The Sydney Morning Herald* (1922, September 9). Henry Lawson. (p. 12).
J. F. Archibald was baptised John Feltham Archibald. As an adult trying to make it on Australia's literary scene, he recast himself as having a French background and changed his name to Jules François Archibald.

2. Henry Lawson (1919, September 18). Archibald's Monument. *The Bulletin.*
It is interesting that Henry here identifies his work as a monument to Archibald. Upon his own death, it was said of Henry's work, 'he has reared for himself a lasting monument.' – *The Sydney Morning Herald* (1922, September 9). Henry Lawson. (p. 12).

3. Bertha Lawson (1931). Memories. In Bertha Lawson & J. Le Gay Brereton (eds.) *Henry Lawson by His Mates*, pp. 79-118. Sydney: Angus & Robertson. (pp. 116-117).
After his return from London, Henry said that Australians in England miss the sun: 'Oh, we don't feel the cold so much, we Australians, in London, but the terrible gloom! Fancy writing by lamplight in the middle of the day, and for three days running, in a London fog! We get heartsick for the sun.' – *The Brisbane Courier* (1902, August 16). Grimy Old Babylon. Henry Lawson's Appreciation. Fraternising With Hodge. A Literary Secret. (p. 15).

4. Colin Roderick (1991). *Henry Lawson: A Life*. Sydney: Angus & Robertson. (e.g. pp. 131, 195, 201, 219, 254, 297).
Archibald also often paid Henry in advance. *The Bulletin* continued to publish original Lawson works after Henry's death but, as one person said, 'We may be morally certain that Lawson still owed *The Bulletin* "copy" on which he had drawn advance payments.' – Walter W. Stone (compiler) (1964). *Henry Lawson: A Chronological Checklist of his Contributions to 'The Bulletin' 1887-1924.* Sydney: Wentworth. (p. 3).

5. Henry wrote, 'They say that self-preservation is the strongest instinct of mankind; it may come with the last gasp, but I think the preservation of the life or liberty of a mate—man or woman—is the first and strongest. It is the instinct that irresistibly impels a thirsty, parched man, out on the burning sands, to pour the last drop of water down the throat of a dying mate, where none save the sun or moon or stars may see.' – Mateship. *Triangles of Life and Other Stories*, 1913.

6. A. G. Stephens (1977). Australian Literature I. In Leon Cantrell (ed.) *A. G. Stephens: Selected Writings*, pp. 76-88. Sydney: Angus & Robertson. (p. 85). (Originally published in *The Commonwealth*, 1901).
Henry's tramping mate was of the opposite opinion. Rather than Henry hating the Bush, Jim Gordon wrote, 'Lawson loved the bush, and mostly all contained therein—its rivers and old homesteads with their straggling bridle-paths and winding tracks. He saw beauty in the plain with its cruel mirages, and found charm in the great silences of the nights of the Never-Never. None of them were lost romances to him.' – Jim Grahame (1931). Amongst My Own People. In Lawson & Brereton, *op. cit.* #3, pp. 210-251. (p. 211).
Stephens believed Henry chose always to look for, and write about, what was bad. He believed Henry's writing was too sombre and that, even when writing about life in the city, Henry also chose

to write gloomily. Stephens said that, while they were true for some people, Henry's portrayals were really only 'an episode of the minority, not a permanent reflection of the majority.' – A. G. Stephens (1922, November 1). Henry Lawson. *Art in Australia (3rd series #2).* (8 pages unnumbered [p. 2]).

7. Henry uses the phrase in 'The Mystery of Dave Regan' (*Short Stories in Prose and Verse*, 1894), 'In a Wet Season' (*While the Billy Boils*, 1896), and 'In Hospital' (Composed 1902-1903. In Roderick *Prose Writings*, pp. 625-627).
8. Henry wrote, 'One of the best cows at the homestead had a calf, about which she made a great deal of fuss. She was ordinarily a quiet, docile creature, and, though somewhat fussy after calving no one ever dreamed that she would injure anyone. It happened one day that the squatter's daughter and her intended husband, a Sydney exquisite, were strolling in a paddock where the cow was. Whether the cow objected to the masher or his lady love's red parasol, or whether she suspected designs upon her progeny, is not certain; anyhow, she went for them. The young man saw the cow coming first, and he gallantly struck a bee-line for the fence, leaving the girl to manage for herself. She wouldn't have managed very well if Malachi hadn't been passing just then. He saw the girl's danger and ran to intercept the cow with no weapon but his hands.
   It didn't last long. There was a roar, a rush, and a cloud of dust, out of which the cow presently emerged, and went scampering back to the bush in which her calf was hidden.' – The Story of Malachi. *While the Billy Boils*, 1896.

Other Day Three Sources

'I wrote it and screwed up courage to go down to the *Bulletin* after hours': Henry Lawson (Composed 1903-1906). A Fragment of Autobiography. In Roderick *Prose Writings*, pp. 714-755.

'Who profit by friends—and forget': Henry Lawson (1896). To an Old Mate. *In the Days When the World Was Wide and Other Verses.*

'You couldn't quarrel with Archibald': Henry Lawson (Composed 1919). Three or Four Archibalds and the Writer. In Roderick *Prose Writings*, pp. 778-784.

'Archibald was there': Roderic Quinn (1931). Glimpses of Henry Lawson. In Lawson & Brereton, *op. cit.* #3, pp. 175-191. (p. 180).

'lost in London gloom': Henry Lawson (1905). A Voice from the City, *When I Was King and Other Verses.*

'held responsible, in a general way, for most of the out-back trouble': Henry Lawson (1900). The Darling River. *Over the Sliprails.*

'No publisher ever did more for one of his clients': Redgum (1933, September 2). George Robertson. Henry Lawson's Tribute. *The Sydney Morning Herald.* (p. 9).

'slight acknowledgement of and small return for his splendid generosity': Henry Lawson (Composed 1910). The Auld Shop and the New. In Colin Roderick (ed.) (1969). *Henry Lawson: Collected Verse: Volume Three 1910-1922.* Sydney: Angus & Robertson. (pp. 47-52, 415-416).

'Nothing that Lawson wrote made [Robertson] laugh so hilariously': Redgum (1933, September 2). George Robertson. Henry Lawson's Tribute. *The Sydney Morning Herald.* (p. 9).

'always insisted on keeping the bag full': Jim Grahame (1925, February 19). Henry Lawson on the Track. *The Bulletin.* (The Red Page).

'To cease walking is to die': Henry Lawson (1896). Stragglers. *While the Billy Boils.*

'no ambition beyond the cricket and football field': Henry Lawson (1893). Our Countrymen. *The Worker* (Sydney). In Roderick *Autobiographical Writings*, p. 20.

'I am back from up the country': Henry Lawson (1896). Up the Country. *In the Days When the World Was Wide and Other Verses.*

'He was not so much flattered as surprised': Madame Rose Soley (1922, September 5). A Reminiscence. *The Sydney Morning Herald.* (p. 10).

'tugging and slipping, and moving by inches': Henry Lawson (1900). Song of the Old Bullock-Driver. *Verses Popular and Humorous.*

'the spokes are turning slow': Henry Lawson (1901, December 14). The Teams. *The Queenslander.*

'It is quite time that our children were taught a little more about their country': Henry Lawson (1888, April 4). A Neglected History. *The Republican.*

'I'd have been riddled like a—like a bushranger': Henry Lawson (1894). The Mystery of Dave Regan. *Short Stories in Prose and Verse.*

Henry grumbled that his mother had made a lot of money from the book: Henry Lawson, letter to Angus & Robertson, Mangamauna, New Zealand, 26 July 1897. In Roderick *Letters* #32.

'My mother's the hardest business man I ever met': J. Le Gay Brereton. In the Gusty Old Weather. In Lawson & Brereton, *op. cit.* #3, pp. 1-16. (p. 7).

DAY FOUR

'We found it was also painful': Henry Lawson (1893). Bush Terms. *The Worker* (Sydney). In Roderick *Autobiographical Writings*, p. 23.

'couldn't say no to a friend': Arthur Parker (1931). Beginnings. In Bertha Lawson & J. Le Gay Brereton (eds.) *Henry Lawson by His Mates*, pp. 17-30. Sydney: Angus & Robertson. (p. 30).

Henry and friend exchanged footwear: Jack Moses. The Clot of Gold. *Ibid.* pp. 37-50. (pp. 42-43).

'You'll catch it hot, you'll see': Henry Lawson (1900). Trouble on the Selection. *Verses Popular and Humorous.*

'swiftly retreats to the safety of its burrow': Stephen Swanson (2007). *Field Guide to Australian Reptiles.* Archerfield, Qld: Steve Parish Publishing.

Henry's sense of humour 'quaint and twisted': Fred J. Broomfield. Recollections of Henry Lawson. In Bertha Lawson & J. Le Gay Brereton (eds.) *Henry Lawson by His Mates*, pp. 61-78. Sydney: Angus & Robertson. (p. 66).

Henry's sense of humour was 'his chief characteristic': J. Le Gay Brereton. In the Gusty Old Weather. *Ibid.* pp. 1-16. (p. 3).

'pranktical jokes': Henry Lawson (1901). The Story of Malachi. *The Country I Come From.*

Rooster story of disturbing the peace: Jack Moses. The Clot of Gold. In Bertha Lawson & J. Le Gay Brereton (eds.) *Henry Lawson by His Mates*, pp. 37-50. Sydney: Angus & Robertson. (pp. 48-50).

'enjoyable unlawfulness': Fred J. Broomfield. Recollections of Henry Lawson. *Ibid.* pp. 61-78. (p. 64).

Tobacco pipe for a revolver: *Ibid.* (p. 67).

'a child with a singing soul in a world of business men': Arthur H. Adams. Stories. *Ibid.* pp. 252-254. (p. 252).

'What kids they were': Bertha Lawson & J. Le Gay Brereton (eds.) *Henry Lawson by His Mates.* Sydney: Angus & Robertson. (editor's footnote, p. 48).

'we aspired to some of the higher branches of the practical joker's art': Henry Lawson (1901). The Story of Malachi. *The Country I Come From.*

'Bushmen are the biggest liars that ever the Lord created': Henry Lawson letter to Aunt Emma, Bourke, New South Wales, 21 September 1892. In Roderick *Letters* #5.

'twice as hard as any of the rest': Henry Lawson (1901). 'Water Them Geraniums'. *Joe Wilson and His Mates.*

'everlasting, maddening sameness of the stunted trees': Henry Lawson (1896). The Drover's Wife. *While the Billy Boils.*

DAY FIVE

1. A. B. ('Banjo') Paterson (1939, February 11). 'Banjo' Paterson Tells His Own Story - 2: Giants of the Paddle, Pen, and Pencil: Henry Lawson at Work: When the Booms Burst. *The Sydney Morning Herald.* (p. 21).

   Henry's wife, Bertha, dismissed Banjo's claim. In a letter to the newspaper where Banjo made his statement, Bertha claimed that Henry did not meet Banjo until after their duelling poems appeared in print. 'So much has been written of Henry Lawson which is not correct,' Bertha wrote. 'Henry felt the wrongs of the people, and the suffering of the women and the children of the bush and the bravery of the men too keenly to stage a mock battle in the Press for paltry gain.' – Bertha Lawson (1939, February 13). Henry Lawson: To the Editor of the *Herald. The Sydney Morning Herald.* (p. 7).

2. Colin Roderick (1991). *Henry Lawson: A Life*. Sydney: Angus & Robertson. (pp. 128 & 131).
   A literary analyst said of Banjo and Henry, 'Over the hot dry country with which the ballad oftenest deals, Paterson rode but Lawson plodded, humping his swag.' – H. M. Green (1951). *Australian Literature 1900-1950*. Melbourne: Melbourne University Press. (p. 14).

Other Day Five Sources

'I was sulky, I was moody (I'm inclined to being broody)': Henry Lawson (1915). A Mixed Battle Song. *My Army, O, My Army! and Other Songs.*

'I think the country's rather more inviting round the coast': Henry Lawson (1892, July 9). Borderland. *The Bulletin.*

'So you're back from up the country, Mister Lawson': Banjo Paterson (1892, July 23). In Defence of the Bush. *The Bulletin.*

'The city seems to suit you, while you rave about the bush': Henry Lawson (1892, August 6). In Answer to 'Banjo', and Otherwise. *The Bulletin.*

'the sad and soulful poet with a graveyard of his own': Banjo Paterson (1892, October 1). In Answer to Various Bards. *The Bulletin.*

'an undignified affair': Paterson, *op. cit.* #1.

'it's quite natural to travel all day without exchanging a word': Henry Lawson (1902). That Pretty Girl in the Army. *Children of the Bush.*

DAY SIX

1. Interestingly, Henry wrote a short story detailing a similar incident. One dark night, Dave Regan has to cross a haunted gully. As he does so, he hears what he believes is the ghost of an old Chinese gold fossicker running up behind him. After a frightful scare, it turns out the approaching footsteps are only the flapping of ribbons fastened to his borrowed hat. See 'The Chinaman's Ghost'. *Joe Wilson and His Mates* (1901).
2. Henry Lawson (1891, May 16). Freedom on the Wallaby. *The Worker* (Brisbane).
   At a meeting of the Queensland Legislative Council, a vote of thanks was extended to the policemen and soldiers who helped to keep the peace and break up the strike camp at Barcaldine, Queensland. Frederick Brentnall read out Henry's poem in disgust, sparking calls for Henry to be arrested for sedition. Henry responded with another poem, entitled 'The Vote of Thanks Debate' (*The Worker* (Brisbane), 25 July 1891). In the poem, Henry referred to 'That foolish speech of Brentnall's in the Vote of Thanks debate.' Henry suggested also that Brentnall had succeeded only in making 'a jackass of one's self.'

Other Day Six Sources

'my spirit revives in the morning breeze, though it died when the sun went down': Henry Lawson (1899, December 16). After All. *The Queenslander.*

'Death and ruin are everywhere': Henry Lawson (1900). The Song of the Darling River. *Verses Popular and Humorous.*

'set your lips and see it through': Henry Lawson (1896). 'Sez You'. *In the Days When the World Was Wide and Other Verses.*

'it's the heat that makes us all a bit ratty at times': Henry Lawson (1900). New Year's Night. *Over the Sliprails.*

'ridges on the floors of hell': Henry Lawson (1893, November 18). Some Popular Australian Mistakes. *The Bulletin.*

'a cramped mind devoted to sport': Henry Lawson (1893). Our Countrymen. *The Worker* (Sydney). In Roderick *Autobiographical Writings*, p. 20.

'the brainless man of muscle has the burial of a god': Henry Lawson (1892, May 28). A Song of Southern Writers. *The Bulletin.*

Brady and Archibald discussing Henry's mood: E. J. Brady quoted in Robyn Burrows & Alan Barton (1996). *Henry Lawson: A Stranger on the Darling*. Sydney: Angus & Robertson. (pp. 13-14).

Squatters' God-given right: Patsy Adam-Smith (1982). *The Shearers*. Melbourne: Nelson. (pp. 88-89).

His father's shock at Henry's drinking: Manning Clark (1978). *In Search of Henry Lawson*. South Melbourne: MacMillan. (pp. 35-36).
'I got £5 and a railway ticket from the *Bulletin* and went to Bourke': Henry Lawson (1899, January 21). 'Pursuing Literature' in Australia. *The Bulletin*.
'took notes all the way up': Henry Lawson letter to Aunt Emma, Bourke, New South Wales, 21 September 1892. In Roderick *Letters* #5.
'In A Dry Season' sketch: Henry Lawson (1896). In A Dry Season. *While the Billy Boils*.
'He had run with Cobb and Co.': Henry Lawson (1900). The Old Bark School. *Verses Popular and Humorous*.
'A hundred miles shall see to-night the lights of Cobb and Co.': Henry Lawson (1900). The Lights of Cobb and Co.. *Verses Popular and Humorous*.

DAY SEVEN

1. Prout suggests a population of 1500, while Burrows and Barton go with 3000:
   – Denton Prout (1963). *Henry Lawson: The Grey Dreamer*. Sydney: Rigby. (p. 107).
   – Robyn Burrows & Alan Barton (1996). *Henry Lawson: A Stranger on the Darling*. Sydney: Angus & Robertson. (p. 30).
   Precise population numbers are difficult given the transient, fluid nature of the town at the time. Henry described Bourke in the following manner:
   'Bourke was just a little camping town in a big land, where free, good-hearted democratic Australians, and the best of black sheep from the old world were constantly passing through; where husband's were often obliged to be away from home for twelve months, and the storekeepers had to trust the people, and mates trusted each other, and the folks were broad-minded.' – That Pretty Girl in the Army. *Children of the Bush*, 1902.
2. Henry said that 'the newspaper reports were enough to frighten Noah's ghost.'
   For instance, note the headlines over four consecutive days of issue (no Sunday issue) in *The Sydney Morning Herald* newspaper. The headings and subheadings reflect some of the excitement and drama of the build up to—and eventuality of—the flood.
   The Sydney Morning Herald (1890, April 17). The Floods in the West: Fighting the Waters. Bourke Still Safe. The Railway Line Endangered. Lord and Lady Carrington on the Spot. (p. 7).
   The Sydney Morning Herald (1890, April 18). The Floods in the West. Great Anxiety at Bourke. A Serious Leakage in the Embankment. The Railway Line Cut to Relieve the Town. Return of Lord and Lady Carrington. (p. 5).
   The Sydney Morning Herald (1890, April 19). The Flood at Bourke. Bursting of the Embankment, The Town Inundated. Boats Being Rowed in the Streets. The Railway Line Swept Away. The Inhabitants Safe. No Danger Apprehended to Life. An Inland Sea for 30 Miles. Great Loss of Property Feared. (p. 9).
   The Sydney Morning Herald (1890, April 21). The Flood at Bourke. Rescue of the Inhabitants. Reported Drowning of a Family. A Scarcity of Boats. Provisions Running Short. The Houses Giving Way. The Water Still Rising. (p. 8).
3. References to Aboriginal legends are derived from the following sources:
   –William Jenkyn Thomas (1923). *Some Myths and Legends of the Australian Aborigines*. Melbourne: Whitcombe & Tombs.
   –K. Langloh Parker (H. Drake-Brockman, selector) (1953). *Australian Legendary Tales*. Sydney: Angus & Robertson.
   –Melva Jean Roberts & Ainslie Roberts (ills.) (1981). *Dreamtime: The Aboriginal Heritage*. Adelaide: Rigby.
4. – Christopher Lee (2002). The Status of the Aborigine in the Writing of Henry Lawson: A Reconsideration. *The La Trobe Journal*, 70, pp. 74-83.
   – W. H. Pearson (1968). *Henry Lawson Among Maoris*. Wellington, N.Z.: A. H. & A. W. Read. (pp. 7-12).

In contrast to revisionist commentaries, a contemporary and friend of Henry said, 'Henry Lawson always had a soft spot for Australian blacks.' – P. J. Cowan quoted by Keith Kennedy (1931, September 12). Lawson and Cowan: Relics of the Last Phase. *The Sydney Morning Herald*. (p. 7).

5. Jack Moses. The Clot of Gold. In Bertha Lawson & J. Le Gay Brereton (eds.) *Henry Lawson by His Mates*, pp. 37-50. Sydney: Angus & Robertson. (p. 37).
   In speaking of her father, Henry's daughter described him as, 'shy [and] retiring,' and said, 'he had little appreciation of any extravagant acclaim.' – Bertha Lawson (1973). *My Father (Abbey's Broadsheet, No. 1)*. Sydney: Henry Lawson's Bookshop. (2 pages unnumbered [p. 1]).
6. Of his time in Hungerford, Henry wrote, 'Hungerford consists of two houses and a humpy in New South Wales, and five houses in Queensland. Characteristically enough, both the pubs are in Queensland. We got a glass of sour yeast at one and paid sixpence for it—we had asked for English ale.' – Hungerford. *While the Billy Boils*, 1896.

Other Day Seven Sources

'Sweeney' poem: Henry Lawson (1896). Sweeney. *In the Days When the World Was Wide and Other Verses*.
'the most drunken town': Henry Lawson (1900). The Darling River. *Over the Sliprails*.
Three churches and nineteen hotels: Burrows & Barton, *op. cit.* #1. (pp. 30-31).
'He stayed drunk for three weeks': Henry Lawson (1901). The Darling River. *The Country I Come From*.
'agreeably disappointed with Bourke': Henry Lawson letter to Aunt Emma, Bourke, New South Wales, 21 September 1892. In Roderick *Letters* #5.
'rescue work was done by a short man with his trousers tucked up': Henry Lawson (n.d.). The Great Flood of '90 [manuscript]. State Library of Victoria, Accession no. MS6026. Lothian Publishing Company, Records 1895-1950, Australian Manuscripts Collection. Box XX1B Folder 3B.
'Went dashing past the camps': Henry Lawson (1918). The Roaring Days. *Selected Poems of Henry Lawson*.
'land of living death': Henry Lawson (1897, February 26). The Bush and the Ideal. *The Bulletin*.
'I preferred to keep dark for a while': Henry Lawson letter to Aunt Emma, Bourke, New South Wales, 27 September 1892. In Roderick *Letters* #6.
'had a great argument with a shearer': Henry Lawson letter to Aunt Emma, Bourke, New South Wales, 21 September 1892. In Roderick *Letters* #5.
Henry's pennames: Burrows & Barton, *op. cit.* #1.
'Henry Lawson told me he had written a poem each for the two local papers': John Hawley (1939, February 20). Letters: Henry Lawson. *The Sydney Morning Herald*. (p. 7).
'a life along the Darling isn't like the life in town': Henry Lawson (1892, October 1). A Stranger on the Darling. *The Western Herald and Darling River Advocate*.
'This is a queer place': Henry Lawson letter to Aunt Emma, Bourke, New South Wales, 27 September 1892. In Roderick *Letters* #6.
'flowing faster in the fear of being late': Henry Lawson (1896). Faces in the Street. *In the Days When the World Was Wide and Other Verses*.
A mate seen as a necessity: Prout, *op. cit.* #1.
'Eager for the bush and wanting a mate': Jim Grahame (1925, February 19). Henry Lawson on the Track. *The Bulletin*. (The Red Page).
'I haven't much time to write, also I am pretty drunk': Henry Lawson letter to Arthur Parker, Bourke, New South Wales, 24 November 1892. In Roderick *Letters* #7.
'He was a stalwart mate': Jim Grahame (1925, February 19). Henry Lawson on the Track. *The Bulletin*. (The Red Page).
'It would take a very long letter to tell you all the news': Henry Lawson letter to Arthur Parker, Bourke, New South Wales, 26 December 1892. In Roderick *Letters* #8.
One of the worst droughts in Australia's history: Sally Sara (2006, July 2). Drought Takes Bourke to the Brink. *ABC Online*. [world wide web].
'you don't see the town till you are quite close to it': Henry Lawson (1896). Hungerford. *While the Billy Boils*.
'The introduction of a few rabbits could do little harm': Government of Western Australia

Department of Agriculture (2001). *The State Barrier Fence of Western Australia, 1901-2001* (Introduction and History). South Perth: Agriculture Protection Board. [world wide web]. www.agric.wa.gov.au/content/pw/vp/barrier_intro_history.pdf (35 pages unnumbered [p. 8]).
'I could hardly eat him for laughing': Henry Lawson (1896). Hungerford. *While the Billy Boils.*
Length of the dingo fence: Roland Breckwoldt (1988). *The Dingo: A Very Elegant Animal.* North Ryde, NSW: Angus & Robertson.
'Half mad with flies and dust and heat': Henry Lawson (1900). The Paroo. *Verses Popular and Humorous.*
'there was peace in Bourke and goodwill towards all men': Henry Lawson (1902). That Pretty Girl in the Army. *Children of the Bush.*
'a sensible cold' Christmas dinner: Henry Lawson (1902). The Ghosts of Many Christmases. *Children of the Bush.*
'their grandmothers used to cook hot Christmas dinners in England': Henry Lawson (1902). That Pretty Girl in the Army. *Children of the Bush.*
Sitting in their underpants drinking beer:
– William Wood (1926, September 24). Reminiscences of Henry Lawson. *Windsor and Richmond Gazette* (Windsor, NSW).
– William Wood (1931). Bourke: A Letter from Paraguay. In Lawson & Brereton, *op. cit.* #5, pp. 34-36. (p. 35).
'free-and-easy costumes': Henry Lawson (1902). The Ghosts of Many Christmases. *Children of the Bush.*
'The blacks may be low and degraded': Henry Lawson (n.d.). King Billy [manuscript]. State Library of Victoria, Accession no. MS6026. Lothian Publishing Company, Records 1895-1950, Australian Manuscripts Collection. Box XX1B Folder 3D.
Henry was 'intensely shy': Arthur Parker. Beginnings. In Lawson & Brereton, *op. cit.* #5, pp. 17-30. (p. 19).
Henry was 'reserved in company': J. Le Gay Brereton. In the Gusty Old Weather. In Lawson & Brereton, *op. cit.* #5, pp. 1-16. (p. 4).
'The average healthy boy's aversion to school': Henry Lawson (Composed 1903-1906). A Fragment of Autobiography. In Roderick *Prose Writings*, pp. 714-755.

DAY EIGHT

1. Bertha Lawson (1931). Memories. In Bertha Lawson & J. Le Gay Brereton (eds.) *Henry Lawson by His Mates*, pp. 79-118. Sydney: Angus & Robertson. (p. 111).
   Henry's daughter so admired the Longstaff portrait of her father that she said it was not only 'his truest likeness' but she said also, 'To see it is to speak to him.' – Bertha Lawson (1973). *My Father (Abbey's Broadsheet, No. 1).* Sydney: Henry Lawson's Bookshop. (2 pages unnumbered [p. 1]).
   Despite how highly Henry's wife and daughter considered the Longstaff portrait, the prickly critic Alfred George Stephens said, 'Longstaff's portrait is exceptionally good, but there is too much painting in it; too much Longstaff.' – A. G. Stephens (1922, November 1). Henry Lawson. *Art in Australia (3rd series #2).* (8 pages unnumbered [p. 7]).
   Interestingly, more than 30 years later and long after Henry was dead, Longstaff won the Archibald Prize for a portrait of Henry's old sparring partner Banjo Paterson. Indeed, the Paterson portrait earned Longstaff his fifth Archibald Prize.

Other Day Eight Sources

Henry grew tired of de Guinney's griping: William Wood (1926, September 24). Reminiscences of Henry Lawson. *Windsor and Richmond Gazette* (Windsor, NSW).
'My Warrego bard was born in St Petersburgh': Henry Lawson (1897, February 26). The Bush and the Ideal. *The Bulletin.*
'I found your letter in the Post Office of this God-Forgotten town': Henry Lawson letter to Aunt Emma, Hungerford, Queensland, 16 January 1893. In Roderick *Letters* #10.
'Being a young dog...': Henry Lawson (1901). The Loaded Dog. *Joe Wilson and His Mates.*
'beautiful dark brown eyes': J. Le Gay Brereton. In the Gusty Old Weather. In Lawson & Brereton, *op. cit.* #1. pp. 1-16. (p. 4).

'woman's eyes, dog's eyes': A. G. Stephens quoted in Colin Roderick (1991). *Henry Lawson: A Life*. Sydney: Angus & Robertson. (p. 217).
'If you know Bourke you know Australia': Henry Lawson quoted in Roderick, *ibid.* (p. 236).
'the man who told me might have been a liar': Henry Lawson (1893, December 14). Hungerford. *The Bulletin.*
'women dry quickly in the bush': Henry Lawson (1907). 'Buckolts' Gate'. *The Romance of the Swag.*
'no place for a woman': Henry Lawson (1900). No Place for a Woman. *On the Track.*
'a person unfit to be at large': Henry Lawson (1893, July 29). Louth, On the Darling. *The Worker* (Sydney).

DAY NINE

1. *The Sydney Morning Herald* (1903, June 5). Divorce Court. (p. 8).
Shortly after the separation became official, Bertha wrote to Henry, 'There is no power on the earth will ever reunite us. You are dead to me as far as affection goes. The suffering I have been through lately has killed any thought of feeling I may have had for you.' – Bertha Lawson to Henry Lawson [15 June 1903] published in John Arnold & Frances Thorn (comps. & eds.). (1981, October). Henry and Bertha Lawson: Some Unpublished Letters and Stories. *La Trobe Library Journal, 28* (7), pp. 73-100. (p. 82).

Other Day Nine Sources

'That's all I ever intend to do with a swag': Henry Lawson letter to Aunt Emma, Bourke, New South Wales, 6 February 1893. In Roderick *Letters* #11.
'Say you, "he's gone to Maoriland, and isn't coming back"': Henry Lawson (1893, June 18). The Emigration to New Zealand. *The Truth.*
'driven out of Australia by the hard times there, and glad, no doubt, to get away': Henry Lawson (1896). Coming Across. *While the Billy Boils.*
'Baby dead and buried when he got back': Henry Lawson letter to Aunt Emma, Wellington, New Zealand, 6 December 1893. In Roderick *Letters* #12.
'the boss said we weren't bushmen': Henry Lawson letter to *The Bulletin*, Sydney, New South Wales, [January 1899]. In Roderick *Letters* #53.
'at the end of a week begged me to marry him': Bertha Lawson (1931). Memories. In Bertha Lawson & J. Le Gay Brereton (eds.) *Henry Lawson by His Mates*, pp. 79-118. Sydney: Angus & Robertson. (p. 80).
'you never realize how innocent you were': Henry Lawson (1902). The Ghosts of Many Christmases. *Children of the Bush.*
'the bones of the dead gleam whitest': Henry Lawson (1896). The Great Grey Plain. *In the Days When the World Was Wide and Other Verses.*
'rather see them dead' and 'deep brown eyes filled with worship of him and his works': George Robertson quoted in Colin Roderick (1991). *Henry Lawson: A Life*. Sydney: Angus & Robertson. (pp. 147 & 155).
Waiver of £10 deposit: Bertha Lawson (1931). Memories. In Bertha Lawson & J. Le Gay Brereton (eds.) *Henry Lawson by His Mates*, pp. 79-118. Sydney: Angus & Robertson. (p. 86).
'surface of the ground was cracked in squares': Henry Lawson (1902). The Shearer's Dream. *Children of the Bush.*
'Where the white man lies dead': Henry Lawson (1891, November 28). Watching the Crows. *Freeman's Journal.*
'getting on each other's nerves': Henry Lawson (1913). Drifting Apart. *Triangles of Life and Other Stories.*
'Old mates seldom quarrel': Henry Lawson (1902). The Sex Problem Again. *Children of the Bush.*
'Past feelin' and despairin'': Henry Lawson (1896). Past Carin'. *In the Days When the World Was Wide and Other Verses.*
'bride of frivolous fashion': Henry Lawson (1913). Australia's Peril: The Warning. *For Australia.*
'She has her dreams': Henry Lawson (1910). A Bush Girl. *Skyline Riders and Other Verses.*
'all her girlish hopes and aspirations have long been dead': Henry Lawson (1896). The Drover's Wife.

While the Billy Boils.
'her heart was nearly broken too': *Ibid.*
'What-did-you-bring-her-here-for? She's only a girl': Henry Lawson (1901). 'Water Them Geraniums': 'Past Carin''. *Joe Wilson and His Mates.*
'some men make fools of themselves then': Henry Lawson (1901). Joe Wilson's Courtship. *Joe Wilson and His Mates.*
'She picked the wrong man': Henry Lawson (1901). Telling Mrs Baker. *Joe Wilson and His Mates.*
'Make the most of your courting days': Henry Lawson (1901). Joe Wilson's Courtship. *Joe Wilson and His Mates.*
'Don't you feel lonely, Mrs Spicer': Henry Lawson (1901). 'Water Them Geraniums': 'Past Carin''. *Joe Wilson and His Mates.*
'Australia! My country! Her very name is music to me': Henry Lawson (1907). The Romance of the Swag. *The Romance of the Swag.*
Henry not able to enlist in war: Colin Roderick (1991). *Henry Lawson: A Life.* Sydney: Angus & Robertson. (p. 345).
'if the cavalry charge again': Henry Lawson (1896). The Star of Australasia. *In the Days When the World Was Wide and Other Verses.*
'no person who could slake the thirst of Henry Lawson': Manning Clark (1978). *In Search of Henry Lawson.* South Melbourne: MacMillan. (p. 56).
'the everlasting friction that most husbands must endure': Henry Lawson (1905). The Secret Whisky Cure. *When I was King and Other Verses.*
Henry complaining of being lent money to get married: Colin Roderick (1991). *Henry Lawson: A Life.* Sydney: Angus & Robertson. (p. 150).
'an ever present temptation': Bertha Lawson (1931). Memories. In Bertha Lawson & J. Le Gay Brereton (eds.) *Henry Lawson by His Mates*, pp. 79-118. Sydney: Angus & Robertson. (p. 93).
''Twas drink and nag—or nag and drink': Henry Lawson (1905). The Secret Whisky Cure. *When I was King and Other Verses.*
Bertha's application for judicial separation: Colin Roderick (1991). *Henry Lawson: A Life.* Sydney: Angus & Robertson. (pp. 246, 253).

DAY TEN

1. Henry's daughter said that her father and her brother were very close. She said of her father, 'He had immense pride in my brother. They were close and lifelong "mates".' Of her own relationship with her father, she said, 'He liked just to be happy, to enjoy being with me, and to know my hopes and dreams.' – Bertha Lawson (1973). *My Father (Abbey's Broadsheet, No. 1).* Sydney: Henry Lawson's Bookshop. (2 pages unnumbered [p. 2])...'special happiness' [p. 1].
2. *Ibid.*
   Henry's friend, Thomas Davies Mutch, Australia's Minister for Education from 1925-1927, unveiled a monument to Henry shortly after the writer's death. The monument was located within the Abbotsford Public School grounds, near where Henry died in Abbotsford, Sydney. At the unveiling, Mutch said, 'It is singularly appropriate that this memorial should be erected in a school ground, for Henry Lawson's love of children was tremendous. Had it been possible to consult Lawson, I feel he would have wished the memorial to be placed where children could see it, in the hope that it would give them some inspiration.' – *The Sydney Morning Herald* (1926, November 1). Henry Lawson. Memorial Tablet. Unveiled at Abbotsford. (p. 10).

Other Day Ten Sources

'Old Black Jimmie lived in a gunyah on the rise at the back of the sheepyards': Henry Lawson (1900). Black Joe. *Over the Sliprails.*
'The land I love above all others': Henry Lawson (1907). The Romance of the Swag. *The Romance of the Swag.*
'I'm meeting Jim': J. & A. Seymour (1931). A Place of Friendly Call. In Bertha Lawson & J. Le Gay Brereton (eds.) *Henry Lawson by His Mates*, pp. 287-300. Sydney: Angus & Robertson. (p. 294).

'I value them very highly': Henry Lawson letter to Dr Frederick Watson, Leeton, New South Wales, 31 March 1916. In Roderick *Letters* #304.
'spending too much on my children': *Ibid.*
'I dot my daddy first': Henry Lawson (1905). The Drunkard's Vision. *When I was King and Other Verses.*
'days will come when you'll be proud': Henry Lawson (1905). To Jim. *When I was King and Other Verses.*
'You break the heart in me!': Henry Lawson (1905). Barta. *When I was King and Other Verses.*
'Here's Mr Lawson! Here's Mr Lawson!': John Barr (1931). Stories. In Bertha Lawson & J. Le Gay Brereton (eds.) *Henry Lawson by His Mates*, pp. 263-264. Sydney: Angus & Robertson. (p. 264).
Henry wrote in children's autograph books: Jim Grahame (1931). Amongst My Own People. *Ibid.* pp. 210-251. (p. 229).
'I loved him, then, for that generous consideration of the boy': *Ibid.* (p. 243).
'I oughter be dead': *Ibid.* (p. 219).
'a crick in my neck and spine for days': Henry Lawson (1907). The Romance of the Swag. *The Romance of the Swag.*

DAY ELEVEN

1. Henry wrote, 'A great bank of rain-clouds is rising in that direction, but no one says he thinks it will rain; neither does anybody think we're going to have some rain. None but the greenest jackeroo would venture that risky and foolish observation.' – Stragglers. *While The Billy Boils*, 1896.
2. Henry writes of 'rats' as an affliction and, elsewhere, as a character. '"You've got rats this mornin', Gentleman Once," growled the Bogan,' writes Henry when Gentleman Once contributes in a manner deemed to be excessively generous to a collection being taken up for the benefit of a sick jackeroo.
   – Send Round the Hat. *Send Round the Hat*, 1907.
   In his well-known short story, 'Rats', Henry tells of an old swagman whose mind has gone because of the hardships of a harsh life on the Track. Although Macquarie, Sunlight, and Milky have fun humouring the old man, as swagmen themselves, they also feel sorry for Rats. It is only with persuasion that Milky agrees to act as time-keeper and referee when a fight is proposed between Rats and a swag. Additionally, Sunlight willingly gives the old man some money and lots of food. Before departing, Sunlight also says, 'You'd best push on to the water before dark, old chap.' Significantly, Henry says that these words are delivered 'kindly'. Sunlight is clearly worried about the old man's well-being. Henry himself—and his characters Macquarie, Sunlight, and Milky—knew enough of the difficulties of the Track to feel sorry for Rats.
   – *While the Billy Boils*, 1896.
   In the story, 'No Place for a Woman,' the character 'Ratty Howlett' is depicted as having 'brooded over [his wife's death] till he went ratty.'
   – *On the Track*, 1900.
3. Bertha wrote to Henry: 'When you have proved yourself a better man and not a low drunkard you shall see your children as often as you like. Until then, I will not let you see them. They have nearly forgotten the home scenes when you were drinking—and I will not let them see you drinking again. I train them to have the same love for you as they have for me.' – Bertha Lawson to Henry Lawson [15 June 1903] published in John Arnold & Frances Thorn (comps. & eds.). (1981, October). Henry and Bertha Lawson: Some Unpublished Letters and Stories. *La Trobe Library Journal, 28* (7), pp. 73-100. (p. 82).
4. Jim Grahame (1931). Amongst My Own People. In Bertha Lawson & J. Le Gay Brereton (eds.) *Henry Lawson by His Mates*, pp. 210-251. Sydney: Angus & Robertson. (p. 219).
   Henry felt that after their separation Bertha proceeded to spread lies about him and his conduct. He felt that many people turned from him as a result of the lies. Henry gave voice to this sentiment in the poem, 'The Old, Old Story and the New Order':
   Have faith, my friends, who stand by me,

In spite of all the lies—
I tell you that a man shall die
On the day that Lawson dies.
– *Skyline Riders and Other Verses*, 1910.

To lose friends in this manner was particularly galling to Henry because, under his code of mateship, a mate was always in the right. He wrote:

We learnt the creed at Hungerford,
We learnt the creed at Bourke;
We learnt it in the good times
And learnt it out of work.
We learnt it by the harbour-side
And on the billabong:
'No matter what a mate may do,
A mate can do no wrong!'
– A Mate Can Do No Wrong. *My Army, O, My Army! And Other Songs*, 1915.

In addition to Bertha's so-called lies, Henry also resented the alimony payments he was obliged to make or be sent to gaol. The poem, 'The Separated Women' suggests Henry felt Bertha was better off than he was. Henry wrote:

The Separated Women
Go lying through the land,
For they have plenty dresses,
And money, too, in hand;
They married brutes and drunkards
And blackguards 'frightful low',
But why are they so eager
For all the world to know?
– Composed 1907 [?], In Colin Roderick (ed.) (1968). *Henry Lawson: Collected Verse: Volume Two 1901-1909*. Sydney: Angus & Robertson. (pp. 234-235).

Bertha certainly disagreed with the assertion that she had 'plenty dresses / and money, too, in hand'. In a letter to her estranged husband, she wrote, 'I had to pawn my wedding ring to pay for a room. And then had to leave the little children shut up in the room, while I sought for work. And when I got work to do I had to leave them all day, rush home to give them their meals. And back to work again. And mind you, I was suffering torture all the time with toothache, and had to tramp the cold wet streets all day, knowing unless I earnt some money that day the children would go hungry to bed.' – Lawson, *op. cit.* #3.

5. In an 1896 newspaper report, it was stated that Charles (Karl) Lind shot himself on 24 March 1896 at Manly. According to the report, 'Henry Lawson, poet and journalist, gave evidence that Lind had been an intimate friend of his.' The report states, 'A letter found on the body of the deceased was addressed to Mr. Lawson.' In part, the suicide note read, 'Harry, my boy, you and I have had many a fine spree and many, to me at least, interesting conversations together. We have lived, loved, and suffered somewhat alike in this world, and have seen the beauties and glories, as well as the sorrows and abomination of existence. You have frequently heard me remark how I looked upon all in this world and all that is connected with it as a farce, more or less hideous, and a horrible farce, and I will simply add that now in the face of the inevitable I have no reason to alter my opinion. Good-bye, old chap. Be good to the girl. Once again be kind to B -Yours, Karl.' – *The West Australian* (1896, March 27). New South Wales. Suicide of a Swede. The Inquest. Letter to Henry Lawson. (p. 5).

   The critic, Alfred George Stephens, of the *Bulletin* reacted strongly to Henry's '"Pursuing Literature" in Australia' ('my advice to any young Australian writer') piece and turned the blame on Henry. 'Plainly, this young man does not know when he is well off,' Stephens wrote about Henry's strong book sales. Stephens said that the only reason Henry had not made a substantial amount of money from his writing was because he was such a poor businessman. 'It seems to me,' Stephens wrote, 'that Lawson made a bad bargain with his publishers; but, after all, it was Lawson who made the bargain.'

Stephens felt that Henry had no one to blame but himself. – A. G. Stephens (1899, February 18). Lawson and Literature. *The Bookfellow,* (2nd number), pp. 21-24. (p. 23).

6. 'Lawson's Fall' is an incomplete manuscript contained within the Lothian papers in the State Library of Victoria, Australian Manuscripts Collection. The poem was published in *The La Trobe Journal, 70* (spring 2002), p. 52. Among other things, Henry suggests the jump may have been the result of 'the torture of the present...or the horror of the future.'
7. A. B. ('Banjo') Paterson (1939, February 11). 'Banjo' Paterson Tells His Own Story - 2: Giants of the Paddle, Pen, and Pencil: Henry Lawson at Work: When the Booms Burst. *The Sydney Morning Herald.* (p. 21).
   Henry echoed this same sentiment in a poem entitled 'Lawson's Dream'. In the poem, Henry dreams of praise and tributes flowing upon the announcement of his death yet, when Henry awakens from the dream, in life things are much different: Henry 'woke to thirst and reek and oaths and spittle.' – *The Bulletin,* 26 August 1915.
   In another work, Henry used this same idea and had one of his characters say, 'I've been dead a few times myself, and found out afterwards that my friends was so sorry about it, and that I was such a good sort of a chap after all, when I was dead that—that I was sorry I didn't stop dead. You see, I was one of them chaps that's better treated by their friends and better thought of when—when they're dead.' – Brummy Usen, *While The Billy Boils,* 1896.
8. Paterson, *ibid.*
   Bertha responded angrily to this claim by Banjo and wrote a letter to the newspaper denying the conversation ever took place. – Bertha Lawson (1939, February 13). Henry Lawson: To the Editor of the *Herald. The Sydney Morning Herald.* (p. 7).
   In a letter to Henry in 1902, Bertha made it clear that she considered Henry's writing to be his work. Bertha wrote, 'Above all get to work and work hard. By so doing, you will forget the past and you will also forget yourself in your work. And I want you to make the most of your chances. You know you are a long way ahead of all Australian writers. But every week you let go by is so much lost. And it makes way for others. Get ahead with your novel. It will be a big success and you must sell your work now, because your name is now before the public.' – Bertha Lawson to Henry Lawson [August 1902?] in Arnold & Thorn, *op. cit.* #3. (p. 76).
   In her biography of Henry, Bertha also points out her attempts to assist Henry with his writing. She says that Henry liked to dictate his stories while she acted as scribe. 'Often he would wake me in the middle of the night with an idea for a story,' Bertha wrote. 'Then I slipped on a dressing gown and sat down for hours, taking it all down, while he paced to and fro.' – Bertha Lawson (1943). *My Henry Lawson.* Sydney: Frank Johnson. (p. 137).

Other Day Eleven Sources

'to what depths a man can sink': Henry Lawson (1905). The Women of the Town. *When I was King and Other Verses.*

'let himself be miserable': Christopher Brennan quoted in Manning Clark (1978). *In Search of Henry Lawson.* South Melbourne: MacMillan. (p. 133).

'the worst side of everything': Fred Davison quoted in Denton Prout (1963). *Henry Lawson: The Grey Dreamer.* Sydney: Rigby. (p. 298).

'shoot himself carefully with the aid of a looking-glass': Henry Lawson (1899, January 21). 'Pursuing Literature' in Australia. *The Bulletin.*

'ACCIDENT TO MR. HENRY LAWSON': *The Sydney Morning Herald* (1902, December 8). Accident to Mr. Henry Lawson. (p. 6).

'I wasn't a success as a flying machine, was I?': Henry Lawson letter to George Robertson, Sydney, New South Wales, 17 December 1902. In Roderick *Letters* #119.

'that mad (?) attempt at suicide': Henry Lawson letter to Dr Frederick Watson, Leeton, New South Wales, 31 March 1916. In Roderick *Letters* #304.

'I scarcely felt it at all after the first night': Henry Lawson (Composed 1902-1903). In Hospital. In Roderick *Prose Writings,* pp. 625-627.

'because of a woman's work': *Ibid.*
'I hope that I shall never be / As dry as Lake Eliza': Henry Lawson (1910). Lake Eliza. *The Rising of the Court and Other Sketches in Prose and Verse.*
'my blessed back seems broke': Henry Lawson (1896). Knocked Up. *In the Days When the World Was Wide and Other Verses.*
'the spirit of a bullock takes the place of the heart of a man': Henry Lawson (1896). 'Some Day'. *While The Billy Boils.*
'dunno if my legs or back or heart is most wore out': Henry Lawson (1896). Knocked Up. *In the Days When the World Was Wide and Other Verses.*
'To cease walking is to die': Henry Lawson (1896). Stragglers. *While the Billy Boils.*

DAY TWELVE

1. *The Advertiser* (1904, December 28). Attempted Suicides: Brave Rescue by Henry Lawson: Exciting Scene at Sydney North Head. (p. 4).
   The Brisbane Courier reported that the baby was taken to the Benevolent Asylum for care. – *The Brisbane Courier* (1904, December 28). Interstate Telegrams. New South Wales. Sydney, Tuesday. (p. 4).
2. Will Lawson (1931). The Meeting of the Lawsons. In Bertha Lawson & J. Le Gay Brereton (eds.) *Henry Lawson by His Mates*, pp. 266-267. Sydney: Angus & Robertson.
   Henry jokingly claimed that beer extended his life. Upon reuniting with Henry after a long separation, his friend, Edwin James Brady, suggested that, with his hard living, Henry should have been long dead. Henry replied that beer saved his life. When Brady asked how that was so, Henry replied, 'If I'd been drinking hard tack [biscuits] I WOULD have been dead long ago.' – Brady recollection quoted by Clive Turnbull (1952, July 26). Clive Turnbull Discusses the Life and Times of E. J. Brady. *The Argus* (Melbourne). (p. 2).
3. Henry Lawson (1896). Baldy Thompson. *While the Billy Boils.*
   In a letter to his aunty when he returned to Bourke, Henry wrote, 'A squatter who knew me gave me as much tucker as I could carry, when I was coming down, and a pound to help me along. Squatters are not all bad.' – Letter to Aunt Emma, Bourke, New South Wales, 6 February 1893. In Roderick *Letters* #11.
   Henry's positive assessment and portrayal of Baldy did not win him friends amongst the Unionists. Squatters were seen as the enemy and there was no good to be found there. In one of Henry's stories, Barcoo-Rot attacks Mitchell for his charity. 'Why, you'd find a white spot on a squatter,' Barcoo-Rot says insultingly. In automatic response 'the chaps half-unconsciously made room on the floor for Barcoo-Rot to fall after Jack Mitchell hit him.' – 'Lord Douglas'. *Children of the Bush*, 1902.
   When his book, *In the Days When the World Was Wide* did not sell as well as he hoped, Henry complained to Banjo Paterson, 'The Labour people are not buying my book. They have declared me bogus for disclosing some good points in a squatter.' – A. B. Paterson (1933, September 20). Some Reminiscences of George Robertson, Australian Publisher. *The Sydney Mail.* (p. 8).
4. Peter FitzSimons (2011, January 12). Plenty to Write Home About. *The Sydney Morning Herald.* (Summer Herald p. 3).
   According to Henry, J. F. Archibald's advice to writers was: "Every man has at least one story; some more. Never write until you have something to write about; then write. Write and re-write. Cut out every word from your copy that you can possibly do without. Never strain after effect; and, above all, always avoid anti-climax.' – Three or Four Archibalds and the Writer, Composed 1919. In Roderick *Prose Writings*, pp. 778-784.

Other Day Twelve Sources

'my body craved...for another hour's sleep': Henry Lawson (Composed 1903-1906). A Fragment of Autobiography. In Roderick *Prose Writings*, pp. 714-755.
'Her mother is also insane': Henry Lawson letter to Bland Holt, [Manly, New South Wales?], [October 1902]. In Roderick *Letters* #118.
'Things always look brighter in the morning': Henry Lawson (1901). 'Water Them Geraniums'. *Joe Wilson and His Mates.*

'a pitiful travesty of life': Denton Prout (1963). *Henry Lawson: The Grey Dreamer*. Sydney: Rigby. (p. 287).
'I can't help her': Henry Lawson letter to Bland Holt, [Sydney], New South Wales, [28 December 1904]. In Roderick *Letters* #132.
'It was a gigantic success and ended in oblivion': Anthony Cashion (1931). Pahiatua. In Lawson & Brereton, *op. cit.* #2, pp. 56-60. (p. 58).
'face half-hid 'neath a broad-brimmed hat': Henry Lawson (1901, December 14). The Teams. *The Queenslander.*
'lie down on a grassy bank in a graceful position in the moonlight and die just by thinking of it': Henry Lawson (1901). Joe Wilson's Courtship. *Joe Wilson and His Mates.*
'I was a good son—one of the best': Henry Lawson letter to Dr Frederick Watson, Leeton, New South Wales, 31 March 1916. In Roderick *Letters* #304.
'That tree haunted my early childhood': Henry Lawson (Composed 1903-1906). A Fragment of Autobiography. In Roderick *Prose Writings*, pp. 714-755.
'it's a terrible thing to die of thirst in the scrub Out Back': Henry Lawson (1896). Out Back. *In the Days When the World Was Wide and Other Verses.*

DAY THIRTEEN

1. In a will written several years earlier (1905), Henry wrote, 'I want to say that Mr Archibald (of the *Bulletin*) and Mr George Robertson (of Angus & Robertson) and my landlady, Mrs Isabel Byers... were my best friends.' – The will was published in *The La Trobe Journal, 70* (spring 2002), p. 18.
   In addition to the high praise for Mrs Byers and J. F. Archibald, the high praise for George Robertson is also noteworthy given that Henry's relationship with his book publishers was not always a good one. Henry had an acrimonious relationship with Thomas Lothian who published two of Henry's books in 1913 under the imprint of the Standard Publishing Company. – See John Arnold (2002, spring). Bringing Lawson to Book: The Lothian Experience. *The La Trobe Journal, 70*, pp. 19-30.
2. Jim Grahame (1931). Amongst My Own People. In Bertha Lawson & J. Le Gay Brereton (eds.) *Henry Lawson by His Mates*, pp. 210-251. Sydney: Angus & Robertson. (p. 231).
   Jim wrote a poem in which he indicated that he and Henry would often go walking together and 'hear the tales / each other had to tell.' Henry would share his dreams of 'books he meant to write.' Even after a long walk, when they got back to the gate to Henry's place, they 'talked and talked' for, as Jim wrote, 'we were loth to part.' – Jim Grahame (1940). When Lawson Walked With Me. *Call of the Bush*. Melbourne: Bread and Cheese Club. (pp. 49-50).
3. In his poem, 'Bourke', Henry suggests he would like to return to Bourke. He certainly leaves no doubt that he thinks with fondness back to his time in and around Bourke, writing:
   I've followed all my tracks and ways, from old bark school to Leicester Square,
       I've been right back to boyhood's days, and found no light or pleasure there.
   But every dream and every track—and there were many that I knew—
       They all lead on, or they lead back, to Bourke in Ninety-one, and two.
   – *When I was King and Other Verses*, 1905.
   In the poem Henry employs poetic license and refers to being in Bourke in [18]91 and '92, whereas he was actually there in 1892 and 1893. Presumably 'two' was an easier or neater rhyme than 'three'.
4. Grahame (1931), *op. cit.* #2. (pp. 241-246).
   In one letter, Henry reported that when he and Jim went to the station and knocked on the door, they received 'a right royal welcome from the cook' and 'got enough tucker to last us a week.' – Letter to Ernest O'Ferrall, Leeton, New South Wales, 1 January 1917. In Roderick *Letters* #326.
   Henry provided more details in another letter, when he said that in walking up to the homestead door and asking for a handout, the 25 years since he and Jim had previously done that seemed to vanish quickly. They 'were just two swagmen interviewing the cook at a station, outback, twenty-five years before, and the Murrumbidgee [River at Leeton] was the Darlin' [River at Bourke].' Henry said that Jim even dropped into 'the old Drought-Haze drawl' when he spoke. This time, Henry said that

they 'got enough grub to last us for three days, and tea and sugar for a week. Jim took it home that night for tea and breakfast for his family.' – Letter to George Robertson, Leeton, New South Wales, [January 1917]. In Roderick *Letters* #335.

Other Day Thirteen Sources

'I'm qualified for the work suggested': Henry Lawson letter to G. J. Evatt (Secretary, Water Conservation and Irrigation Commission), North Sydney, 16 December 1915. In Roderick *Letters* #294.

'*lo and behold! the pub is not here*': Henry Lawson (1916). Amongst My Own People (New Series) (First Impressions of Leeton), *The Murrumbidgee Irrigator*. In Roderick *Autobiographical Writings*, pp. 312-315.

'She has all her vegetables up and the place well stocked': Henry Lawson letter to G. J. Evatt, Leeton, New South Wales, 4 August 1916. In Roderick *Letters* #316.

'I lost him with the aid of two mates': Henry Lawson letter to George Robertson, Leeton, New South Wales, [January 1917]. In Roderick *Letters* #335.

'Lost, a Dorg named "Charley"': Henry Lawson letter to J. W. Gordon, Leeton, New South Wales, [1916?] In Roderick *Letters* #319.

'the applause of all of Australia and other parts of the world': Grahame (1931), *op. cit.* #2. (pp. 215-216).

'Drop in any day you're in and we'll get acquaint again': Henry Lawson letter to J. W. Gordon, Leeton, New South Wales, [18 January 1916]. In Roderick *Letters* #300.

'What a hearty, silent handshake we had!': Grahame (1931), *op. cit.* #2. (p. 216).

'we seem to have changed not at all. Not to each other': Henry Lawson (1916, May 13). By the Banks of the Murrumbidgee. *The Bulletin*.

'the colour came back to Lawson's cheeks': Grahame (1931), *op. cit.* #2. (p. 218).

'written by the same hand that made the name of Henry Lawson famous': James W. Gordon, letter to Henry Lawson, (1916, May 23). Quoted in Colin Roderick (ed.) (1970). *Henry Lawson Letters 1890-1922*. Sydney: Angus & Robertson. (p. 477).

Henry fussy about the appearance of his swag: Jim Grahame (1925, February 19). Henry Lawson on the Track. *The Bulletin*. (The Red Page).

'The swag is usually composed of a tent "fly" or strip of calico': Henry Lawson (1907). The Romance of the Swag. *The Romance of the Swag*.

'That pleased him immensely. It was the right way, he said—the bushman's way': Grahame (1931), *op. cit.* #2. (p. 219).

'nothing matters between mates': *Ibid.* (pp. 238-239).

'There were many other Mitchells': Henry Lawson letter to Frank Beaumont Smith, Leeton, New South Wales, [June 1916]. In Roderick *Letters* #312.

'There used to be two young fellas knockin' about Bourke and west-o'-Bourke named Joe Swallow and Jack Mitchell': Henry Lawson (1916, May 13). By the Banks of the Murrumbidgee. *The Bulletin*.

'the bush had an idea that I might have done away with him': Henry Lawson (1902). That Pretty Girl in the Army. *Children of the Bush*.

'Christ didn't believe in prohibition': Henry Lawson (1916). A Letter From Leeton. *The Australian Soldiers' Gift Book*. In Roderick *Autobiographical Writings*, pp. 319-323.

'a Place is not natural without the Pub': Henry Lawson (1916). Amongst My Own People (New Series) (First Impressions of Leeton). *The Murrumbidgee Irrigator*. In Roderick *Autobiographical Writings*, pp. 312-315.

'I'm the Commander of the Army of the Fed-ups in Leeton': Henry Lawson letter to J. W. Gordon, Leeton, New South Wales, [August 1917]. In Roderick *Letters* #424.

'interested in every human being whom he met': Bertha Lawson. Memories. In Lawson & Brereton, *op. cit.* #2. pp. 79-118. (p. 94).

Henry referred to himself as 'painfully shy': Henry Lawson (Composed 1903-1906). A Fragment of Autobiography. In Roderick *Prose Writings*, pp. 714-755.

'There were times when I really wished in my heart I was on my own': Henry Lawson (1913). Drifting Apart. *Triangles of Life and Other Stories.*

DAY FOURTEEN

1. Henry wrote, 'Travelling with the swag in Australia is variously and picturesquely described as "humping bluey," "walking Matilda," [and] "humping Matilda".'
   – The Romance of the Swag. *The Romance of the Swag*, 1907.
2. Colin Roderick (1991). *Henry Lawson: A Life*. Sydney: Angus & Robertson. (e.g. pp. 265, 269). For recollections from one who worked at Angus & Robertson and witnessed Robertson's attempts to hide and Henry's determination to talk with him, see Harry G. Hodges, (2002, spring). Lawson Memories. *The La Trobe Journal, 70*, pp. 72-73.
3. One reviewer said of Jim's writing, '[He] has done and seen and suffered much. He writes at a horseman's speed; he writes in vigorous rhythms what bushman think and feel.' This reviewer, however, felt that Jim's writing was too gloomy. This is a fascinating perspective given that, five years earlier, the same reviewer wrote the same thing about Henry's writing. Of Jim's descriptions of the Bush, the reviewer wrote, 'His verses are all true, yet typically they are too sombre.' The reviewer explained, 'The picture is true, yet partial; there are good seasons as well as bad seasons.' – A. G. Stephens (1927, September 14). A Survey of Some Recent Australian Poetry. *The Sydney Mail*, pp. 8-9. (p. 9). As a criticism of what he saw as Henry's gloomy, one-sided descriptions of the Bush, the reviewer wrote, 'There are good seasons as well as bad; there are wet times as well as dry times; there is grass as well as sand; and joy as well as pain.' – A. G. Stephens (1922, November 1). Henry Lawson. *Art in Australia (3rd series #2)*. (8 pages unnumbered [p. 4]).
4. Henry Lawson letter to J. W. Gordon, Coolac, New South Wales, 22 March 1920. In Roderick *Letters* #468.
   A month later, Henry wrote to Angus & Robertson to further push Jim's cause and to remind them that Henry had left some of Jim's work with them to peruse. – Letter to George Robertson, [Sydney, New South Wales?], 19 April 1920. In Roderick *Letters* #474.

Other Day Fourteen Sources

'the solemn immensities of the sunburnt plain': Arthur H. Adams (1931). Stories. In Bertha Lawson & J. Le Gay Brereton (eds.) *Henry Lawson by His Mates*, pp. 252-254. Sydney: Angus & Robertson. (p. 252).

'Lawson thanked her for the privilege of being allowed to read it': 'Kelby', *ibid.* p. 259.

'Mary had a little lamb': R. J. Cassidy. Lawson as I Knew Him. *Ibid.* pp. 198-203. (p. 201).

'I get drunk because...': Henry Lawson letter to *The Bulletin*, No address, [December 1903?]. In Roderick *Letters* #130.

'some indefinable honour': Bartlett Adamson quoted in Roderick, *op. cit.* #2. (p. ix).

Robertson paid Mrs Byers to keep Henry away: *Ibid.* (e.g. pp. 272, 276).

'It's them that's mad, not us': E. J. Brady. Mallacoota Days and Other Things. In Bertha Lawson & J. Le Gay Brereton (eds.) *Henry Lawson by His Mates*, pp. 125-150. Sydney: Angus & Robertson. (p. 130).

'I went there to escape from the damned lunatics outside': R. J. Cassidy quoted in Roderick, *op. cit.* #2. (p. 358).

'shamefully, appallingly overcrowded': Henry Lawson (n.d.). The Old Men's Home [manuscript]. State Library of Victoria, Accession no. MS6026. Lothian Publishing Company, Records 1895-1950, Australian Manuscripts Collection. Box XX1B Folder 3D.

'You know how I love bananas': Roderic Quinn. Glimpses of Henry Lawson. In Bertha Lawson & J. Le Gay Brereton (eds.) *Henry Lawson by His Mates*, pp. 175-191. Sydney: Angus & Robertson. (p. 183).

Henry gave children bananas in the hospital: Walter Jago. Stories. *Ibid.* pp. 272-275. (p. 273).

'The heat is bad, the water's bad, the flies a crimson curse': Henry Lawson (1896). Knocked Up. *In the Days When the World Was Wide and Other Verses.*

Henry's reaction to Jago telling him of the journalist's struggles: Walter Jago. Stories. In Bertha Lawson & J. Le Gay Brereton (eds.) *Henry Lawson by His Mates*, pp. 272-275. Sydney: Angus & Robertson. (p. 275).
Donald Macdonell tribute: Henry Lawson (Composed 1911). Donald Macdonell. In Colin Roderick (ed.) (1969). *Henry Lawson: Collected Verse: Volume Three 1910-1922.* Sydney: Angus & Robertson. (pp. 332-334).
'this was his best reward—to be loved and appreciated by his bushmen': Jack Moses. The Clot of Gold. In Bertha Lawson & J. Le Gay Brereton (eds.) *Henry Lawson by His Mates*, pp. 37-50. Sydney: Angus & Robertson. (p. 41).
'we sits an' thinks beside the fire': Henry Lawson (1910). Down the River. *Skyline Riders and Other Verses.*

DAY FIFTEEN

1. J. T. Lang (1956). *I Remember.* Sydney: Invincible. (pp. 191-192).
   John Thomas Lang and Henry Lawson married sisters. Lang became the Premier of New South Wales from 1925-1927 and 1930-1932.
2. In this respect, and with all due respect, Jane is perhaps a bit like Henry's wife. Bertha said, 'I would not live in Bourke with its dust, heat, flies, mosquitoes, and bad weather for £1000.' – Quoted in Colin Roderick (1991). *Henry Lawson: A Life.* Sydney: Angus & Robertson. (p. 248).

Other Day Fifteen Sources

Henry looked 'white and shaken': J. & A. Seymour (1931). A Place of Friendly Call. In Bertha Lawson & J. Le Gay Brereton (eds.) *Henry Lawson by His Mates*, pp. 287-300. Sydney: Angus & Robertson. (p. 299).
Henry suffered a seizure and was assisted to bed:
– *The Argus* (1922, September 4). Death of Henry Lawson. Poet and Author. State Funeral Arranged. (p. 8).
– *The Sydney Morning Herald,* (1922, September 4). Henry Lawson: Death Announced: Poet and Prose Writer. (p. 8).
Henry's death: Roderick, *op. cit.* #2. (pp. 393-394).
'it can look more like rain without raining...than in most other places': Henry Lawson (1896). Stragglers. *While The Billy Boils.*
'if things don't alter, they'll be the same': Roderic Quinn. Glimpses of Henry Lawson. In Bertha Lawson & J. Le Gay Brereton (eds.) *Henry Lawson by His Mates*, pp. 175-191. Sydney: Angus & Robertson. (p. 176).
'Their snaky heads well up': Henry Lawson (1900). But What's the Use. *Verses Popular and Humorous.*
'The mighty bush with iron rails / Is tethered to the world': Henry Lawson (1896). The Roaring Days. *In the Days When the World Was Wide and Other Verses.*
'To see "The Death of Henry Lawson" printed / In letters tall and black': Henry Lawson (1915, August 26). Lawson's Dream. *The Bulletin.*
Coffin adornment: *The Sydney Morning Herald* (1922, September 5). Henry Lawson: State Funeral: Impressive Service. (p. 10).
Prime Minister's eulogy:
– *The Argus* (1922, September 4). Death of Henry Lawson. Poet and Author. State Funeral Arranged. (p. 8).
– *The Brisbane Courier* (1922, September 4). Henry Lawson. Death in Sydney. A Great Australian Writer. (p. 7).
– *The Sydney Morning Herald,* (1922). Henry Lawson: Death Announced: Poet and Prose Writer. (4 September) (p. 8).
'The characteristic I most distinctly recall is Henry Lawson's modesty': Madame Rose Soley (1922, September 5). A Reminiscence. *The Sydney Morning Herald.* (p. 10).
Henry rarely read his own published work: Henry Lawson, letter to Frank Beaumont Smith, Leeton, New South Wales, [June 1916]. In Roderick *Letters* #312.

Henry 'never quoted his own lines': Roderic Quinn. Glimpses of Henry Lawson. In Bertha Lawson & J. Le Gay Brereton (eds.) *Henry Lawson by His Mates*, pp. 175-191. Sydney: Angus & Robertson. (p. 183).
Henry wouldn't recite poem at Christmas gathering: J. & A. Seymour (1931). A Place of Friendly Call. *Ibid.* pp. 287-300. Sydney: Angus & Robertson. (p. 294).
'sorrow came as it only came to me once before': Jim Grahame. Amongst My Own People. *Ibid.* pp. 210-251. (pp. 248-250).
'The Bush Mourns': Jim Grahame (1947). *Under Wide Skies*, p. 195. Leeton, NSW: Citizens of Leeton.
'The stars have never seemed so bright': Jim Grahame (1940). When Lawson Walked With Me. *Call of the Bush*, pp. 49-50. Melbourne: Bread and Cheese Club. (p. 49).
'I don't wanter; I've been there': Henry Lawson (1896). In A Dry Season. *While the Billy Boils*.
Henry's bath story: Henry Lawson (1910). The Bath. *The Rising of the Court and Other Sketches in Prose and Verse*.
Dean Karnazes quote: Quoted by Bryn Swartz, The Greatest Athlete You've Never Heard Of. *Bleacher Report*, 19 April 2009. http://bleacherreport.com/articles/158847-the-greatest-athlete-youve-never-heard-of

APPENDIX

1. Mrs Swaffer has in her possession a copy of a Bill of Fare from the Royal Mail Hotel from 1894. As this was just one year after Henry's visit, it provides an interesting insight into the type of menu Henry may have seen when he visited the Royal Mail (although this menu was prepared for a special occasion). For such a small and remote location, the surprisingly extensive beer and wine list is noteworthy. It is particularly interesting given that in his 'Hungerford' sketch (*While the Billy Boils*, 1896), Henry wrote that at the pub he 'had asked for English ale.'
   The Bill of Fare is annotated, 'Menu used by T. G. [Thomas Goodman] Foster at the Hotel 1894. For the Hungerford Race Dinner.'
   On one side, the card reads:
   1894
   ROYAL MAIL HOTEL
   RACE CLUB DINNER
   HUNGERFORD

   On the other side, it reads:
   BILL OF FARE

   Clear and Oyster Soup.
   Roast and Boiled Fowl, Turkey, and Duck.
   Roast, Corned, and Spiced Beef
   Sucking Pig, Ham.
   Baked and Mashed Potatoes, Green Peas,
   French Beans, Pumpkin.
   Plum Pudding. Custard. Jellies.
   Mince Pies. Blanc Mange.
   Compotes of Apricots, Peaches, and Pears.
   Snow Egg. Lemon Sponge.
   Fruit in Season and Preserves.
   Salad and Cheese.

   WINE AND BEERS

Claret V.D.O.	Foster's English
Muscat	Aitchison's English
Burgundy	St. Louis Lager

Chablis	Brandt's Lager
Hock	Milwaukee Lager
Port and Sherry	Munchausen Lager

2. Mrs Swaffer has in her possession a poem written by Anita Bernich, a granddaughter of Thomas and Agnes. The poem relates some of the family's Outback history. In one verse, the poem reads:
   When Lawson came by Cobb & Co.
   He stayed and ate his fill,
   But his pockets were so empty
   He wrote a poem to pay his bill.
3. Colin Roderick (1991). *Henry Lawson: A Life*. Sydney: Angus & Robertson. (e.g. p. 295). Roderick describes some of these compositions as 'rough and uncomplicated'.
4. One contemporary newspaper report read: 'The embankment burst below the Hospital, 200 yards above the railway line, yesterday afternoon [April 18] at 4 o'clock. The town is now submerged to an avenge depth of 3 feet, the maximum being 5 feet and the minimum 2 feet. It appears that the water was commencing to run over the banks and a man dug his spade into the base of the dam for the purpose of stopping the slight overflow, when 50 ft. of the structure burst away in one piece, carrying several men a distance of about fifty yards before they could recover themselves.' – *The Morning Bulletin* (Rockhampton, Qld) (1890, April 21). The Darling in Flood: Bourke Ruined. The River Still Rising. (p. 5).
5. *Launceston Examiner* (1890, April 24). The Flood at Bourke. Official Bungling, Disorderly Scenes. Deaths from Drowning. Flood Stationary. (p. 2).
   The Sydney Morning Herald (1890, April 21). The Flood at Bourke. Rescue of the Inhabitants. Reported Drowning of a Family. A Scarcity of Boats. Provisions Running Short. The Houses Giving Way. The Water Still Rising. (p. 8).
   The *Sydney Morning Herald* report included the information, 'It is rumoured that a woman and three children have been drowned at a place called Gowtown, on the north-east side of the town.'
6. Information on hotel and publican licences obtained from a list compiled by Rusheen Craig (last updated 2006) entitled *Aus-NSW-West Mailing List: Hotels and Publicans Licences. Western NSW 1865-1900, Part of Central NSW 1865-1870.* The list was compiled from Publicans listed in the NSW Government Gazette and appears on the world wide web @ http://homepages.rootsweb.ancestry.com/~surreal/NSWW/Hotels/index.html
   Another source lists seven hotels between Bourke and Hungerford and identifies their locations as: North Bourke; Lauradale; Fords Bridge; Youngerina; Lake Eliza; Yantabulla; and, Paragundy. The date of the existence of these establishments is not recorded. The original source of this information is unknown but was provided by Sheree Parker, current publican of the Royal Mail Hotel (Hungerford) on a map entitled, 'Hotels Between Bourke & Hungerford.' Copy in possession of the author.
7. Roderick, *op. cit.* #3. (p. viii).
   The critic, A. G. Stephens, said of Henry's writing, 'Sometimes the prose is merely egotistic maundering, and the verse underdone doggerel.' – A. G. Stephens (1922, November 1). Henry Lawson. *Art in Australia (3rd series #2).* (8 pages unnumbered [p. 8]).
8. One scholar wrote of Henry, 'His spelling, always shaky, got worse when he was under stress; and his use of punctuation and capitals was not always consistent.' – John Barnes (2002). Lawson Manuscripts in the Lothian Papers. *The La Trobe Journal, 70*, pp. 31-32. (p. 31).
   Noted historian, Manning Clark, wrote that Henry's spelling 'left much to be desired' (p. 12). Clark said also of Henry's poor spelling, punctuation and grammar, 'To the end of his days everything he wrote needed generous use of the "grammarian's" pencil.' (p. 24). – Manning Clark (1978). *In Search of Henry Lawson.* South Melbourne: MacMillan.
   In his authoritative biography of Henry's life, Lawson scholar Colin Roderick makes several references to Henry's poor spelling and poor punctuation (e.g. pp. 236, 367, and 382) and to the heavy involvement of

editors in the preparation of his work for publication (e.g. p. 365). – Roderick, *op. cit.* #3.
A revealing illustration of the jumbled and disorganised nature of some of Henry's submissions to his publishers is provided in John Barnes (2002). From the Editorial Chair. *The La Trobe Journal, 70*, pp. 2-3. (p. 2).

9. *The Sydney Morning Herald* (1908, October 14). Deaths. (p. 12).
The death notice reads:
DEATHS.
BUCKLEY-At his home, Yantabulla, Bourke, on October 5, John Buckley. In his 65th year. R.I P.

Other Appendix Sources

Reputation for begging and willingness to scribble verses for drinking money: Roderick, *op. cit.* #3 (see e.g. p. 317).
Henry's manner of writing poems: *Ibid.* (e.g. pp. 264-265).
'a dirty mark around the doorsteps where the water had been': Henry Lawson (n.d.). The Great Flood of '90 [manuscript]. State Library of Victoria, Accession no. MS6026. Lothian Publishing Company, Records 1895-1950, Australian Manuscripts Collection. Box XX1B Folder 3B.
Henry's poems only making it into print because of his reputation: *The Sunday Times* (Perth) (1907, February 3). The writer's reward. (p. 9).
George Mure Black comments about 'His Father's Mate': George Black quoted in Denton Prout (1963). *Henry Lawson: The Grey Dreamer*. Sydney: Rigby. (p. 73).
Henry felt 'an editorial cut in his copy as keenly as if it were a cut in his flesh': A. G. Stephens (1899, February 18). Lawson and Literature. *The Bookfellow*, (2nd number), pp. 21-24. (p. 21).
'misspelt, ungrammatical but forcible copy of his early days was edited into something like decent form': A. G. Stephens (1977). Lawson's Last Book—A Temporary Adjustment. In Leon Cantrell (ed.) *A. G. Stephens: Selected Writings*, pp. 231-237. Sydney: Angus & Robertson. (p. 233). (Originally published in *The Bulletin*, 12 January 1901).
Concession that Henry could only guarantee the quality of his ideas: Barnes, *op. cit.* #8.
'I don't spell—never could': Henry Lawson letter to Jack McCausland, Leeton, New South Wales, 17 February 1916. In Roderick *Letters* #299.
'disguised but not exaggerated': Henry Lawson (1899, January 21). 'Pursuing Literature' in Australia. *The Bulletin*.

ENDNOTES

ABOUT THE AUTHOR

Gregory Bryan is a professor in literacy education who has co-authored a number of books and articles and presented at conferences across Australia and the world. His knowledge of Henry Lawson's life and appreciation of the quality and significance of his writing are borne of a lifetime of study of his life and works. Gregory is passionate about Australia's natural and social history. Together with his brother, in 2011, he became the first person ever to recreate Henry Lawson's 1893 walk from Bourke to Hungerford and back.